Systematic Hydrology

Systematic Hydrology

JOHN C. RODDA, BSc, PhD, FRMet Soc

RICHARD A. DOWNING, BSc, MIWES, FGS

FRANK M. LAW, BSc, Dip Hydrol, MICE, MIWES

Newnes–Butterworths
LONDON BOSTON
Sydney Wellington Durban Toronto

THE BUTTERWORTH GROUP

UNITED KINGDOM	Butterworth & Co (Publishers) Ltd London: 88 Kingsway, WC2B 6AB
AUSTRALIA	Butterworths Pty Ltd Sydney: 586 Pacific Highway, NSW 2067 Also at Melbourne, Brisbane, Adelaide and Perth
CANADA	Butterworth & Co (Canada) Ltd Toronto: 2265 Midland Avenue, Scarborough, Ontario, M1P 4S1
NEW ZEALAND	Butterworths of New Zealand Ltd Wellington: 26-28 Waring Taylor Street, 1
SOUTH AFRICA	Butterworth & Co (South Africa) (Pty) Ltd Durban: 152-154 Gale Street
USA	Butterworth (Publishers) Inc Boston: 19 Cummings Park, Woburn, Mass. 01801

First published in 1976

ISBN 0 408 00234 4

Photoset by Amos Typesetters, Hockley, Essex
Printed in England by Butler and Tanner Ltd,
Frome and London

Preface

Modern society faces a series of challenges. One of the most important concerns the uses and abuses of water. Few natural resources are as universally and as constantly in demand, and few are employed in so many ways. Water is needed for drinking, for power generation and for cooling purposes; it is required for the carriage of wastes and for irrigation; it is the basis of many leisure activities. Too much water causes destruction and leads to disease; too little results in shortages and famine. Intense erosion can lay waste fertile regions; serious pollution destroys the life of rivers and can bring about epidemics. To combat these extremes large sums of money are spent in many countries on the construction of dams and flood alleviation schemes, on the development of water distribution systems and on the provision of purification works. Basic to many water resource projects is hydrology, a science with its foundations in ancient Egypt and Mesopotamia and one that can be approached in a number of different ways. Engineers, chemists, physicists, geologists, meteorologists, geographers and biologists have all made significant contributions to the science of hydrology. Application of the principles of hydrology creates a potential for improving the lives of those in the world without an adequate water supply. In countries where water resources are almost fully developed and there are the conflicting demands of water supply, effluent disposal, recreation, amenity and environment, hydrology provides a framework for planning and development.

The theme of this book is hydrology in the United Kingdom, its advancement and role in the solution of questions of water use and control. These small islands present a range of problems resulting from the impact of a dense population and intense industrialisation on a landscape varied in its topography, geology and climate. But the basic problems are common to many parts of the world, and it is hoped that their treatment here will be of value to hydrologists and others concerned with water outside the UK.

The text is designed to form a convenient source of reference to basic scientific studies and to indicate how such studies have been applied, in the hope that it will be helpful to those involved in the policy, management and implementation of national, regional and local water programmes in the UK and abroad. The volume is also aimed at the 'students' of hydrology on whom the future of the science rests.

In preparing the book we have drawn upon many sources for assistance,

Preface

advice and reference and we wish to express our gratitude to all those concerned. We are especially indebted to the Institute of Hydrology, the Water Resources Board, Binnie and Partners, the Water Data Unit and the Central Water Planning Unit for the information, illustrations and help they have provided during our preparation of the text. For their constructive comments and suggestions on the manuscript we are very grateful to: Mr. A. Bleasdale, Mr L. Ellis, Mr F. O. C. Hodges, Dr J. S. G. McCulloch, Mr H. R. Oliver, Mr J. M. A. Pontin, Mr D. H. A. Price, Mr J. B. Stewart, Mr L. M. J. Standon-Batt, Dr W. B. Wilkinson and Mr F. Young; to Mr A. Lambert, Mr H. R. Taylor and Dr G. E. Hollis for their help with material for Chapter 6; and to Mr S. W. Smith for his help with illustrations. A large part of the typing was undertaken by Mrs D. Downing and Mrs M. Upton, while Mrs E. Dharzive, Mrs D. Sherwood, Mrs P. Derbyshire and Mrs G. Kinnane were responsible for the remainder. Our thanks are due to these ladies. Finally we wish to express our thanks to our families for their tolerance, encouragement and help.

John C. Rodda
Richard A. Downing
Frank M. Law

Contents

Contents

Chapter 1

Introduction

Hydrology the science

Water is one of the natural resources in greatest demand and it is a demand that is constantly increasing. Use of water by industry and agriculture is accelerating rapidly, while requirements in the home are growing apace. In the UK, allowing for an increase in consumption per head and the natural growth in population, early in the 21st century the total national demand is expected to be more than twice that of the present day.

Population increase and urban expansion have a considerable impact on the problems of flooding and erosion. Pressure on the land, and competition between differing forms of land use, often bring about developments in flood plains at sites that have not been occupied previously. These and other riparian works cause a general rise in the amount of property at risk from floods. Pressure on the land also results in intensification of its use, and this applies to agriculture and forestry as well as to other forms of usage. Marginal land is ploughed under subsidy for improvement, while steep slopes are furrowed prior to tree planting. There is some argument about whether these measures increase the incidence of floods, but there is little doubt that the ploughing initiates a rise in the rate of erosion.

A greater population and an expansion of industrial activities also increase the incidence and risk of pollution. Most rivers in the UK are cleaner today than for many years, but at times the flow in some is composed very largely of treated sewage effluent, often undesirably high in nitrates and phosphates and lacking dissolved oxygen.

Of course water is important in a number of other ways. Whether it quenches thirst, produces power, carries away waste or provides enjoyment, the value of water will continue to increase, until a substitute is discovered—a very unlikely event!

Water resources and pollution on the one hand, and flooding and erosion on the other, are the chief concerns of the hydrologist. There are many other topics that involve him, but none so important from the practical point of view. Indeed, although hydrology is still an applied science to a considerable extent, the time has passed when it grew simply to meet the demands of the engineer undertaking some project or other. Now it has an increasing pure science content; in fact, for some time it has been acknowledged as a science in its own right, a part of geophysics. As an earth science it is closely allied to

1

those other sciences that deal with the movement of waters about the globe. With them it is concerned in the study of global patterns of water circulation, the hydrologist's interest naturally being focused on the land-based section of the 'hydrological cycle'—the omnibus term for the passage of water from the sea, to the land and back to the sea, by many and devious routes. The description of the paths in the hydrological cycle that involve a land route might be sufficient for some as an indication of the domain of hydrology. Others[1] prefer a succinct definition such as 'the study of what happens to the rain'. In fact, there are a large number of definitions from which to choose, one[2] of the most comprehensive being 'the science of the world's waters, the different forms in which they exist, their circulation and distribution over the globe, their physical and chemical properties, their interactions with the environment, including their response to human activity'.

This definition and the majority of its alternatives demonstrate the very extensive nature of the science. The scope of hydrology is extremely broad, as is evident from the way in which water enters into so many of man's activities. It follows that hydrology is linked by common content to a large number of other scientific disciplines, particularly to environmental sciences such as geomorphology, climatology and ecology. Evidence of especially strong points of connection is given by the existence of the more specialised branches of hydrology, namely hydrogeology, hydrometeorology, glaciology, limnology and forest hydrology. Then, as Horton[3] recognised, there is the unique relationship between hydrology and hydraulics. Because of the earlier use of the name 'hydraulics' for the study of the mechanics of fluids, but particularly water, the term hydrology is limited to the study of the natural occurrence and circulation of water. Nevertheless hydraulics is essentially a part of hydrology, in spite of the latter's growth from the former. As Valentine[4] puts it, 'rather have the Frankensteins of hydraulics produced an amiable monster that bids fair to subordinate its creators'.

Floods and droughts result from distortion of the normal pattern of the hydrological cycle. Man too can alter and interrupt it locally. Yet these variations in the distribution of water, though significant in the short term, are infinitely small by comparison with the total amount of fresh water flowing into the seas. The fresh-water phase is itself one of the minor modes of occurrence of terrestrial water, the salt-water mode being by far the most abundant. A global study of the modes of occurrence of water, though informative, is not very rewarding scientifically. On the other hand, studies of the exchanges and transports between modes began centuries ago and continue to absorb the efforts of a variety of scientists. Of these fluxes two are particularly important to the hydrologist, namely precipitation and evaporation. These, occurring at the surface of the globe, are complicated by the boundary conditions that arise at this interface and by heterogeneity both in the atmosphere and at and below the ground surface. In fact, in terms of the global cycle, the hydrologist confines his interest to these two fluxes, to storage changes on and within the land mass, and to the water that is residual to evaporation and storage changes, namely runoff. Furthermore he directs his attention to individual catchment areas, where an attempt can be made to define the input and output from this subsystem of the hydrological cycle. For a particular basin, over a time t, the relationship between these components, or in other words the water balance, is usually expressed as:

$$P = R + E + \Delta W \qquad (1.1)$$

where P = mean basin precipitation
      R = mean basin runoff
      E = mean basin actual evaporation and transpiration
     ΔW = mean storage change over the basin

Alternatively, a systems representation of a basin water balance might be employed *(Figure 1.1),* the one shown here adopting the classic division of the hydrograph into storm runoff (Q_o), interflow (Q_i) and baseflow components (Q_s). No two basins are alike, however, in either climate or

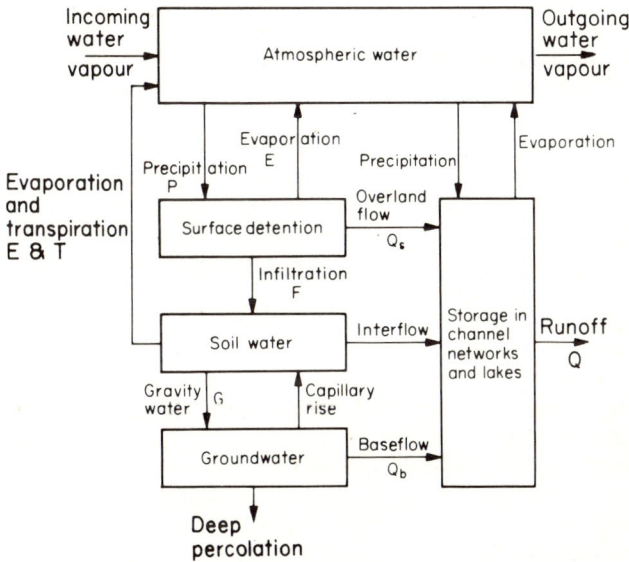

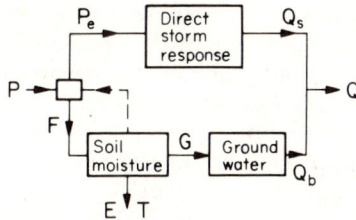

Figure 1.1   A systems representation of a basin water balance (after Dooge)

terrain features. Thus the quantity of water in each component and the residence time of water in storage and runoff vary enormously from one basin to another. Some controls of the water balance, such as geology, are of a permanent nature. Others, like land use, and size and path of storms, are transient. Whether involved in a practical project or in a 'pure' study, the hydrologist is concerned in some way or other with the relationships in the water balance and their adjustments to the local surface and subsurface characteristics.

United Kingdom, the study area

One large island, part of a second and a host of smaller ones comprise the territory of the United Kingdom of Great Britain and Northern Ireland, more commonly but less properly known as Britain and sometimes even as England. The islands occupy an area of some 241 000 km² and span the ten degrees of latitude north of 50°N. No place is further than 130 km from the sea and even the highest summit reaches only 1343 m above sea level. Moorlands and mountains dominate the north and west, in contrast to the flatter and more undulating landscape of the south and east. This convenient division[5] is important hydrologically and in terms of most other spatially distributed variables. Scale is another factor that must be given special consideration, for within these islands diversity is not related to distance and variability is not governed by extent. Indeed, region-to-region and district-to-district differences must be unrivalled in any other area of similar size elsewhere in the world.

Because representatives of almost the entire stratigraphical sequence occur within the UK, the geology is varied in the extreme. The north and west are characterised by ancient compact rocks, which have been extensively folded and faulted. They include representatives of the Precambrian and the Palaeozoic sequence from the Cambrian to the Devonian. In the south and east of England the more recent Mesozoic and Cainozoic sediments form relatively low ground. A dominant feature of the landscape is the series of escarpments with intervening clay vales formed by the Mesozoic rocks.

The nature and form of the superficial deposits have been strongly influenced by glacial action. They range from outwash sands and gravels to boulder clays and alluvial deposits. This variety adds considerably to the spectrum of soil types that are present. Immature podsols occur in the more elevated regions, while well developed brown earths are characteristic of the lower areas. Of course, the landscape represents the balance achieved locally between geology, structure and the agents of erosion. Running water is the main agent, but ice has been an important factor over nearly all the country except the south. In geomorphological terms, most of the land surface is in the mature and post-mature stage of evolution, but in many regions several changes in sea level have added the complications of regeneration and multiple surfaces to those caused by glaciation. Hence, although the topography lacks grandeur, it is very heterogeneous.

The hydrological implications of these differences in geology and geomorphology are considerable. Lithology is extremely varied, so that high yielding aquifers can give way to impermeable deposits over very small distances. Surface deposits, such as peat and glacial clays, may effectively isolate groundwaters from the circulation of surface waters. Steep-sided valleys, cut in impermeable rock by short swift rivers, give rise to problems of flash flooding, while drainage is difficult in clay vales and other low-lying regions. In the case of the large areas of chalk and limestone in the south and east, little or no surface runoff is produced, but groundwater occurs in abundance. In contrast, there is generally little groundwater storage in the north and west and not even a great deal of surface storage in natural lakes.

The British weather and climate are virtually free from the extremes experienced in other parts of the world. Prolonged and extensive drought, storms that devastate wide areas and periods of intense heat or cold are phenomena that are foreign to these islands; yet, in contrast to many other parts of the globe, the day-to-day weather is characterised by a series of frequent and sometimes dramatic changes. On the other hand, the climate of the UK as a whole is lacking in variety, one season being difficult to distinguish from another.

The position of the UK is the most important control of its weather and climate. The islands are situated within the belt of extra-tropical Westerlies, where the prevalent sequence of variable disturbances, with their oceanic routes, can be interrupted by contrasting systems evolved over the continental land mass. This juxtaposition of land and sea causes further modifications, including amelioration, particularly along the western margins, where the maritime influences are greatest. The Atlantic fringe is narrowed by the presence close to the west coast of the bulk of the mountainous areas. These relief differences add considerable local diversity to the weather and climate[6], and they are largely the cause of countrywide patterns in the distribution of most hydrological elements. Precipitation is greatest in the north and west and least in the south and east, while the reverse is true for evaporation. In England and Wales, because the main watersheds are nearest the west coast, east flowing rivers are generally the longest; this also applies in Scotland to a certain extent. In general the soils of the south and east have the greatest moisture capacity, reflecting the nature of the strata beneath.

Until the middle of the eighteenth century, the UK was an agricultural country with a population of less than 7 million and few large towns. By 1801 it had become very industrialised and the population of nearly 10 million was becoming increasingly concentrated in towns and cities. The Industrial Revolution altered the distribution and character of the population, brought about its increase and caused a most drastic transformation of the landscape in many regions. Most of the towns lacked a piped water supply and drainage facilities were rudimentary. Many of the rivers became polluted and enteric disease was rife. This state continued well into the latter half of the nineteenth century, but improvements followed the increasing use of unpolluted sources of water supply (usually based on impounding reservoirs), the establishment of procedures for chemical and bacteriological examination of water, the introduction of water-treatment processes and the development of more satisfactory means of distribution[7]. Now nearly all aspects of the supply of water are covered by a series of Parliamentary Acts embracing a wide range of topics from the prevention of pollution to the safety of reservoirs. These ensure a constant and potable piped water supply to 99 per cent of the population[8] of 56 million (1974) whose average consumption was well over 310 litres per head per day by 1974 *(Table 1.1)*.

The distribution of population, hence the demand for water, reflects the

Table 1.1 AVERAGE TOTAL AND PER CAPITA DAILY WATER CONSUMPTION IN ENGLAND AND WALES

Year	*Average total consumption*			*Average per capita consumption**		
	(million m^3/d)	(million gal/d)	*Percentage change from previous year*	(m^3/d)	(gal/d)	*Percentage change from previous year*
1955	8.90	1958	—	0.200	44.1	—
1960	10.00	2200	+2.5	0.218	48.0	+1.7
1961	10.47	2303	+4.7	0.226	49.8	+3.8
1962	10.79	2373	+3.0	0.231	50.8	+2.0
1963	11.26	2479	+4.5	0.240	52.8	+3.7
1964	11.52	2533	+2.2	0.243	53.5	+1.3
1965	11.57	2545	+0.5	0.243	53.4	−0.2
1966	11.95	2629	+3.3	0.249	54.8	+2.6
1967	12.26	2696	+2.5	0.254	55.8	+1.8
1968	12.62	2777	+3.0	0.260	57.1	+2.3
1969	13.18	2899	+4.4	0.269	59.1	+3.5
1970	13.47	2963	+2.2	0.275	60.5	+2.4
1971	13.61	2994	+1.0	0.279	61.3	+1.3
1972	13.90	3057	+2.1	0.284	62.4	+1.8
1973	14.19	3125	+2.1	0.289	63.6	+1.8

*Per capita figures have been calculated using mid-year population estimates as published in *Annual abstract of statistics 1972*

basic division of Britain into two. The bulk of the population is concentrated in the area between the Humber and Mersey estuaries in the north and Southampton Water and the Thames estuary in the south. In the south of this quadrilateral, water supplies are largely based on groundwater and river abstraction, particularly from the River Thames. In the north, many of the cities obtain their supplies by aqueduct from reservoirs situated some distance away in the neighbouring highland areas[9]. At the present time, proposals for reservoirs to meet the increasing demands for water face stiffening opposition from farming, amenity and conservation interests, and they have aroused nationalist sentiment in Wales. The construction of estuarial barrages in Morecambe Bay, across the Dee and the Solway Firth and also in the Wash would solve some of the water supply problems and those associated with them, but others would be created. Bunded reservoirs within estuaries rather than barrages across them should limit these fresh problems, such as lack of control of water quality and interruption of the movement of fish. Obviously the basic question is one of land use policy and the contrasting claims of the water engineer, the sheep-farmer, the forester and those seeking recreation. Patterns of water use in the south east *(Figure 1.2)* indicate the trends that the rest of the country is likely to follow. Domestic requirements will rise as the numbers of washing machines, dish washers, garbage disposal units and garden hose-pipes increase. The demand from industry, including water to circulate for cooling purposes, will grow. More water will be used in agriculture, particularly for irrigation, in which case the water cannot be employed again. There are also demands for water for hydro-electric power generation and from thermal and nuclear power stations. The first applies in certain limited regions and the water is only stored and diverted. In the case of thermal power stations, however, high quality water is needed for the turbines and a much larger quantity is required for cooling purposes, some being evaporated while the remainder is returned to the river at temperatures 6 to 9°C higher than when it was abstracted. Because the water requirements of nuclear power stations are so large, all except one are sited on the coast. Water is also used for navigation, and then there are fishing, amenity and recreational interests to consider.

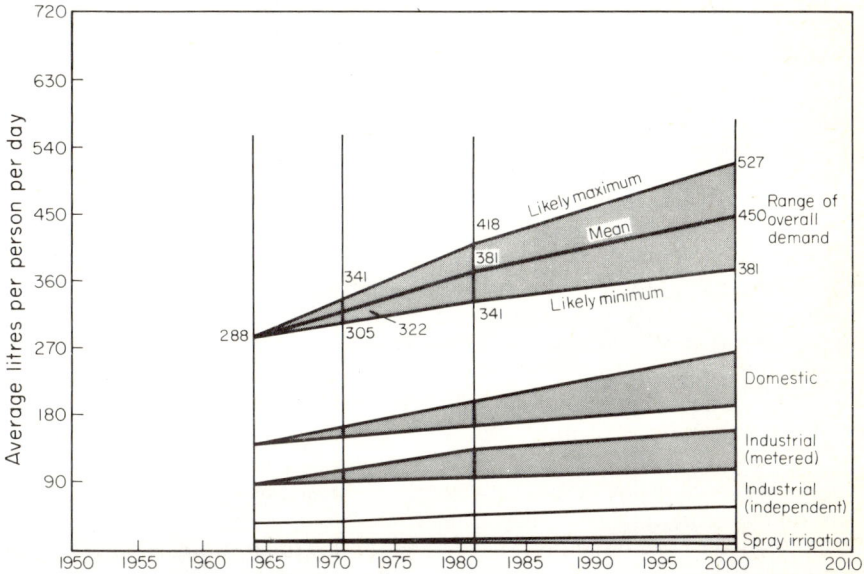

Figure 1.2   *Patterns of water demand in south-east England: total per capita demand for all purposes (from Sharp [10], courtesy Institution of Water Engineers)*

Most parts of the UK have suffered at one time or another from flooding. Alleviation of flooding is dependent mainly on embankments and river training works, but such measures have often stimulated the growth of high-value property behind them, as a result of the misconception that the flood risk has been eliminated. More recently a number of river-regulating reservoirs have been constructed, and these can provide storage for the runoff that would otherwise cause floods. The operation of reservoirs of this type is coupled to programmes that treat a whole river basin as a single system, flood control being one aspect of the scheme. Unfortunately, many rivers cannot be controlled in this way, the best hope of flood warning systems for them being through the extension of weather-radar coverage, linked to very efficient local warning systems. Obviously, the value of schemes for flood warning and mitigation has to be balanced against their cost, but the damage and inconvenience caused by floods also have to be taken into account. In most years losses run into millions of pounds, tens of millions occasionally; fortunately loss of life is a rare occurrence.

Severe and fairly widespread drought has been experienced on a number of occasions during the present century. Notable drought years were 1921, 1933 and 1934, 1944, 1959, 1965 and 1968, but there have been differences between the years in the timing and distribution of the drought. The 1968 drought, for example, was largely confined to the north and west, whereas in 1959 the drought affected most of the country. The drought in 1965 was the culmination of the driest three successive winters since records of rainfall commenced in 1729[11], dry winters being particularly serious in areas that rely on the winter recharge of aquifers for water supply. With an increasing population and a bigger demand for water the effects of drought could be far more serious in the future than at present[12]. Obviously the consequences of droughts are more widespread and prolonged than those of floods, but there is also the very great difficulty of obtaining a meaningful measure of drought severity.

Of course reservoir storage capacity should be sufficient to prevent draw-down below the lowest outflow level during extreme drought. To prevent this occurrence, approaches like Hawksley's Rules, the Deacon Diagram and the Lapworth Chart were formulated along with their more recent counterparts. There is also the concept of minimum acceptable flow in rivers, the statutory minimum below which discharge cannot be reduced. This is a particularly important matter to consider in relation to dry-weather abstractions, when demand for water is usually at a peak and pollution is greatest.

River pollution is attendant on most of man's activities. Originally, the main hazards were from chemically simple industrial wastes carried in suspension and solution, but now complex effluents, both industrial and domestic, and the residues of agricultural pesticides and herbicides are becoming more and more important. There is also the so-called thermal enrichment by power stations to consider and, what is more important, the radio-active wastes from thermo-nuclear explosions and in certain local discharges. One of the greatest problems facing man is how to deal with the increasingly complicated nature of the waste he produces; this is particularly vital in a sophisticated industrial society, such as exists in the UK. Water quality is monitored continuously at several points on a number of rivers, and many individual samples are analysed at frequent intervals to study the changes in the characteristics of river chemistry. Measurements of erosion are rarely as comprehensive and certainly not as widespread. Rates of erosion are being assessed in a few representative and experimental basins, but greater attention is given to measurements of the suspended load in certain rivers and even those of bed load. Information has also been collected about the silting of a number of reservoirs.

Development

The beginnings of hydrology are lost at the start of history: the science is at least as old as the ancient riverine civilisations of Egypt, the Indus and Mesopotamia. Nile flood records date from before 3000 B.C., while the early irrigation systems reached peaks of development about 2000 B.C. In Britain, some of the first works concerned with water seem to have been the sea defences built by the Romans. One of the Roman governors of Britain, Julius Frontinus (A.D. 35–103), later became Rome's water engineer and wrote a lengthy treatise on the city's water supply[13]. No other major works appear to have been undertaken until Myddleton brought the 'New River' to London in 1613 and the Fens were drained later in the same century. On the other hand several British philosophers, such as Geraldus Cambrensis[14] (1146–1222), entered into the discussion on the origin of springs and rivers that continued from the time of Ancient Greece well into the Middle Ages. In fact, it was the renowned Edmund Halley (1656–1742) who, following the measurements of rainfall and flow made in the Basin of the Seine by Perrault (1608–1680) and Mariotte (1620–1684), showed that evaporation from the Mediterranean Sea was greater than the river flow into it[15,16]. These three men demonstrated conclusively that the landward phase of the hydrological cycle is essentially atmospheric. In so doing, they more or less put an end to theories that the hydrological cycle delivered water to the land surface by subterranean routes, lifting it and ridding it of salt in the process.

Of course, this progress was made possible by the development of the raingauge and other instruments, together with rudimentary methods of flow measurement. The first automatic raingauge was devised by Sir Christopher Wren (1632–1723). It operated on the tipping-bucket principle and was displayed to the Royal Society in January 1663[17]. It is likely, however, that this gauge was never used for routine observations of rainfall[18]. On the other hand it was incorporated with other instruments in Wren's ideas for a 'weather-cock', which were later developed by Robert Hooke (1635–1703) into the first automatic weather station. Hooke[19] was able to show his 'weather wiser' to the Royal Society in 1679. It consisted of a barometer, a thermometer, a hygrometer, a wind vane and the tipping-bucket raingauge, together with a clock that unwound a paper strip from a cylinder. Records from each instrument were punched on the paper strip every 15 minutes by the action of the clock hammer, the mode of recording being similar to that of some modern weather stations.

Hooke also devised a non-recording raingauge[20], but the first regular measurement of rainfall in Britain was commenced by Richard Townley in 1677, at Townley Hall near Burnley[21]. Halley carried out some further evaporation experiments[22] and for the year of 1693 recorded an annual loss of 203 mm (8 in) from an evaporimeter with a surface diameter of 203 mm (8 in). At about the same time Robert Plot (1640–1696) made one of the first classifications of rivers[23] and a few years later John Keill[24,25] put perhaps the final nails in the coffin of the subterranean theory.

The number of sites where rainfall and other elements were observed increased appreciably during the first half of the 18th century, but after Halley there is no evidence of evaporation being recorded until 1772. These observations were made at Liverpool over a period of four years by Dr Dobson[26], who used a 305 mm (12 in) diameter tin cylinder 152 mm (6 in) deep alongside a raingauge of the same dimensions. He found the average annual total for the four years was almost 940 mm (37 in), which is rather high by comparison with modern records[27]. A little earlier than this, in 1765, Benjamin Franklin (1706–1790) used a pond on Clapham Common for an experiment in spreading oil on water surfaces[28]. This may have been the forerunner of the modern efforts to suppress evaporation using monomolecular films. In the late 1760s, Heberden[29] was carrying out experiments with raingauges on and around Westminster Abbey, his results being some of the earliest showing that catch decreases as height of gauge increases. These conclusions stimulated an argument, over the origin and mode of formation of rainfall, that continued for over a hundred years. This was due largely to other investigators finding that more elevated gauges caught less rain, but ignoring the effects of wind. The canal building era was commencing about the time these experiments were being conducted, with obvious effects on hydrology. Starting in 1761, when the Duke of Bridgewater's canal was opened to Manchester, some 4800 km (3000 miles) of canals were constructed for barge traffic, a traffic which was to decline rapidly with the coming of the railways.

By the end of the 18th century, the observational beginnings of hydrology and meteorology were well under way and it is fitting that, at about this time, as great a scientist as John Dalton (1766–1844) was commencing his work on evaporation. With the help of Thomas Hoyle, Dalton recorded evaporation for the three years commencing 1796, at a site near Manchester[30]. His 'cylindrical vessel of tinned iron 10 inches in diameter and 3 feet deep' was fitted with two drains, one at the top and the other at the bottom, filled with soil and sunk into the ground. It seems to have been operated in the same way as the present-day irrigated lysimeter, in that 'water was poured on to sodden the earth' and the amount draining through was measured. For the first year Dalton kept the soil surface bare, but then he grassed it over and found that the evaporation averaged 635 mm (25 in) for the three years and 595 mm (23.5 in) over six years. He concluded that evaporation increased with rainfall, but not proportionally, and that there was little difference between evaporation from bare soil and that from grass. From these evaporation observations, records of rainfall and the estimates of runoff he made from discharge measurements carried out in the Thames, Dalton went on to assess the water balance for the whole of England and Wales. The value of 894 mm (35.2 in) that he used for the mean rainfall is remarkably close to the 1881–1915 mean of 894.1 mm (35.23 in) for the two countries, but in spite of this it was 178 mm (7 in) *less* than his total of runoff and evaporation! In a subsequent paper[31], Dalton put forward what must be his most important contribution to hydrology, namely a statement of the relationship between the rate of evaporation and the vapour-pressure gradient above a water surface:

$$E = C (e_s - e_d) \tag{1.2}$$

where E = rate of evaporation per unit area
         C = a constant
         e_s = vapour pressure at the water surface
         e_d = vapour pressure in the air above

The Dalton formula, or one of its variants, forms the basis of any number of modern methods for estimating evaporation (see equation 3.12, page 101).

Whereas the 18th century was almost entirely a period of innovation for British hydrology, the 19th was as much concerned with organisation. The amateur working singly and in groups continued to be the main motivating force, rather than any State institution, but as the century progressed governmental participation increased. In 1815, the records of rainfall that continue today were commenced at the Greenwich and Radcliffe Observatories. Interest was developing in the field of groundwater through the efforts of William Smith (1769–1839) and the Geological Society, which was founded in 1807. The Geological Survey was formed in 1835, and as the early surveys extended across the country information about wells and springs was collected and published in the memoirs. The British Association appointed a committee in 1874 to investigate groundwater in the New Red Sandstone and Permian, and in 1882 this was extended to cover all permeable deposits. From 1875 to 1895 the results of this work were published in the committee's annual reports. In 1899 the Geological Survey introduced the series of county water-supply memoirs[32]. Of those associated with this early work on groundwater, the name of W. Whitaker deserves mention.

The Meteorological Department was established in 1854 as part of the Board of Trade. This followed the display of the world's first synoptic weather maps at the Great Exhibition in 1851, an enterprise that owed much to the initiative of James Glaisher (1809–1903), the Superintendent of Meteorological Observations at Greenwich. In fact it was at Greenwich in 1841 that Glaisher had commenced the first standard meteorological observations and encouraged others to follow the same practice. Standardisation became easier to achieve from 1866 onwards with the use of the Stevenson screen, a practice fostered by the British Meteorological Society, which had been founded in 1850, and by the Scottish Meteorological Society, which was set up five years later. Nathaniel Beardmore's *Manual of Hydrology* was published in 1850, one of the first modern books on the subject.

The value of standardisation was also recognised for rainfall measurements. This manifested itself in 1860, when the British Rainfall Organisation was established by G. J. Symons (1838–1900) from the large number of volunteer observers who were recording rainfall at that time. Symons ensured uniformity of gauge and observer practice and published the collected records annually in *British Rainfall* from 1860 onwards. A description of the present standard raingauge was given in *British Rainfall 1866,* and from that time onwards the proportion of instruments with a height of 305 mm (1 ft) and a diameter of 127 mm (5 in) increased rapidly. Symons's successors, Mill and then Glasspoole, had a considerable share in bringing about the increase, which was such that when the British Rainfall Organisation became part of the Meteorological Office in 1919 there were over 3400 raingauges in use. By 1971 the total had risen to 6831[33] with a good prospect of the growth continuing.

About the time of the inception of the British Rainfall Organisation, there were a number of investigations in progress into the performance of raingauges of different sizes, shapes and heights. The Calne, Rotherham and Aldershot experiments determined the size and height of the standard gauge, and it is interesting to note that a ground-level gauge was among those tested. The value of installing a raingauge with its rim at ground level had been discussed by Stevenson in 1842[34], but the earliest records from this type of gauge are from Calne, no indication being given of the nature of the surface surrounding the instrument[35]. From these experiments, the work of

Jevons[36] and an earlier study of the variation of rainfall with altitude made by Miller[37], it became obvious that the loss of catch in an elevated gauge was due to wind action. This explanation finally discredited the ideas about rain condensing in the lower layers of the atmosphere that had been current since Heberden's time.

The second half of the 19th century saw considerable developments relating to flowing water, but no organisation emerged comparable with those concerned with rainfall and groundwater. The Thames Conservancy was created in 1857, and over the next thirty years there were a number of Royal Commissions and subsequent Acts of Parliament dealing with water supply and river pollution. In 1881 work was started on the Vyrnwy Dam, which was to form what was for a long time Britain's largest artificial reservoir. This was one of the first high masonry dams and its success prompted the building of a number of others of similar design. The systematic recording of the flow of the Thames commenced at Teddington in 1883. At about the same time, progress was being made in the study of the flow of fluids. Osborne Reynolds published the results of his classic experiments on the resistance to flow of fluids[38]; while in 1889 Manning proposed an equation[39] for determining the velocity of flow *(V)* in channels or pipes, which was developed later into:

$$V = \frac{1.5}{n} R^{2/3} S^{1/2} \tag{1.3}$$

Subsequently the terms 'Reynolds number' and 'Manning's *n*', the roughness coefficient, became part of the normal parlance of both hydrology and hydraulics. Some of the first attempts to estimate reservoir yield and flood discharges were made before the turn of the century, while attention had also been given to the problems of assessing percolation and evaporation. Percolation gauges were installed at Rothamsted in 1870, and in 1885 regular measurements of evaporation were commenced at Camden Square.

During the 19th century the state of rivers in industrial areas deteriorated. This led to the setting up of Royal Commissions on prevention of river pollution in 1865 and 1868, from which stemmed the River Pollution Prevention Acts of 1876 and 1890. A Royal Commission on Sewage Disposal was appointed in 1898 and published a series of reports between that date and 1915. Many of the Commission's recommendations are still accepted today, for example the well known requirement that, where an effluent is diluted in fresh water in the ratio of 8:1, discharges should be satisfactory if they contain no more than 30 mg/l of suspended solids and have a biological oxygen demand no greater than 20 mg/l.

There was sufficient interest in pollution problems for the Water Pollution Research Board to be set up in 1927, followed by the Freshwater Biological Association in 1932, but surprisingly no official body was initiated in the early part of the 20th century to organise the collection of river flow records on a national basis. Capt. W. N. McClean established his own River Flow Records organisation and installed and maintained gauging stations on a number of Scottish rivers from the late 1920s onwards. In 1932 the British Association meeting at York appointed a committee to examine the need for an Inland Water Survey. The committee recommended that such an organisation was indeed required, this requirement being emphasised by the drought of 1933 and 1934. The Inland Water Survey Committee was appointed in 1935, responsible to the Minister of Health and the Secretary of State for Scotland. The duty of the committee was to collect and collate records of river flows and underground waters, the latter being the responsibility of the Geological Survey. The Inland Water Survey published records of river flow in the *Surface Water Year Book of Great Britain,* commencing in 1935 with records from 21 gauging stations. The River Boards Act of 1948 and the efforts of individuals such as W. Allard helped to increase the number of gauging stations and the development of the Survey, so that when the Surface Water Survey was integrated with the Water Resources Board in 1965 records from 331 stations were being published. Over the years the Geological Survey amassed a unique collection of well and borehole records, and well catalogues containing summaries of these data have been published. This work is being continued by the Institute of Geological Sciences, which now incorporates the Geological Survey.

There has been considerable rationalisation in the fields of water conservation and water supply since 1945. Amalgamation reduced the number of water undertakings in England and Wales from 1100 at that time to 286 by the end of 1966 and to 189 by 1973. The Water Resources Act of 1963 placed the main responsibilities for conserving and managing water resources in England and Wales on the 29 river authorities. In Scotland the river purification boards and in Northern Ireland the Ministry of Development fulfilled some of the functions of the river authorities. The river authorities were required among other things to:

1. survey resources in their areas, and estimate future demands;
2. prepare hydrometric schemes, and investigate underground sources;
3. determine the minimum acceptable flows in their rivers;
4. plan and execute measures to conserve water, e.g. major reservoir projects.

In fulfilling these obligations and those relating to pollution prevention, navigation, flood mitigation and land drainage, the river authorities had a considerable role to play in relation to hydrological matters. On a countrywide scale, the Water Resources Board, a Government-appointed body, advised both the river authorities and the central Government on water conservation. The Board too had far-reaching responsibilities relating to hydrology, because it was charged particularly with:

1. reviewing water demands and resources, and conducting regional surveys designed to establish demand and possible measures needed;
2. scrutinising the hydrometric schemes of river authorities, exercising an informal supervision of their more detailed survey work and advising them on planning;
3. collecting information on water resources and engaging in research projects.

Of these measures, the most significant in hydrological terms was the one concerned with hydrometric schemes. This required measurements of rainfall, evaporation, stream flow and other factors to be made, so that it became possible to think for the first time of a national hydrological network.

In 1973 a Bill was introduced in Parliament[40] for reorganising the water industry in England and Wales. (See Appendix 2 for details of the Water Act.) Ten regional water authorities (RWAs) were proposed to take over from 1 April 1974 the functions of the river authorities, the water supply undertakings and the sewerage and sewage-disposal authorities *(Figures 1.3 and 1.4)*. The RWAs have the responsibility for meeting the various

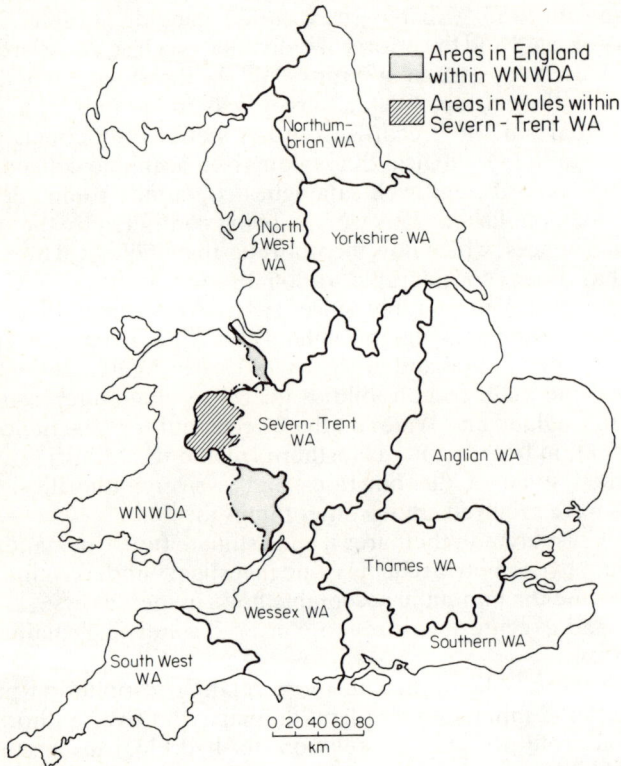

	Population, 1974 (millions)	Rateable value (£ millions)	Water supplied per capita, 1974 (l/d)
Anglian WA	4.6	190	323
North West WA	7.1	290	352
Northumbrian WA	2.7	110	370
Severn-Trent WA	8.2	350	261
South West WA	1.3	50	284
Southern WA	3.8	200	282
Thames WA	11.8	950	272
Wessex WA	2.2	110	339
Yorkshire WA	4.5	170	288
Welsh National Water Development Authority	3.0	120	390

Figure 1.3   Regional water authorities in England and Wales (from ref. 41, reproduced by permission of the Controller, Her Majesty's Stationery Office)

Before 1 April 1974	After 1 April 1974
River authorities	Regional water authorities
Local authorities Statutory water undertakings Sewerage & sewage disposal authorities	National Water Council
	Water Space Amenity Commission
Statutory water companies	Statutory water companies

Department of the Environment
Directorate General of Water
  Engineering
Development Division, DGWE
Water Pollution Research Lab.
Hydraulics Research Station
 	
Water Resources Board
	Department of the Environment
	Directorate General of Water
	  Engineering
	Central Water Planning Unit
	Water Data Unit
	Hydraulics Research Station

Water Research Association	Water Research Centre
British Hydromechanics RA	British Hydromechanics RA
British Waterways Board	British Waterways Board
Meteorological Office	Meteorological Office
Ministry of Agriculture, Fisheries & Food	Ministry of Agriculture, Fisheries & Food
Forestry Commission	Forestry Commission
Central Electricity Generating Board	Central Electricity Generating Board
UK Atomic Energy Authority	UK Atomic Energy Authority
Natural Environment Research Council Institute of Hydrology Institute of Geological Sciences Freshwater Biological Association	*Natural Environment Research Council* Institute of Hydrology Institute of Geological Sciences Freshwater Biological Association
Agricultural Research Council Rothamsted Experimental Station Other ARC stations	*Agricultural Research Council* Rothamsted Experimental Station Other ARC stations
Scottish Development Department Local authorities River purification boards	**After 16 May 1975** Scottish Development Department Regional councils | A smaller River purification boards | number
Department of Agriculture & Fisheries, Scotland	Department of Agriculture & Fisheries, Scotland
Electricity boards	Electricity boards
Ministry of Development, NI Local authorities, water companies	**After October 1973** Department of the Environment, NI
Ministry of Agriculture, NI	Ministry of Agriculture, NI
Universities	Universities

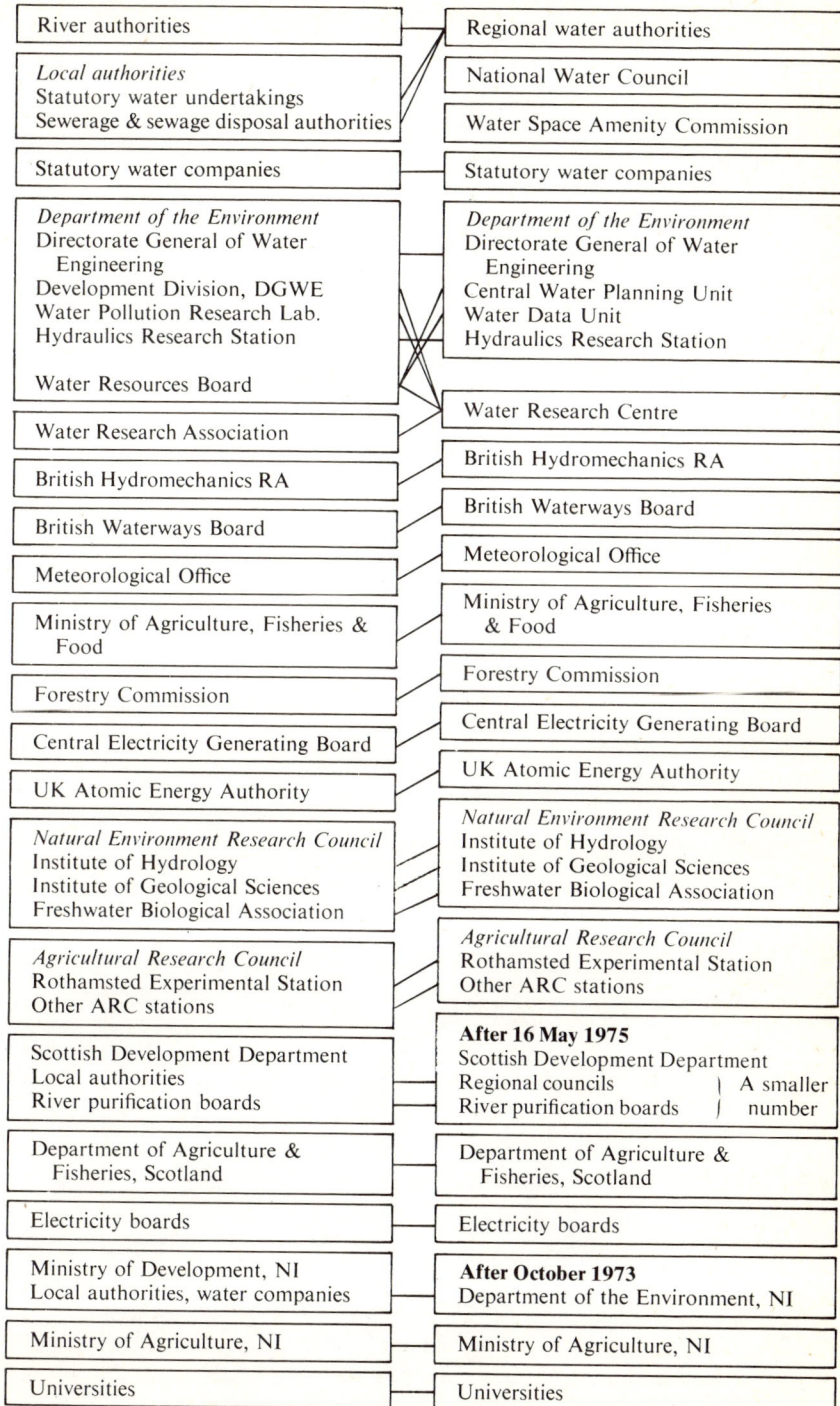

Figure 1.4   Bodies concerned with water in the UK

demands on river systems and on water resources[41]. Each authority is responsible for water supply, sewerage and sewage disposal, prevention of pollution, land drainage, flood protection, fisheries, recreation and in some cases navigation. Together the water authorities have an annual expenditure of between £600 and £700 million and a staff of about 55 000. Similar legislation came into force in Northern Ireland in October 1973, a single water authority being formed. In Scotland reorganisation dates from 16 May 1975. On that date the regional councils were formed and they assumed responsibility for water supply and sewerage, while seven new river purification boards *(Figure 1.5)*, covering the whole country, took over control of pollution in rivers, estuaries and coastal waters. Amongst their other duties the water authorities in England, Wales and Northern Ireland and the similar bodies in Scotland are required to survey the water resources and water usage in their areas, estimate future demand, plan the more

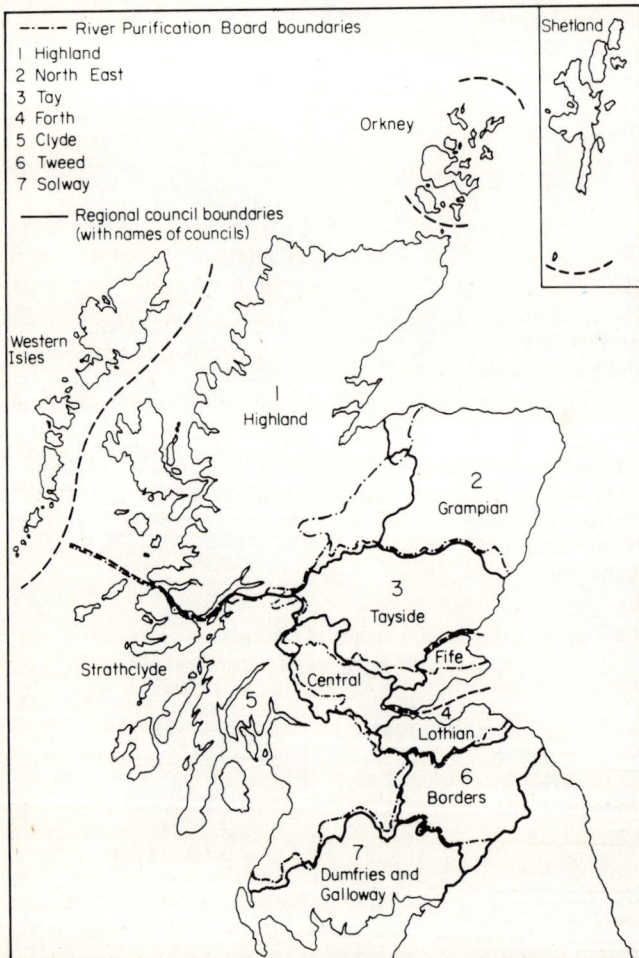

Figure 1.5   Regional council and river purification board boundaries in Scotland

efficient management of the water resources and take measures for restoring and maintaining the wholesomeness of inland and coastal waters.

At the national level a consultative and advisory body was established—the National Water Council. The NWC consists of the chief executives of the water authorities and other members, and acts on matters of interest common to authorities and as a link between them and the Government on general issues. Also at the national level, but within the government service, are the Central Water Planning Unit and the Water Data Unit. The former provides planning advice to the Government, the NWC and the RWAs on augmentation of water resources, on water quality and on pollution prevention. The latter's task is to acquire and archive data concerned with water quantity and quality, supplies and finance. These two bodies were formed from parts of the Water Resources Board, which ceased to function on 1 April 1974. On the same date the Water Research Centre was established by combining the Water Research Association, the Water Pollution Research Laboratory, the remaining part of the Water Resources Board and some of the functions of the Department of the Environment's Directorate General of Water Engineering. The aim was to meet the research needs of the whole water industry by bringing together in one organisation the research on water supply, sewage treatment and water resources. Each RWA is a member of the Water Research Centre, together with their Scottish and Northern Irish equivalents, manufacturers of equipment, consulting engineers, universities and researchers. The Water Space Amenity Commission is the body that advises the Secretary of State and the RWAs on recreation and amenity matters relating to water.

In March 1976 the government made proposals, in a consultative document, for further changes in the organisation of the water industry in England and Wales. The main proposal is that there should be a strong National Water Authority, which would replace the National Water Council and incorporate the Central Water Planning Unit and the Water Research Centre. It would have considerable powers including the responsibility for strategic planning, research and the overall financial direction of the industry. Proposals are also made to merge the British Waterways Board with the National Water Authority and to integrate the water companies with the water authorities. The document argues against further radical changes in the water authorities' powers or in the areas for which they are responsible.

Conservation of water resources and the urge to establish exactly what happens to the water in a river basin led McClean to carry out his studies of catchments in Scotland[42,43] and later in the Lake District[44]. These first British studies of catchment water balances did not involve a change in land management; in fact it was not until the 1950s that the hydrological significance of different land uses was investigated[45]. In one of his studies, Law[46] found that the evaporation from a small plantation at Stocks Reservoir was some 330–355 mm (13–14 in) a year greater than that from grass, a result that created some doubt among hydrologists and concern among water engineers operating extensively forested gathering grounds. One consequence of these feelings was the formation, under the guidance of the Department of Scientific and Industrial Research, of several committees on hydrology. These were later merged into the Committee on Hydrological

Research, with the task of looking at current hydrological problems and the research aimed at their solution. The Committee pointed to gaps in the research and particularly to the fact that no organisation had the responsibility for studying the hydrological consequences of land management. This led to the formation of an establishment with this role, namely the Hydrological Research Unit, which was set up in 1962 as part of the Hydraulics Research Station at Wallingford. Subsequently the Unit embarked upon a number of hydrological studies of land-use contrasts and changes in land use, employing both the single and paired catchment approach. Particular attention was given to the effects of afforestation of upland areas, one of these studies being concerned with the adjoining headwater catchments of the rivers Severn and Wye, on the east flank of Plynlimon in Mid-Wales. There, the Wye in sheep pasture provided an excellent opportunity for a hydrological comparison with the forested Severn, which was similar in other aspects. In 1965 the Hydrological Research Unit was placed within the Natural Environment Research Council, and in 1968 it was given the title of the Institute of Hydrology, with a much wider remit than it had previously.

Of course, during recent years, there have been a number of other important developments of both a scientific and an organisational nature. For example, Penman's well-known work on evaporation[47,48] provided a method for estimating that flux, a method that has come to be widely employed by hydrologists. Perhaps one of its most important practical uses is in the assessment of irrigation need, irrigation and also land drainage being the responsibility of the Ministry of Agriculture. Urban drainage has been given considerable attention by Watkins[49] at the Transport and Road Research Laboratory. This work has culminated in a scientific approach to the design of stormwater systems known as the RRL Hydrograph Method that is now widely used. One of the bodies that cooperated in these studies was the Meteorological Office, particularly over the question of intense rainfall. Two important investigations were conducted into intensity–frequency relationship over an area, one at Cardington and the other at Winchcombe, where relief was a complicating factor. In addition, the studies that Bilham[50] carried out into intense rainfall in the 1930s were updated, but little change was made to Bilham's original method[51]. Work of this sort, and that undertaken for the Water Resources Board, has brought about an appreciable expansion of the Meteorological Office's hydrological activities. One study of particular interest is the use of radar to measure the rainfall over the catchment of the River Dee.

The Hydraulics Research Station was one of the bodies cooperating in the Transport and Road Research Laboratory's urban drainage work and has also undertaken research on flood routing and channel morphometry. The design of structures for stream gauging has been given considerable attention at the HRS, where Nash carried out much of his early work on flood prediction in natural catchments[52,53]. These studies contained some of the most original research material incorporating the use of the unit hydrograph. They represented a major advance over previous work, such as that of Richards[54] in the early 1950s and the Institution of Civil Engineers 1933 *Interim report on floods in relation to reservoir practice*[55]. It was recommended that this report should be revised[56] and to undertake this revision a 'Floods Team' was set up at the Institute of Hydrology and a small group at the Meteorological Office. The intensive programme of work was begun in 1970 and completed in 1974[57].

In the groundwater field Ineson[58] successfully applied to British aquifers the analysis of steady and non-steady flow of water to wells. Boulton[59] made significant contributions to well hydraulic theory, particularly in relation to water table conditions, and Childs[60] analysed in detail the problems associated with soil drainage.

In cooperation with a number of river authorities, the Water Resources Board carried out a number of investigations into the feasibility of large-scale development of groundwater in conjunction with surface resources. Attention was also given to the regulation of river flows by groundwater. Major experiments have been carried out in the Thames, Severn and Great Ouse basins and in the Fylde area of north-west England. In addition to continuing the collection and publication of surface water records, the Water Resources Board published, in the *Groundwater Yearbooks,* records of water levels in selected observation wells.

Research, education and related activities

Lack of strong centralised control of hydrology and its common content with so many other sciences are perhaps the factors that have contributed most to make hydrological research so very fragmented. Overall responsibility for hydrological research is vested in the Natural Environment Research Council (NERC), as opposed to the planning, service and operations roles which are shared between the Meteorological Office, the Central Water Planning Unit, the Water Data Unit and the RWAs and their Scottish and Northern Irish equivalents, together with the Ministry of Agriculture. In practice, coordination of hydrological research is extremely difficult because of the large number of organisations involved, to say nothing of the universities. Committees such as the former NERC Hydrology Committee and its working groups, the Hydrology Sub-committee of the Royal Society's Geophysics Committee and several representative meetings of NERC have been and are the formal agents of coordination. Financial support is provided by the NERC to universities for research in hydrology and, of course, to the Institute of Hydrology and to the Institute of Geological Sciences for its Hydrogeology Department. In addition, the Institute of Terrestrial Ecology, the British Antarctic Survey and the Freshwater Biological Association allocate a proportion of their budgets to researches in hydrology, but the other organisations involved in hydrological research are not funded by the NERC. The Water Research Centre is such a body; its funds come partly from the water industry and partly from government. The Science Research Council provides some support for university research in hydrology.

A wide range of research topics is covered[61], particular attention being given to flood prediction and groundwater studies. Subjects of research extend from the applications of radar in hydrological forecasting to the development of hydraulic weighing lysimeters, and from assessment of flood risk for insurance purposes to airborne radio echo-sounding of the Antarctic ice sheet. In the catalogue of UK projects prepared for the International Hydrological Decade, some 500 topics were covered by the government and semi-government organisations, universities and other bodies known to be involved in hydrological research. In addition, there are a considerable number of *ad hoc* investigations being undertaken in the UK by civil engineering consultants, who also conduct similar studies abroad. The Ministry of Overseas Development also carries out projects in developing countries, particularly former British Colonies.

A number of universities and colleges teach some hydrology as part of their undergraduate courses in a variety of subjects, but there is no degree course in hydrology itself. Courses where hydrology features as part of the syllabus include agriculture, botany, civil engineering, environmental science, forestry, geography, geology and meteorology. In most universities individual postgraduate research in hydrology can lead to a higher degree, but there are several where formal one-year courses are available in departments of agricultural engineering, civil engineering, geology and forestry. In addition, a number of research institutes provide training in specific hydrological techniques. There are also professional institutions that award qualifications by examination, but these are mostly concerned with aspects of engineering related to hydrology.

The oldest and foremost international scientific association concerned with hydrology, and the only one dealing with it exclusively, is the International Association of Hydrological Sciences, until 1971 the International Association of Scientific Hydrology. IAHS was established in 1922, as part of the International Union of Geodesy and Geophysics (IUGG). Organisational matters concerning the IAHS in the UK are dealt with by the Royal Society's Hydrology Sub-committee, but there are several other international scientific gatherings that have interests in some part of hydrology. These bodies include the International Association of Hydraulic Research (IAHR), the International Association of Hydrogeologists (IAH), the International Geographical Union (IGU), the International Association of Meteorology and Atmospheric Physics (IAMAP), the International Association of Theoretical and Applied Limnology (IAL), the International Commission on Irrigation and Drainage (ICID) and the International Commission on Large Dams (ICOLD) of the World Power Conferences. Most of these bodies are members of the International Council of Scientific Unions (ICSU), which itself has a Scientific Committee on Water Research (COWAR) concerned with the coordination of its members' activities. The International Biological Programme (IBP) also involves a certain amount of research and cooperation in hydrology, as does the UNESCO 'Man and the Biosphere' programme.

There is no one governmental international agency that has sole responsibility for hydrology. However, the World Meteorological Organisation (WMO) in Geneva, continuing the interests of its predecessor, the International Meteorological Organisation (IMO), has been concerned with hydrology for a longer period than the other agencies. Now, through its Commission for Hydrology and its Advisory Committee on Operational Hydrology, its publications, such as the *Guide to Hydrological Practices,* its support for projects in developing countries and its organisation of symposia and seminars, WMO makes an effective contribution to operational and research aspects of hydrology.

A ten-year programme designed to stimulate the scientific, research and educational facets of international collaboration in hydrology was inaugurated in 1965. The International Hydrological Decade (IHD) had its Secretariat at UNESCO in Paris and pursued the proposals adopted by its Coordinating Council. Each nation had its own IHD Committee to organise its particular share of the global programme, while the other international agencies added their special aptitudes. Thus the UN Food and Agriculture

Table 1.2 UNITS FOR HYDROMETEOROLOGICAL ELEMENTS (BASED ON A TABLE IN THE WMO GUIDE TO HYDROLOGICAL PRACTICES)

Element	Recommended unit	Alternative units	Factor for conversion from alternative unit to recommended unit
Water-level (stage)	mm	ft	305
Stream discharge	m³/s	ft³/s	0.0283
Unit discharge	m³ s⁻¹ km⁻²	ft³ s⁻¹ mile⁻²	0.0103
Volume (storage)	m³	ft³	0.0283
		acre-ft	1230
		ft³/s-day	2450
Well discharge	l/s	gal/h	0.01263
Runoff depth	mm	in	25.4
Precipitation	mm	in	25.4
Precipitation intensity	mm/h	in/h	25.4
Snow depth	mm	in	25.4
Snow cover, area	%		
Water equivalent of snowpack	mm	in	25.4
Ice thickness	mm	in	25.4
Evaporation	mm	in	25.4
Evapotranspiration	mm	in	25.4
Soil moisture	%, volume	%, weight (conversion depends on density)	
Soil-moisture deficiency	mm	in	25.4
Sediment discharge	tonne/day	ton/day	0.907
Sediment concentration	kg/m³	ppm (conversion depends on density)	
Chemical quality	mg/l	ppm	
Energy (heat)	cal (gram)	Btu	252
Radiation	cal/cm²	ly	
Radiation intensity	cal cm⁻² min⁻¹	ly/min	
Sunshine	% possible	h (conversion depends on possible sunshine)	
Temperature	°C	°F	5/9 (°F — 32)
Wind speed	knot	mile/h	0.868
		m/s	1.95
Relative humidity	%		
Vapour pressure	mb	mm Hg	1.333
		in Hg	33.86
Atmospheric pressure	mb	mm Hg	1.333
		in Hg	33.86
Area	km²	mile²	2.59
		acre	0.00405
		hectare	0.01

Note. Abbreviations used in the table are as follows:

Btu	British thermal unit	Hg	mercury	mb	millibar
°C	degrees Celsius	h	hour	mg	milligram
cal	calorie	in	inch	min	minute
cm	centimetre	kg	kilogram	mm	millimetre
°F	degrees Fahrenheit	km	kilometre	ppm	parts per million
ft	foot	ly	langley		by weight
l	litre	m	metre	s	second

Organisation (FAO), which utilises hydrological skills in many of its development projects, was especially interested in man's influence on the hydrological cycle. The International Atomic Energy Agency (IAEA) also collaborated in the IHD, as did the World Health Organisation (WHO) and of course WMO. In 1975 UNESCO and WMO launched the International Hydrological Programme (IHP) to continue the stimulus to worldwide cooperation in hydrology that was given by the IHD.

In addition to IHP activities, there is cooperation and collaboration between the UN specialised agencies in a variety of development projects sponsored by the UN. Each continent has its own UN Economic Commission, and these concern themselves with an immense range of topics including affairs concerning water and water resources.

Units, definitions and standards

The units, definitions and standards employed in a number of branches of science are still in a state of flux, and hydrology is no exception. Hydrology in the UK faces the added problem that many of the measurements, factors, constants and other 'tools' in daily use are in the Imperial system of units, when international scientific practice is becoming more and more committed to the SI system (Système International d'Unités) rather than the Imperial system. Although the aim in the UK is to employ metric units generally in the near future, it is likely that for some years to come British hydrologists will continue to think and work to a certain extent in Imperial units.

The reader will find that this text contains a mixture of units, and to help convert one system to the other *Table 1.2* has been included. This table, which is based on one prepared by WMO, lists the common hydrological elements and their conversion factors. In fact WMO has made a number of contributions to standardisation[62] and in collaboration with UNESCO has prepared a multilingual glossary of hydrological terms, their definitions and equivalents in Spanish, Russian and French. A more comprehensive list of conversion factors is given in Appendix 1.

REFERENCES

1. Bruce, J. P. and Clark, R. H., *Introduction to hydrometeorology*, Pergamon Press, 319 (1966)
2. Batisse, M., 'The International Hydrological Decade—a world-wide programme of scientific research', in *Water and Life, UNESCO Courier*, 4–9 (July-Aug 1964)
3. Horton, R. E., 'The field, scope, and status of the science of hydrology', *Trans. Am. Geophys. Union*, **12**, 189–202 (1931)
4. Valentine, H. R., *Water in the service of man*, Pelican Books, 223 (1967)
5. Mackinder, H. J., *Britain and the British seas*, Heinemann (1902)
6. Manley, G., *Climate and the British scene*, Collins, London, 314 (1952)
7. Institution of Water Engineers, *Manual of British water supply practice*, 2nd edn, Cambridge, 983 (1958)
8. Department of the Environment, *Water data 1974*, Water Data Unit, Reading, 55 (1975)
9. Gregory, S., 'Contribution of the uplands to the public water supply of England and Wales', *Trans. Inst. Brit. Geographers*, **25**, 153–165 (1958)
10. Sharp, R. G., 'Estimation of future demands on water resources in Britain', *J. Instn Water Engrs*, **21** No. 3, 232–249 (1967)

11. Grindley, J., 'Rainfall over Great Britain and Northern Ireland during 1965', *Water and Water Engng,* **70,** 158–161 (1966)
12. Allen, R. G., 'Water supplies and conservation', *J. R. Soc. Arts,* 384–402 (April 1963)
13. Frontinus, J., *The water supply of the city of Rome,* transl. C. Herchel, Longman & Co. (1913)
14. Geraldus Cambrensis, *Topographia Hiberniae,* Rolls Series No. 21, Translation 1.6 (1861)
15. Halley, E., 'An estimate of the quantity of vapours raised out of the sea by the warmth of the sun', *Phil. Trans. R. Soc.,* **15–16,** 366–370 (1685–87)
16. Dooge, J. C. I., 'The development of hydrological concepts in Britain and Ireland between 1674 and 1874', *Hydrol. Sci. Bull.,* **19,** 279–302 (1974)
17. Birch, Thomas, *The history of the Royal Society of London,* 1, 74 (1757)
18. Biswas, A. K., 'The automatic raingauge of Sir Christopher Wren', *Notes and Records of the Royal Society,* **22** No. 1 (1967)
19. Gunther, R. T., *Early science in Oxford,* Oxford (1930)
20. Hooke, R., 'An account of the quantities of rain fallen in one year in Gresham College, London, begun August 12, 1695', *Phil. Trans. R. Soc.,* **19** No. 223, 357 (1695)
21. Townley, R., 'A letter from Richard Townley, of Townley in Lancashire, containing observations on the quantity of rain falling monthly for several years successively', *Phil. Trans. R. Soc.,* **18** No. 208, 52 (1964)
22. Halley, E., 'An account of the evaporation of water', *Phil. Trans. R. Soc.,* **18** No. 212, 183–190 (1694)
23. Plot, R., *De origine fontium tentamen philosophicum,* Oxonii (1685)
24. Keill, J., *An examination of the reflections on the theory of the earth,* Oxford (1699)
25. Keill, J., *An examination of Dr Burnet's theory of the earth, together with some remarks on Mr Whiston's new theory of the earth,* London (1734)
26. Dobson, D., 'Observations on the annual evaporation at Liverpool in Lancashire; and on evaporation considered as a test of moisture or dryness of the atmosphere', *Phil. Trans. R. Soc.,* **67,** 244–259 (1777)
27. Rodda, J. C., 'Eighteenth-century evaporation experiments', *Weather,* **18** No. 9, 264–269 (1963)
28. Biswas, A. K., 'Experiments on atmospheric evaporation until the end of the eighteenth century', *Technology and Culture,* **10** No. 1, 49–58 (1969)
29. Heberden, W., 'Of the quantities of rain which appear to fall at different heights over the same spot of ground', *Phil. Trans. R. Soc.,* **59,** 359 (1769)
30. Dalton, J., 'Experiments and observations to determine whether the quantity of rain and dew is equal to the quantity of water carried off by the rivers and raised by evaporation; with an enquiry into the origin of springs', *Mem. Lit. and Phil. Soc. Manchester,* **5** Pt II, 346–372 (1802)
31. Dalton, J., 'Experimental essays on the constitution of mixed gases; on the force of steam or vapour from water and other liquids in different temperatures, both in a Torricellian vacuum and in air, on evaporation; and on the expansion of gases by heat', *Mem. Lit. and Phil. Soc. Manchester,* **5** Pt II, 536–602 (1802)
32. Bailey, Sir Edward, *Geological Survey of Great Britain,* T. Murby & Co., London (1952)
33. Meteorological Office, *Annual Report 1971* (1972)
34. Stevenson, T., 'On the defects of raingauges with description of an improved form', *Edinburgh New Phil. J.,* **33,** 12–21 (1842)
35. *British Rainfall 1865,* **6,** 14 (1866)
36. Jevons, W. S., 'On the deficiency of rain in an elevated raingauge as caused by wind', *London, Edinburgh and Dublin Phil. Mag.,* **22,** 421–433 (1861)
37. Miller, J. F., 'On the meteorology of the Lake District of Cumberland and Westmorland; including the results of experiments on the fall of rain at various heights, up to 3166 ft above the sea level', *Phil. Trans. R. Soc.,* **139,** 318–329 (1849)
38. Reynolds, O., 'The motion of water and the law of resistance in parallel channels', *Phil. Trans. R. Soc.,* **174,** 935–982 (1883)
39. Manning, R., 'On the flow of water in open channels and pipes', *Trans. Instn Civil Engrs of Ireland,* **20,** 161–207 (1889)
40. *A Bill to make provision for a national policy for water* (23 Jan 1973)
41. *A background to water reorganisation in England and Wales,* HMSO, 36 (1973)
42. McClean, W. N., 'Rainfall and flow-off in the River Garry, Inverness-shire', *Trans. Instn Water Engrs,* **32,** 110–146 (1927)

43. McClean, W. N., 'River flow records: River Dee, Aberdeenshire', *Q. J. R. Met. Soc.*, **61**, 433–435 (1935)
44. McClean, W. N., 'Windermere Basin, rainfall, runoff and storage', *Q. J. R. Met. Soc.*, **66**, 337–362 (1940)
45. Conway, V. M. and Millar, A., 'The hydrology of some small peat-covered catchments in the northern Pennines', *J. Instn Water Engrs*, **14**, 415–424 (1960)
46. Law, F., 'The effect of afforestation upon the yield of water by catchment areas', *J. Instn Water Engrs*, **11**, 269–276 (1957)
47. Penman, H. L., 'Natural evaporation from open water, bare soil and grass', *Proc. R. Soc. Arts*, **193**, 120–145 (1948)
48. Penman, H. L., 'Evaporation over the British Isles', *Q. J. R. Met. Soc.*, **76**, 372–383 (1950)
49. Watkins, L. H., 'Research on surface water drainage', *Proc. Instn Civil Engrs*, **24**, 305–330 (1963)
50. Bilham, E. G., 'Classification of heavy falls in short periods', *British Rainfall 1935*, **75**, 262–280 (1936)
51. Holland, D. J., 'Rain intensity frequency relations in Britain', *Met. Office Hydrological Memo No. 33* (1964)
52. Nash, J. E., 'Determining runoff from rainfall', *Proc. Instn Civil Engrs*, **10**, 163–184 (1958)
53. Nash, J. E., 'A unit hydrograph study with particular reference to British catchments', *Proc. Instn Civil Engrs*, **17**, 249–282 (1960)
54. Richards, B. D., *Flood estimation and control*, 2nd edn, Chapman and Hall, London (1950)
55. Institution of Civil Engineers, *Interim Report of the Committee on Floods in Relation to Reservoir Practice* (1933)
56. Institution of Civil Engineers, *Flood Studies for the United Kingdom*, Report of the Committee on Floods in the United Kingdom (1967)
57. *Natural Environment Research Council, Flood Studies Report* (1975)
58. Ineson, J., 'Some observations on pumping tests carried out on Chalk wells', *J. Instn Water Engrs*, **7**, 215–225 (1953)
59. Boulton, N. S., 'The drawdown of the water table under non-steady conditions near a pumped well in an unconfined formation', *Proc. Instn Civil Engrs*, **3**, 564–579 (1954)
60. Childs, E. C., 'The water table, equipotentials, and streamlines in drained land', *Soil Sci.*, **59**, 313–327 (1945)
61. Natural Environment Research Council, *Hydrological research in the United Kingdom 1965–1970* (1970) and *1970–1975* (1976)
62. World Meteorological Organisation, *Standards in hydrology and related fields*, WMO/IHD Report No. 18 (1973)

Chapter 2

Precipitation

Instruments and instrument techniques

Precipitation is the term employed for all forms of atmospheric moisture deposited on the ground. It includes drizzle, rain, sleet, snow and hail, together with dew, hoar frost, rime and similar phenomena, of which fogdrip from trees and glazed frost sometimes demand attention. Rain is the most common form and it is the only one which is gauged extensively and with any degree of certainty. In favourable circumstances estimates of the other forms can be attempted. From time to time snow can be as important as rain, but the other components are usually effective only temporarily and locally. All precipitation is measured as if it were liquid, the measurement being expressed as the depth to which the ground would be covered if the water remained where it fell and there was no loss.

The formation of precipitation requires uplift in order to produce cooling of the body of air and prime the other necessary mechanisms, condensation and growth of cloud droplets. There are three main ways in which uplift can occur and these are employed to classify precipitation into convective, orographic and cyclonic types, the last being associated with depressions and frontal systems. The occurrence of precipitation has been studied according to its origins[1,2], but more important from the hydrological point of view are the spatial and temporal characteristics of each type. Thus precipitation of a convective origin usually occurs at a rate greater than 4 mm per hour, as showers or heavy downpours that are generally brief and localised. On the other hand, cyclonic precipitation is usually more widespread and moderate (2.5 mm/h), characteristically lasting between six and twelve hours. These features vary with season, the type of weather system and its movement, and one detailed study of the Midlands[3] has shown that each rainfall of a different origin has a characteristic mean intensity *(Figure 2.1)*. Pure orographic precipitation is unusual in the UK, the effect of orography being to reinforce the convective or cyclonic situations so that the precipitation is more frequent, extensive and lengthy. Other characteristics of precipitation change according to its origin, particularly drop size *(Table 2.1)*. There is no single drop size, however, that is unique to a particular mechanism, each fall being comprised of drops over a range of different sizes. Nevertheless, several workers[4,5] have established empirically relations between pairs of the following factors: median-volume drop diameter (D_n), rainfall intensity

Precipitation

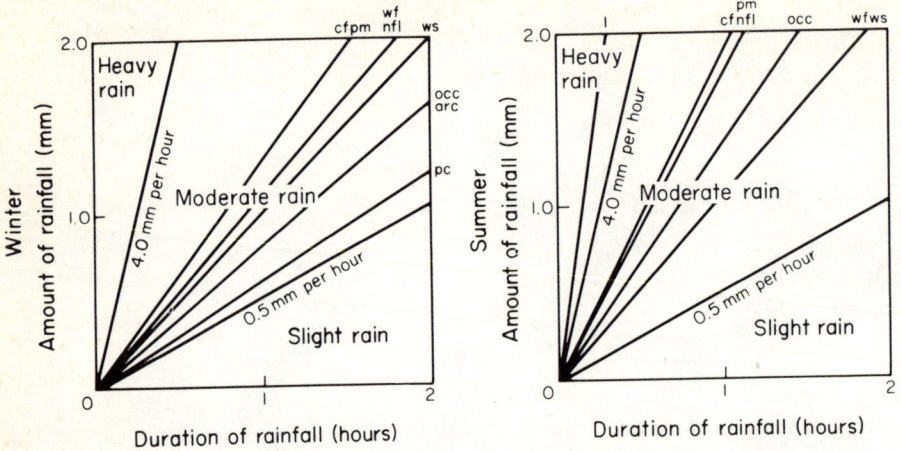

Figure 2.1   Intensity-origin relationships for rainfall at stations in the Midlands (from Matthews[3], courtesy Weather)

Table 2.1   DROP CHARACTERISTICS

Form	Radius (μm)	Terminal speed (mm/s)	Terminal speed (m/s)	Remarks
Cloud	1	0.1		Aerosols
	5	2.5		Sea fog
	10	10		
	50	250		Mist
Drizzle	100	700		Mist
	250		2.0	
Rain	500		3.9	
	1000		6.5	
	1500		8.1	Typical UK drop size
	2000		8.8	
	2500		9.1	Thunderstorm

(*I*), volume of rainwater per unit volume of air *(W)*, and the sum of the sixth powers of the drop diameters $(\Sigma n D^6)$. One of the most commonly used of these formulae, and the one that is important in the radar measurement of rainfall, has the form:

$$\Sigma n D^6 = aI^b \tag{2.1}$$

where *a* and *b* are constants.

Precipitation characteristics other than those of a purely physical nature can be important. Certain chemical constituents may be significant, for example, in promoting weathering, but greater attention is usually given to the concentration of radioactive isotopes, particularly those produced by the detonation of thermo-nuclear devices. At present a large proportion of the atmospheric tritium (^{3}H), the radioactive isotope of hydrogen, has this origin. Tritium has been injected into the atmosphere in a series of pulses that commenced in 1954, reaching a maximum in 1963. The naturally occurring tritium, which has a concentration of about 3 to 10 tritium units has been swamped by this artificial production, but precipitation is now labelled with sufficient tritium to make its measurement relatively easy. Carbon-14 (^{14}C) is another product of nuclear tests that has supplemented the naturally occurring atmospheric isotope and has a considerable potential for dating groundwater.

The measurement of precipitation

Non-recording raingauges

The vast majority of non-recording raingauges in the UK network have collecting funnels with an aperture 127 mm (5 in) in diameter and stand with

(a)

(b)

(c)

Figure 2.2   Raingauges: (a) Met. Office Mk2 (standard gauge) within turf wall; (b) Met. Office Mk3; (c) Met. Office Mk4

(d)　　　　　　　　　　　　　　　　　　　(e)

(f)

(g)

Figure 2.2   (d) Octapent; (e) modified Dines tilting-siphon rain recorder; (f) magnetic-tape rain recorder; (g) ground-level gauge

their rims 305 mm (12 in) above ground level. The standard gauge (Meteorological Office Mk2), made of brass and copper, has an aperture formed to within ±0.4 per cent of the required dimension. The deep-set collecting funnel conducts the rain water to a glass bottle of 100 mm (4 in) capacity, housed in an overflow can that can accommodate a further 75 mm (3 in) of rain *(Figure 2.2a)*. Measurements of the collected rain are made daily with a glass measuring cylinder at 0900 GMT at most sites. Less frequent measurements are made at isolated sites, where storage gauges of the Bradford, Seathwaite or Octapent pattern are employed, the last two types having a capacity of 685 mm (27 in) *(Figure 2.2d)*. Some gauges with an aperture of diameter 203 mm (8 in) are in use, while since 1965 the Meteorological Office has been testing a new interchangeable system of gauges with collecting funnels 150 and 750 cm² in area, as opposed to the 127 cm² funnels of the standard gauge[6] *(Figure 2.2b* and *c)*. These funnels are also used with the new rain recorder, all the large components being made of glass-fibre. Several different shapes of funnel are employed, although their rim height remains the same for the current standard installation.

Rain recorders

Most of the 1000 or so rain recorders[7] in the UK *(Figure 2.3)* register the cumulative amount of rain, although there are some, such as the Jardi, that record the instantaneous rate. There are three main kinds of rain recorder, the weighing, the tipping bucket and the siphon types; the last of these is the most common, particularly the Meteorological Office float chamber tilting-siphon instrument developed by Dines[8] *(Figure 2.2e)*. Strip-chart mechanisms have been developed at the Transport and Road Research Laboratory and by the Meteorological Office. These may be substituted for the normal drum chart. Frost protection may be afforded by insulating the outer casing and warming the collecting chamber, using a short length of battery-powered heating tape[9].

Few gauges that record the rain by weighing are employed in the UK, and until the introduction of the Meteorological Office glass-fibre gauges there were not many that operated on the tipping-bucket principle. Now the 150 cm² and 750 cm² collecting funnels may be fitted to base members including a tipping-bucket unit, each bucket having a capacity of 15 grams[10]. Every tip of a bucket corresponds to a 1.0 mm increment of rain when the 150 cm² funnel is used, and 0.2 mm for the larger funnel. The movement of the buckets is employed to close a contact momentarily in a circuit, which may include a simple impulse counter or alternatively an impulse recorder. The first gives the total fall over a period, while the second provides duration and a step-wise measure of rate as well. Tipping-bucket gauges have also been developed that record on magnetic tape. A simple twin-track battery-operated recorder logs the number of tips on one track and time on the other. The recorder is housed within the gauge in some models, or contained in a separate watertight box in others *(Figure 2.2f)*. A tipping-bucket gauge is employed as one sensor in most types of automatic weather station and gauges of a similar kind are used to telemeter

Figure 2.3   Location of autographic raingauges, excluding dense network operated by Greater London Council and London boroughs (from Lowing and Newson[7], courtesy Water Services)

information on rainfall to a distant base[11]. Transmission of the signal can be by telephone line or UHF radio, the system being arranged to provide an answer when the gauge is interrogated or to activate an alarm if a preselected value is exceeded. Alternatively, the signal from the gauge can be recorded continuously.

Snow and ice

Measurements of solid precipitation can be obtained by attempting to catch the falling snow or ice, or by determining the depth of lying snow, its density and water equivalent. If the fall of snow has been small and only light winds have occurred, the snow caught in the raingauge funnel can be melted and its liquid equivalents determined. Large amounts of snow bury the gauge and strong winds cause drifting and considerable redistribution brings about increased spatial variability. Under these conditions, measurements of depth of freshly fallen snow and its conversion to a water equivalent by density measurements[12], or use of approximate relationships (e.g. 10 mm fresh snow = 1 mm water), are usually resorted to. Identifying the depth where each new fall of snow starts is facilitated by the use of snow boards made of a light, coloured material, the boards being placed on top of the snow after each reading. The standard raingauge funnel is often used to obtain snow core samples, by inverting and pressing it into the snow until the ground is reached. The sample is melted indoors to obtain its water equivalent, but great care has to be taken to obtain representative sampling sites and when the snow is greater than approximately 150 mm in depth, more than one core has to be obtained. The use of a network of snow poles or stakes is an aid to securing more representative measurements of depth. The poles, which usually carry 50 mm graduations, are set up in a predetermined pattern. The form of network is favoured that permits the largest number of poles to be read from the smallest number of points by means of a telescope. Snow poles can be associated with snow courses. These are permanently established lines where measurements of depth and water content are made at fixed intervals of space and time. Such a course might consist of ten or more sampling points, spaced 10–100 m apart, where depth measurements and core samples are obtained. In countries where snow is a normal feature of the winter scene, more attention has been given to snow than in the UK. Snow pits are employed for density studies, radio isotope snow gauges have been developed, and instruments have been devised that indicate the depth of snow by registering the attenuation of the natural radioactivity of the soil due to the snow. The use of snow pillows and remote sensing techniques from aircraft and satellites has grown during recent years and should add much to knowledge of snow and its distribution.

Radar measurement of rainfall

Radars designed specifically to detect rainfall are being used at a number of sites. Most operate by transmitting a continuous beam of pulses of electromagnetic radiation, which scans around a few degrees above the horizon. The pulses are emitted in very rapid succession and, for the most commonly used radars, are intercepted by particles in the atmosphere larger than 200 μm in diameter. The particles reflect the beam and are indicative that precipitation is taking place. Echoes received back at the radar between pulses are converted into a visual display of the distribution of precipitation on a plan position indicator (PPI).

The transmitted power, the reflected power, characteristics of the radar and other factors are related in the 'radar equation'[13], given here in simplified form as:

$$P_r = \frac{P_t \pi^6 A_e h |K|^2}{256 \log_e 2r^2 \lambda^4} \Sigma n D^6 \qquad (2.2)$$

where P_r = average power (watts) received from a series of echoes
        P_t = peak transmitted power (watts)
        A_e = effective antenna area (m²)
        h = pulse length (m)
        r^2 = distance from radar to target (m) (as large as 200 km)
        λ = wavelength of radiation (m) (usually 0.03–0.1 m)
        n = number of drops
        D = drop diameter (mm)
        K = $(\gamma-1)/(\gamma-2)$, where γ is the dielectric constant

   Hence the intensity of the echo is directly proportional to $\Sigma n D^6$, but as the number and size of droplets varies from storm to storm it is difficult to provide good quantitative estimates of rainfall from radar measurements alone. To overcome this, drop size distributions have been measured at ground level and related to rainfall intensity, the most common form found for this relationship being:

$$\Sigma n D^6 = 200\, I^{1.6} \qquad (2.3)$$

where I = rainfall intensity (mm/h). In addition, comparisons have been made between radar estimates of rainfalls and those obtained from dense raingauge networks. The results from both approaches have been less successful than might be expected. The constants in equation 2.3 have been observed to vary from 80 to 660 and from 1.2 to 1.9, and it has been concluded by one worker that the mean error arising from drop-size variations in assessing continuous point rainfall would be about 25 per cent[14]. On the other hand, errors ranging from 25 to over 60 per cent have been quoted elsewhere[15], in contrast to errors of from 15 to 50 per cent when the radar has been calibrated against a single gauge or small cluster of gauges. This is the technique being used in the Dee Weather Radar Project *(Figure 2.4)*, where the radar near Llandegla scans over a network of seventy tipping-bucket magnetic tape recording gauges but is calibrated against a gauge site near Bala[16]. The mean percentage difference between radar-measured and network-measured three-hourly rainfalls over sub-catchments has been found to vary between −15% close to Bala, and +20% at distances over 20 km[17]. These differences decrease as the time of integration increases and if more than one calibration site is employed.
   Radars have an enormous potential for use in hydrology, for they offer the possibility of providing rainfall measurements for large areas from a single point[18,19]. Yet there are a number of problems that have to be overcome before their full potential can be realised. Refraction of the beam and its attenuation, the range of the radar and its use in areas of diverse topography, snow and the radar equation itself, all present difficulties. Then there are the improvements in the display and analysis of the data that are required in order to make the radar an integral part of an automated data-acquisition

system. Nevertheless, a complete coverage of the UK by radars of this type would provide a valuable addition to existing flood warning and other surveillance systems. It would also provide information about rainfall over the sea that is unobtainable by other means.

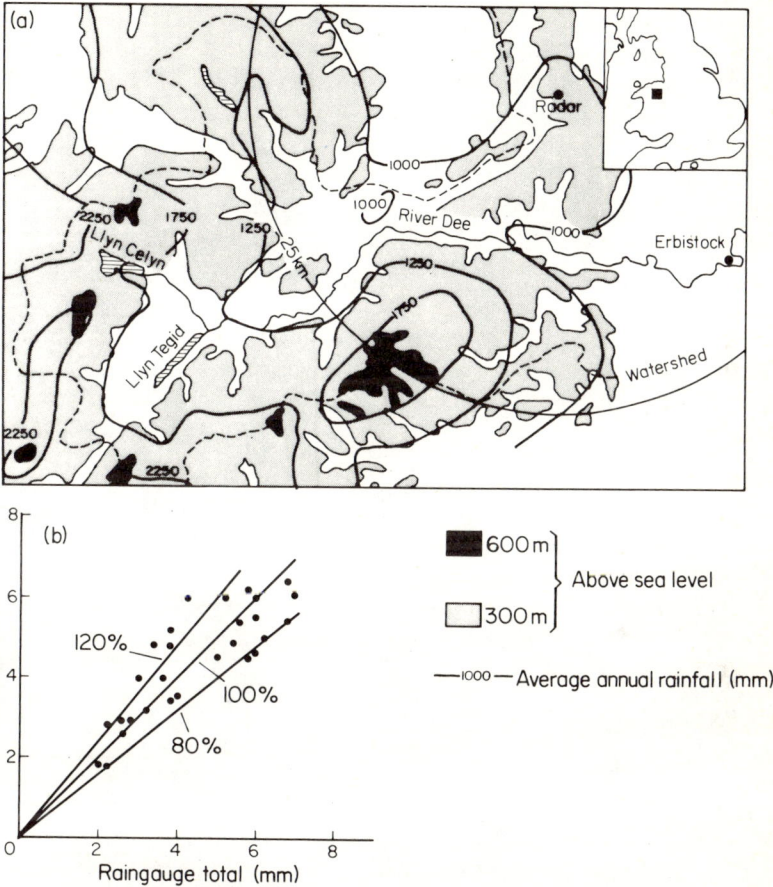

*Figure 2.4   Dee weather radar: (a) the area of the experiment; (b) comparison between gauge-
and radar-measured rainfalls (courtesy Harrold, Bussell and Grinsted[15])*

Errors of measurement

The measurements produced by a well-exposed and well-maintained standard raingauge are, for many purposes, sufficiently near the true rainfall. It is as well to recognise, however, that the catch of a standard gauge is not the true rainfall, namely the amount of rain which would have reached the ground if the gauge had not been there. For there is no method of measuring the true rainfall, no absolute standard of rainfall measurement having yet been devised. Hence all rainfall measurements are relative[20], a fact not generally recognised. Nevertheless, studies have been made that demonstrate what are the main sources of error *(Figure 2.5)*, errors that for the most part cause a gauge to under-register in a systematic fashion. As a consequence, it is possible to design a gauge and devise a method of installing it that is most free from the sources of these errors. The design of the standard British gauge largely eliminates the possibility of evaporation and splash out, two quite important faults, while observer practices reduce some of the others. Wind, however, is the factor having the most serious consequences and these are not avoided by the usual method of installation,

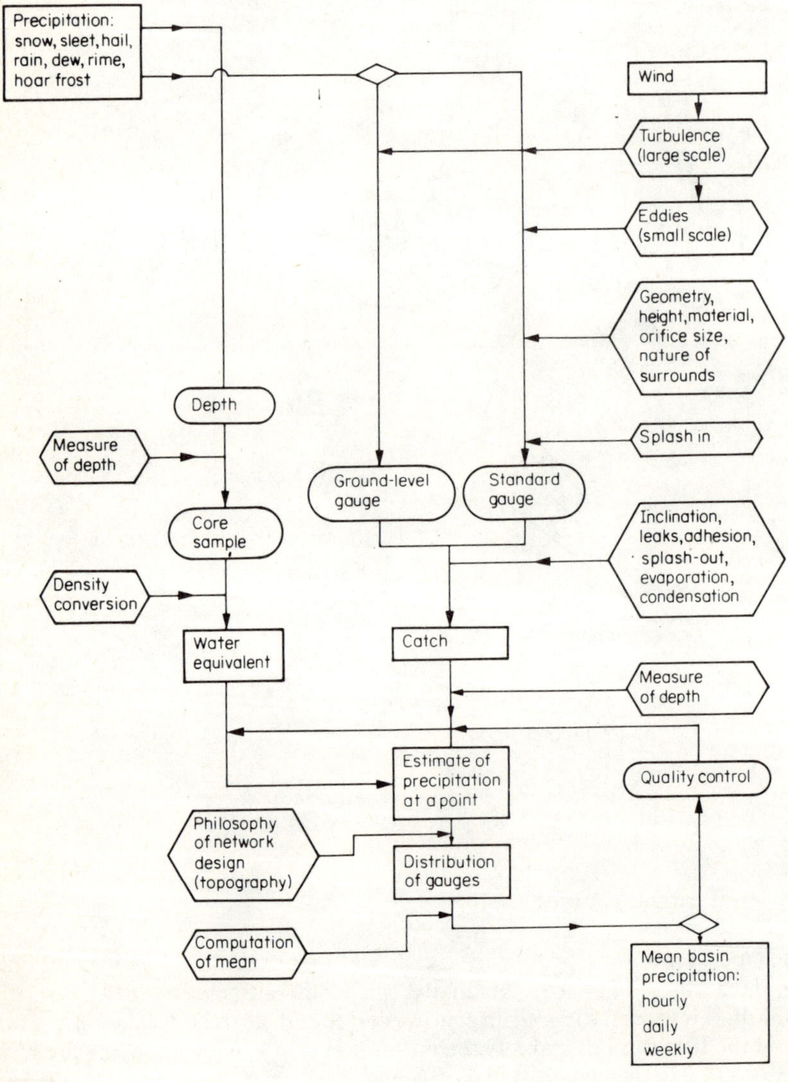

Figure 2.5   Sources of error in determining precipitation (from Edwards and Rodda[113], courtesy Hydrological Sciences Bulletin)

particularly at windy sites. Wind causes raindrops to fall obliquely, while turbulence and eddies produced by the gauge and the features of its site have a further effect on the drops, especially in the region immediately above the instrument[21,22]. Drop trajectories are distorted even more, especially those of the smaller drops, which can be carried completely over the gauge[23]. Loss of catch is greatest in storms with small drops and high wind-speeds, while tall gauges are more susceptible to loss from wind action than short ones. In this respect, it is fortunate that the British gauge is only 305 mm (1 ft) high, most other national gauges being much taller. On the other hand, there is evidence to show that the British standard gauge is likely to suffer from splash in, particularly when it is surrounded by a wet surface. Under such conditions, splash from large diameter drops could reach a height of 600 mm[4] and even 1.2 m in exceptional cases[24]. A poorly maintained lawn appears to produce least splash[25].

Various attempts have been made to overcome the effects of wind, such as by fitting the gauge with a shield[26,27] or by surrounding it with a turf wall *(Figure 2.2a)*[28]. The most satisfactory solution, however, is to instal the gauge in a shallow pit so that its rim is level with the surrounding ground surface. The space between the gauge and the sides of the pit is spanned by some material that inhibits splash and minimises the effect of the pit on the air flow about the gauge. The most widely used anti-splash surface is a grid made of metal or plastic *(Figure 2.2g)*, but wire mesh, angled louvres[29] and door-mats[30] have also been employed. Among a number of tests he conducted on rain gauges at a wind-swept site, Green[31,32] investigated the size of the surround needed to prevent splash into a ground-level gauge. Gravel has been used to surround the ground level gravimetric rain gauge at Kew Observatory[33], the gauge itself filled with gravel on the basis that splash in will equal splash out.

Comparisons of ground-level and standard gauges have been made at a number of sites in the UK. Invariably the gauge at ground level has caught most, the difference being greatest for gauges in the west and least for those in the south-east. This might be expected in view of higher wind speeds in the mountainous west, but the annual catch deficit of over 20 per cent for some of those areas is surprisingly large. An indication of the likely distribution of this deficit over the whole country can be obtained from the map of the annual driving rain index *(Figure 2.6)*, an index combining the effects of wind and rain[34]. This shows a value proportional to the total amount of rain that would be driven on to a vertical surface always facing the wind. Of course, for monthly totals the catch deficit is larger during the winter than the summer, while for certain individual storms the systematic error is two or three times larger than that for annual totals.

Network design

Most raingauge networks have grown in a haphazard fashion: few have been planned scientifically. Often the only design criteria were the availability of observers and the presence of well exposed sites. There are, however, a number of scientific methods that can be used, but there are few grounds for establishing that one particular method should be given preference.

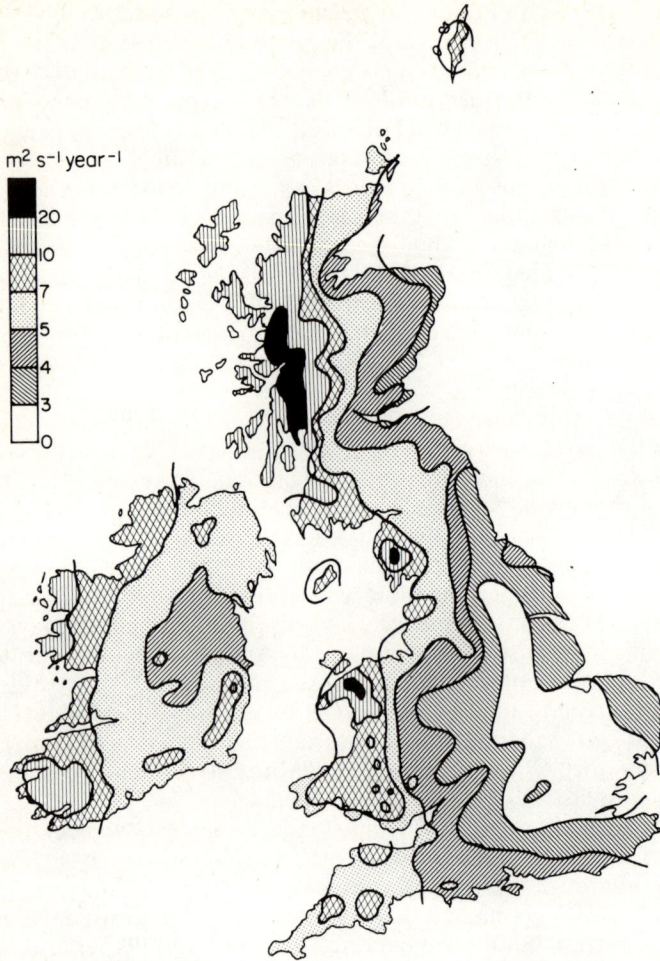

Figure 2.6   Guide to the distribution of losses in standard raingauges, based on a driving rain index (from Lacy and Shellard[34], reproduced by permission of the Controller, Her Majesty's Stationery Office)

There are three basic questions involved in network design[35]:
1. How many sites need to be sampled?
2. Where are these sites to be located?
3. How long is the network to operate?
It is usual to find that the third question is avoided, because once a network is established for a long time its continuance is usually assured. The second is often answered in terms of the ideal site for a gauge. This is supposed to be similar to an open suburban garden, where there are two wind breaks, one afforded by fences and shrubs some 5 m away and the other by houses at a greater distance[36]. In practice, such an exposure is quite difficult to find and of course it could be argued that the selection of ideal sites introduces a bias into the design of the network. Sometimes it is even hard to comply with the

'rule' that a gauge should be surrounded by short grass and sited at a distance at least twice the height of an obstacle away from that obstacle. None of these standards for exposure is particularly scientific. What is needed is some measure of site characteristics that would express, in simple terms, its physical features combined with those resulting from the presence of buildings, trees and like objects. Physical features such as the altitude, slope and orientation of sites have been used for the design of networks in small catchment experiments, but not for general network design.

There are a number of scientific methods of network design, but these really apply to individual basins rather than to a whole country. For example, raingauges or any other type of instrument could be sited at random over an area; this method has the advantage that it provides statistically valid estimates of the mean rainfall. Nevertheless, there are several objections to it, one being the practical difficulties it entails. These objections also apply to a variant of the method—the use of gauges moved at random within a specified area[37]. An alternative is to adopt the systematic approach, so that instruments would be installed at predetermined intervals over an area, intervals that are fixed in both horizontal and vertical planes. An approach of this type was employed by the Devon River Authority[38] for upgrading the network in the Authority's area; a density of one gauge per 23 km² being adopted as desirable, actual gauge locations being determined by the availability of observers. This method may not be as valid statistically as when gauges are sited at random, but it provides more realistic estimates of the mean basin rainfall. Stratified random sampling is an approach possessing some of the advantages of both the other techniques, but one difficulty is to derive a satisfactory means for delimiting strata. A simple method would be to divide a basin into a number of areas of equal size and to select a gauge site within each area by superimposing a grid over it, choosing an intersection as the gauge site from a throw of dice or a table of random numbers.

Several other methods of network design are of considerable interest; for example, the operation of a two-level network incorporating primary and secondary stations. Primary stations are those that will be operated continuously for a long time ahead; secondary stations are only maintained for long enough to establish a relationship with a primary station before the gauge is moved to another site and the process repeated. Another and separate approach, which offers a useful way of determining the number and spacing of gauges in a network, is through the use of correlation analysis[39]. Correlation coefficients are determined between every pair of stations in a network, using the observations made at the sites, so that the decay in correlation with increasing gauge separation can be determined. Records from the Cardington network[40] were employed in this type of analysis *(Figure 2.7)* and the results could have been used to indicate whether or not improvement of the network was necessary. The correlation between one year of observations in a network and those for the next year[41] has also been suggested for use in network design, through the separation of the random and persistent patterns of rainfall variation.

In another approach, Stephenson[42] employed the idea that for monthly totals recorded in a network of n gauges, uniformly distributed, the coefficient of variation of the mean (CV_m) for each month could be employed in determining the adequacy of the network. Values of CV_m are

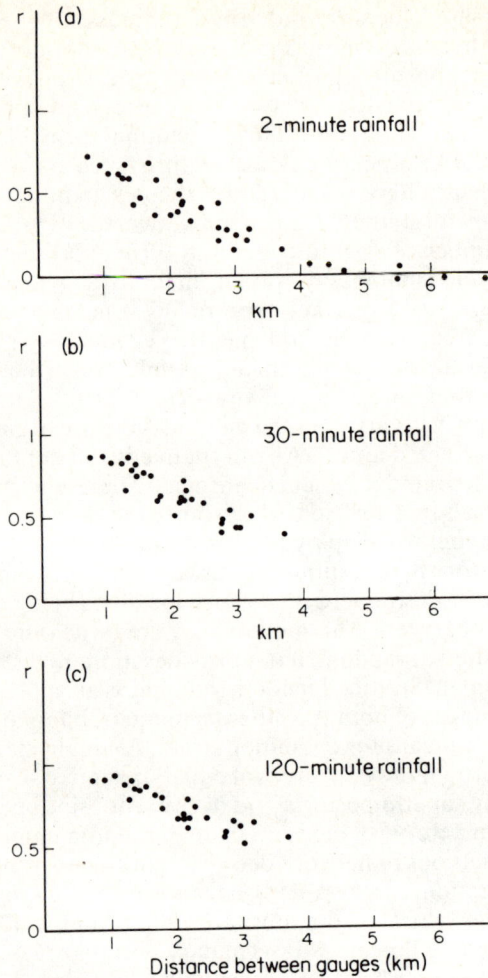

Figure 2.7 Correlations between rainfalls at Cardington, Beds., at various spacings and intervals of time (from Holland[40], reproduced by permission of the Controller, Her Majesty's Stationery Office)

calculated for the network using the monthly amounts expressed as a percentage of average annual rainfall. This calculation was performed for 120 months and the values used to construct a cumulative frequency curve from which the value (C') of CV_m exceeded on 5 per cent of occasions was determined. If C' is less than 10, the number of gauges is considered adequate. When C' is more than 10 then the number of gauges can be calculated from:

$$N = (C'/10)^2 n \qquad\qquad (2.4)$$

Other methods of network design have been put forward; for example, use of the reduced standard error of the mean[43] and analysis of variance[44].

In terms of number of instruments and length of records, the British raingauge network is one of the world's best[45]. Yet, like most other national networks, its present form is a result of intermittent and haphazard growth. The availability of an observer rather than a requirement for knowledge of the rainfall at a particular site has been the prime reason for its growth. Networks in reservoired catchments are an exception, as many of these have been planned and fulfil the recommendations[46] for minimum numbers of raingauges *(Table 2.2)*. There have been similar recommendations made by the Meteorological Office for minimum densities in the countrywide network *(Table 2.3)*, but even so the distribution of instruments is very uneven. Most are situated where the rainfall is least and vice versa, access to sites being an important factor.

Table 2.2   RECOMMENDED MINIMUM DENSITIES FOR RAINGAUGE NETWORKS IN RESERVOIRED BASINS (AFTER BLEASDALE[46])

Area		Raingauge type		
(mile²)	(km²)	*Daily*	*Monthly*	*Total*
0.8	2	1	2	3
1.6	4	2	4	6
7.8	20	3	7	10
15.7	41	4	11	15
31.3	81	5	15	20
46.9	122	6	19	25
62.5	162	8	22	30

Table 2.3   RECOMMENDED MINIMUM DENSITIES FOR COUNTRYWIDE RAINGAUGE NETWORKS GIVING MONTHLY PERCENTAGE OF AVERAGE ESTIMATES (AFTER BLEASDALE[46])

Area		Number of
(mile²)	(km²)	*raingauges*
10	26	2
100	260	6
500	1300	12
1000	2600	15
2000	5200	20
3000	7800	24

Averages and departures from average

It has been claimed that the end product of virtually all important rainfall work lies in the analysis and presentation of averages, variability and extremes[47]. Such an analysis and presentation is aimed at here, this section dealing with averages and variability and the next section with extremes. A comprehensive display of maps of rainfall and other elements of climate is contained in the *Climatological atlas of the British Isles*[48] to which reference should be made.

Where rainfall is concerned, it is generally assumed that use of the words *average* or *mean* implies annual or sometimes monthly amounts, rather than those for any shorter period. Yet information about weekly, daily or even hourly averages could be as useful in a number of circumstances. Another assumption, now less prevalent, is that a long record, for example of fifty or even a hundred years, is necessary in order to establish a true average.

Table 2.4 ESTIMATES OF GENERAL ANNUAL RAINFALL (mm) OVER ENGLAND AND WALES, 1727–1975, PREPARED BY AVERAGING ANNUAL TOTALS MEASURED AT SELECTED STATIONS (AFTER WALES-SMITH[49])

Year	mm	Year	mm	Year	mm	Year	mm	Year	mm
1727	876	1777	846	1827	937	1877	1133	1927	1100
1728	983	1778	940	1828	1024	1878	986	1928	1026
1729	909	1779	843	1829	866	1879	975	1929	894
1730	734	1780	658	1830	932	1880	1008	1930	1052
1731	582	1781	737	1831	1021	1881	975	1931	975
1732	820	1782	1021	1832	894	1882	1135	1932	922
1733	744	1783	843	1833	937	1883	958	1933	726
1734	1039	1784	800	1834	790	1884	787	1934	851
1735	884	1785	765	1835	861	1885	904	1935	1011
1736	983	1786	871	1836	1008	1886	1046	1936	975
1737	851	1787	935	1837	787	1887	663	1937	986
1738	660	1788	584	1838	798	1888	876	1938	886
1739	876	1789	1031	1839	1024	1889	838	1939	1013
1740	734	1790	831	1840	782	1890	805	1940	904
1741	607	1791	947	1841	1123	1891	993	1941	859
1742	711	1792	1057	1842	841	1892	841	1942	841
1743	630	1793	780	1843	940	1893	742	1943	833
1744	869	1794	902	1844	775	1894	965	1944	897
1745	907	1795	831	1845	864	1895	859	1945	833
1746	800	1796	770	1846	937	1896	833	1946	1057
1747	914	1797	973	1847	853	1897	904	1947	823
1748	754	1798	798	1848	1130	1898	777	1948	953
1749	775	1799	935	1849	846	1899	841	1949	785
1750	660	1800	833	1850	798	1900	975	1950	1021
1751	993	1801	884	1851	772	1901	790	1951	1110
1752	815	1802	792	1852	1265	1902	752	1952	902
1753	932	1803	757	1853	874	1903	1146	1953	757
1754	803	1804	851	1854	686	1904	798	1954	1085
1755	853	1805	739	1855	790	1905	770	1955	785
1756	879	1806	945	1856	810	1906	904	1956	869
1757	876	1807	777	1857	848	1907	886	1957	899
1758	874	1808	810	1858	704	1908	813	1958	1029
1759	780	1809	851	1859	899	1909	940	1959	805
1760	907	1810	917	1860	1069	1910	1011	1960	1171
1761	851	1811	851	1861	810	1911	841	1961	874
1762	810	1812	889	1862	925	1912	1118	1962	790
1763	1092	1813	798	1863	815	1913	876	1963	851
1764	950	1814	904	1864	683	1914	968	1964	706
1765	787	1815	856	1865	991	1915	986	1965	993
1766	732	1816	947	1866	991	1916	1019	1966	1024
1767	864	1817	914	1867	859	1917	876	1967	983
1768	1191	1818	869	1868	907	1918	958	1968	980
1769	787	1819	894	1869	907	1919	940	1969	909
1770	958	1820	792	1870	734	1920	975	1970	912
1771	734	1821	1006	1871	864	1921	627	1971	798
1772	963	1822	879	1872	1288	1922	942	1972	848
1773	980	1823	1021	1873	795	1923	1011	1973	739
1774	897	1824	1057	1874	846	1924	1074	1974	991
1775	996	1825	856	1875	1024	1925	947	1975	752
1776	859	1826	737	1876	1046	1926	912		

Thirty-five year averages were selected for British rainfalls in view of the cycle of that period suggested by Bruckner, but thirty year averages have also been employed. In addition, attempts have been made to evaluate how much short period averages deviate from long-term values. However, present opinion on climatic change seems to indicate that no one period of a particular length beyond about ten years has any peculiar advantage. In fact it is possible that long-term averages could even be misleading, when used to anticipate future rainfalls.

Long unbroken sequences of records are available for a few sites (e.g. Oxford Radcliffe Observatory since 1815) and for some others records have been 'reconstructed' for longer periods. For instance, monthly rainfalls representative of Kew Observatory have been estimated[49] for the years 1697–1871 when records commenced; values for Manchester have been reconstructed in a similar fashion[50]. Similarly, monthly estimates of the general rainfall over England and Wales have been made[51] and recently updated *(Table 2.4)*.

Average annual rainfall

Maps of average annual rainfall exist for a number of different periods, but the ones most commonly used span the years 1881–1915 and 1916–1950 *(Figure 2.8)*. A map for the period 1901–1930 was included in the *Climatological atlas of the British Isles* and a similar map has been prepared as a contribution to the WMO *Climatic atlas of Europe,* but for the years 1931–1960. More recently a preliminary map for the years 1941–1970 has been produced.

As might be expected there are differences between these maps, the most noticeable feature being a general tendency for rainfall amounts to increase, an increase that may not be continuing in the 1970s. Annual averages are greater for 1916–1950 than for 1881–1915, except for some small areas along eastern and western coasts, where some slight decreases are even to be found. The most striking increases are for the counties of Lanark and Argyll, where averages are up to 14 per cent higher. Increases of over 5 per cent occur over much of northern England[52], most of Wales, parts of the Midlands, Dartmoor and part of Scotland's east coast. Whether these differences are due to genuine climatic change or to an improved network and methods of measurement is not entirely clear, but it seems likely that they are related in some way to the progressive decline of the westerlies and the increasing continentality which has followed the so-called 'climatic optimum' of the early 1940s.

For the country as a whole, there has been an increase of about 6 per cent in the estimated general rainfall *(Table 2.5)* during the more recent thirty-five year period. On the other hand, the differences between the 1916–1950, the 1931–1960 and the 1941–1970 rainfalls are not so distinct, although there seems to be a tendency for the increases to continue in a number of areas. Yet the general pattern of rainfall remains much the same whatever period is selected. Average annual totals range from about 500 mm (20 in) for the Thames Estuary, to nearly 5000 mm (200 in) on Snowdon, at Styhead Tarn (Lake District) and near Loch Quoich in

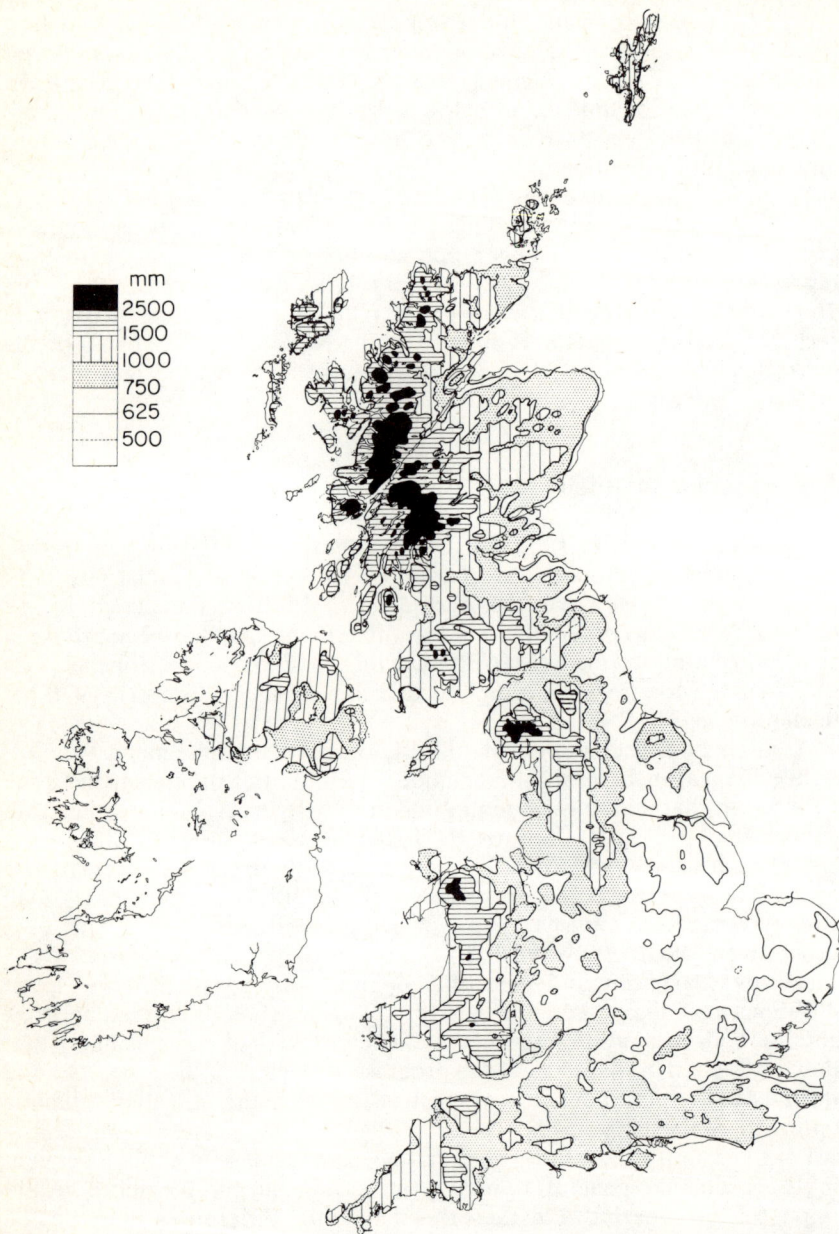

Figure 2.8   Average annual rainfall, 1916–1950 (reproduced by permission of the Controller, Her Majesty's Stationery Office)

Inverness-shire. Over most of the country rainfall amounts show a strong correlation with relief. Detailed rainfall maps have been constructed for certain areas from knowledge of altitude–rainfall relationship alone[53,54]. There are regions where such relationships differ greatly from those that apply generally, and this is particularly the case for mountainous areas away from west coasts, notably the Grampians[55]. On the other hand, over some of the flatter eastern parts of the country, rainfall of a convectional origin makes up for lack of orographic rainfall.

Table 2.5 COMPARISON OF GENERAL RAINFALL AVERAGES (mm) 1881–1915 AND 1916–1950

	England 1881–1915	England 1916–1950	Wales 1881–1915	Wales 1916–1950	Scotland 1881–1915	Scotland 1916–1950	United Kingdom 1881–1915	United Kingdom 1916–1950	% change
January	69	86	122	150	124	145	91	112	+22
February	61	64	102	104	107	102	79	79	0
March	64	56	99	86	104	84	79	69	−13
April	51	58	76	81	76	84	61	71	+17
May	56	64	76	86	79	81	64	71	+12
June	58	53	79	79	74	81	66	66	0
July	71	79	91	107	97	107	81	91	+13
August	81	79	122	117	117	114	94	94	0
September	61	74	91	114	102	122	76	94	+23
October	94	86	145	147	124	150	109	114	+5
November	81	91	135	145	135	135	104	112	+7
December	91	84	152	142	150	135	117	107	−9
Year	838	874	1290	1358	1289	1340	1021	1080	+6

The average or mean $(\bar{X})$ is the most frequently used parameter, but knowledge of the scatter or dispersion of the individual observations X_1, X_2 . . . X_N) about the mean is no less valuable. There are a number of measures of this variability; for example, the range between highest and lowest value, and the proportion of observations in each tercile or quartile. Two more useful parameters are the relative variability and the coefficient of variation (C_v). Both of these have been employed to study the variability of annual rainfall[56–58], the second being preferred because it is based on the standard deviation (σ). From the definition:

$$\sigma = \sqrt{\left(\frac{\Sigma(X_i - \bar{X})^2}{N}\right)} = \sqrt{\left(\frac{\Sigma X_i^2}{N} - (\bar{X})^2\right)} \qquad (2.5)$$

it follows that the coefficient of variation is:

$$C_v(\%) = 100 \, \sigma/\bar{X} \qquad (2.6)$$

Maps displaying the distribution of C_v have been constructed for different periods *(Figure 2.9)* and again there are differences from one period to another. Values of C_v range from 8 to 20 per cent, with a marked tendency for the variability to increase towards the south and east in all cases. This pattern seems to have been reinforced for the most recent periods but it appears to have been accompanied by some increases in variability in the high rainfall areas, particularly in western Scotland.

The foregoing measures of dispersion apply to populations that are normally distributed. For some purposes it is expedient to assume that annual and seasonal rainfall totals are normally distributed, and this makes it possible to apply the normal probability theory in order to relate amount to the probability of occurrence. This method has been applied to show the probability of the annual rainfall exceeding or falling short of a specified amount *(Figure 2.10).* Alternatively, the amount of rainfall can be calculated for a given probability *(Figure 2.11):* in this case the technique has been applied to totals for the winter months November to March. Both maps brings out the contrasts between the east and west of the country, the second being of particular interest in relation to water supply in areas that depend upon winter rainfall for the recharge of aquifers.

Figure 2.9 (a) Distribution of the coefficient of variation of annual rainfall (per cent), 1881–1915 (from Bleasdale et al.[59], courtesy Institution of Civil Engineers)

Figure 2.9 (b) Distribution of the coefficient of variation of annual rainfall (per cent), 1916–1950 (from Bleasdale et al.[59], courtesy Institution of Civil Engineers)

Seasonal and monthly averages

Maps of average rainfalls for durations of less than one year exhibit contrasts for the different periods of record, for example 1881–1915 as opposed to 1916–1950, that are more marked than the contrasts between annual averages for these different periods. Even these differences, however, are usually not as large as those between one part of the year and another. Taking the country as a whole, there have been noteworthy increases in the 1916–1950 amounts for January and September *(Table 2.5)* compared with totals for the earlier period, while the summer half of the year has experienced a larger increase than the winter. Some of these changes are much more evident for particular areas. In Wales, for example, the average January rainfall has increased by about 28 mm (1.1 in). Some of these trends

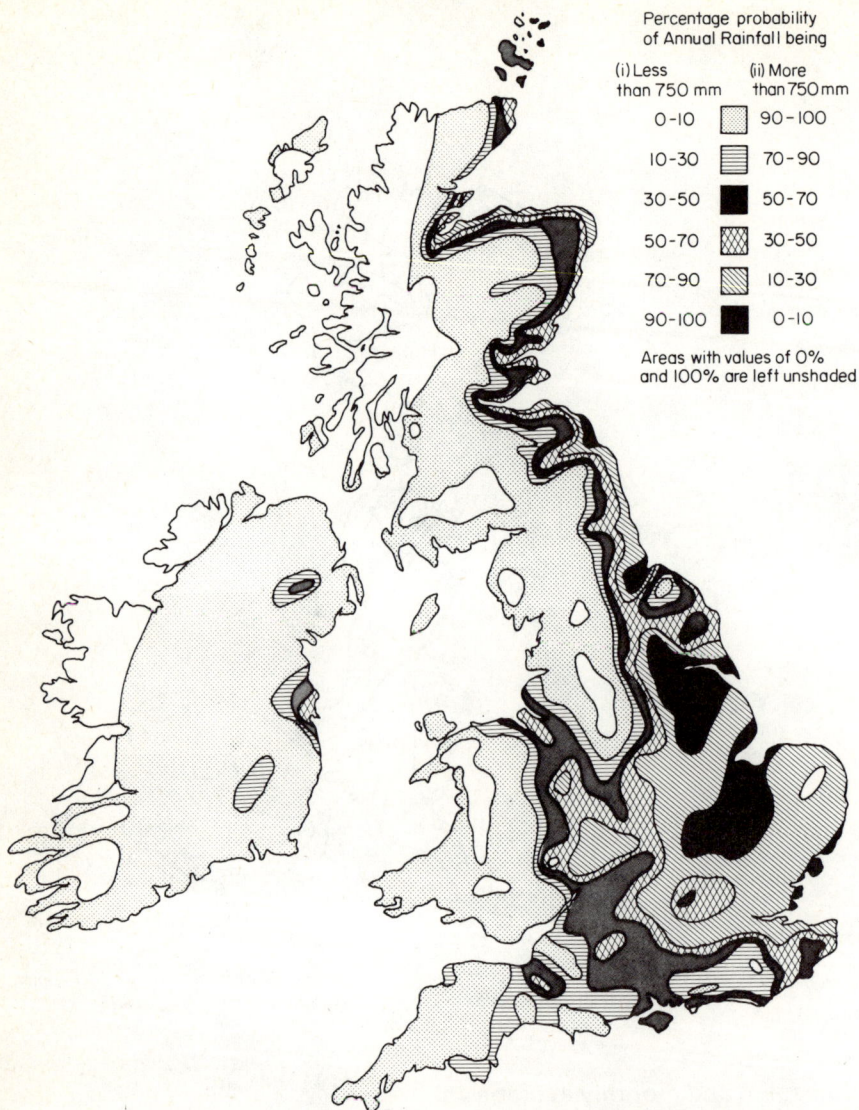

Percentage probability
of Annual Rainfall being

(i) Less than 750 mm		(ii) More than 750 mm
0-10		90-100
10-30		70-90
30-50		50-70
50-70		30-50
70-90		10-30
90-100		0-10

Areas with values of 0%
and 100% are left unshaded

Figure 2.10   Probability of receiving more or less than 750 mm annual rainfall (from Gregory[57], courtesy Royal Meteorological Society)

seem to continue in the 1931–1960 period; for instance Julys tend to be wetter in the north and west. On the other hand, some winter months, December for example, may be becoming slightly drier in the south-east. In contrast, a study[60] of monthly rainfalls for England and Wales since 1727, in which totals were divided into three equal classes, showed that rainfalls of summer months have not changed over the 240 years, but that winters are wetter since 1873, particularly Januarys.

Seasonal and monthly rainfall averages can be shown as actual amounts, or in terms of percentages of the annual average. The latter method is used here, first to display the proportion of rainfall in the summer half of the year *(Figure 2.12)*. This map shows that many of the drier areas have a summer maximum, the time of the year when the wetter areas are experiencing a minimum. This was a characteristic brought out by Crowe[61] in his study of seasonal incidence of rainfall, one that is again apparent in the isomeric maps for January and July *(Figure 2.13)*. The first of these demonstrates the importance of orography, a feature of lesser significance in the second map where there is a tendency for rainfall of a convectional origin to attain some dominance in eastern areas.

— — — — —5   Frequency of irrigation need
5 years out of 10 (after Penman)

Figure 2.11   Total winter rainfall (November–March, mm) likely to be exceeded on 98 per cent of occasions (courtesy Institute of Hydrology)

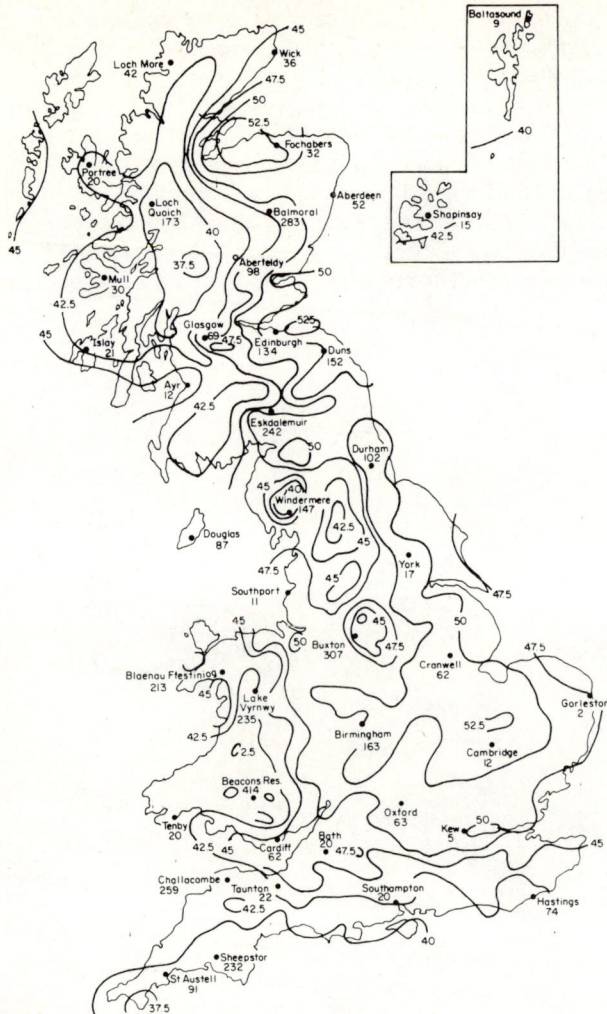

Figure 2.12 Summer rainfall (April–September) as a percentage of annual average, 1916–1950; positions of representative rainfall stations are shown, with heights in metres above sea level (reproduced by permission of the Controller, Her Majesty's Stationery Office)

Analyses of the dispersion of seasonal and monthly totals have not been carried out to the same extent as for annual amounts. Probability studies for periods of less than a year face the fact of the distribution becoming increasingly skewed as the period shortens. For some purposes and sites the error arising from assuming normality is small, however, and consequently seasonal and even monthly[62] rainfalls have been studied in this way *(Table 2.6)*, their distribution having been assumed to be normal. In one study for the Thames Conservancy area, the probability was determined of the rainfall for *n* consecutive months not reaching a specified percentage of the

January

Figure 2.13 *(a) January rainfalls as percentages of annual averages, 1916–1950 (reproduced by permission of the Controller, Her Majesty's Stationery Office)*

July

Figure 2.13 (b) July rainfalls as percentages of annual averages, 1916–1950 (reproduced by permission of the Controller, Her Majesty's Stationery Office)

Table 2.6 PROBABILITY OF SEASONAL RAINFALL TOTALS (mm) AT BLAENAU FFESTINIOG, 1911–1961

Probability, per cent	Winter (mean 789)		Spring (mean 477)		Summer (mean 657)		Autumn (mean 820)		Probability, per cent
	Low	High	Low	High	Low	High	Low	High	
0.5	269	1309	130	824	165	1149	201	1439	99.5
1.0	320	1259	163	791	213	1102	262	1379	99.0
2.5	393	1185	213	741	283	1032	350	1291	97.5
5.0	457	1121	255	699	343	972	425	1216	95.0
10.0	530	1048	304	650	412	902	512	1128	90.0
15.0	580	998	337	617	459	855	572	1069	85.0
20.0	619	959	363	591	496	818	618	1023	80.0
25.0	653	925	386	568	528	786	658	982	75.0

average *(Figure 2.14)*, but these results were arrived at empirically with no assumption about the normality of the totals.

The dependence of sequential events is an obvious characteristic of many hydrological phenomena. There are also the various popular beliefs about the weather of one season having some control over that of a season following. Should such a control be distinguishable, it could have far-reaching implications for long-range weather forecasting, water conservation and like activities. Hence it is not surprising that seasonal weather

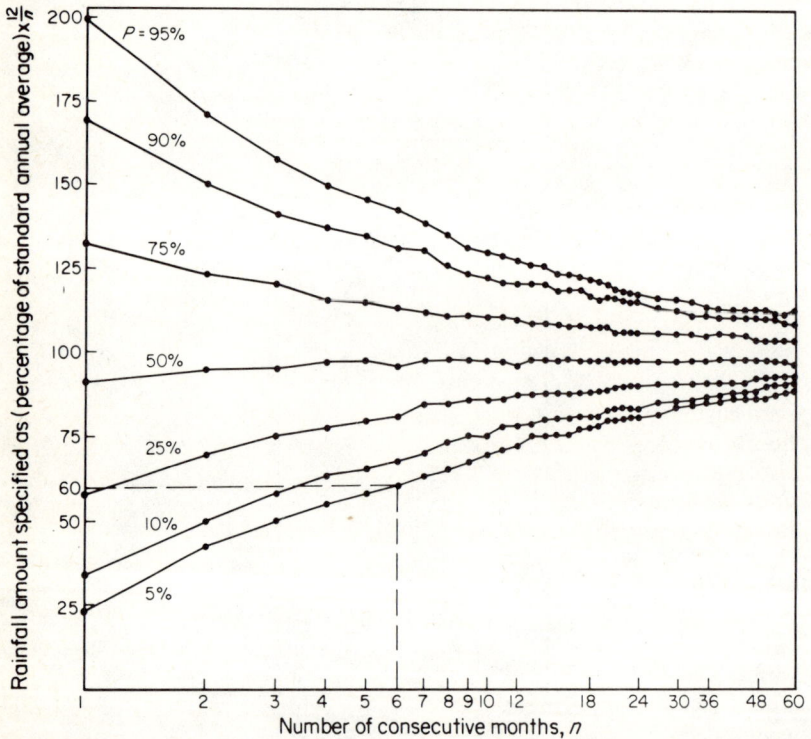

Figure 2.14   Probability of consecutive months' rainfalls not attaining certain percentages of average for the Thames catchment (9950 km²). This is one form of rainfall variability graph. The 50% (median) curve lies below the value of the ordinate marked 100 (average or mean) mainly because the frequency distribution of rainfall amounts is skew. The wettest months (or other periods) depart more from the average than do the driest (decreasingly so as the period lengthens). Also the average rainfall for the whole period 1888–1966 was slightly less (nearly 2%) than the value of 734 mm for the standard period 1916–50 now in use. Such graphs, or alternative forms, can be used to estimate the probabilities of occurrence of severe conditions for water supply (without of course any information on when they may occur). For instance, for rainfall over the Thames Valley as a whole, using 1888–1966 as a guide, there is a 5% (1:20) chance that in a 6-month period there may be no more than 60% of the 6-month average, that is 30% of the annual average, or 221 mm. The graphs can also be used as an aid in assessing whether the variations of rainfall during an important investigation, such as the Dee investigation and the Lambourn valley pilot scheme, differ significantly from expected variability over longer periods (though other forms of graph could be more suitable for this purpose) (from ref. 63, by permission of the Controller, Her Majesty's Stationery Office)

patterns have been investigated for sequences of like events, rainfall records in particular being examined for wet or dry spells. Results from some of the earlier studies[64,65] showed that the likelihood of any season or month being wetter than average, or alternatively drier than average, is not very different from random expectation. Recent analyses of general rainfall amounts by Murray[66] and Stephenson[67] have confirmed these earlier findings, little evidence of persistence being discovered in runs of dry, average and wet periods, when examined by means of contingency tables or when compared with runs predicted from geometric progressions. On the other hand, there are several examples where some evidence of persistence was discovered in records from individual sites[68]. For example, Beer *et al.*[69] studied rainfall records for seven stations in order to distinguish sequences of wetter-than-average months, or alternatively drier-than-average months. They found a close relationship between the number of successive wet (or dry) months *(m)* and the frequency of occurrence *(F)* of the series of months, in the form:

$$\log F = Rm + S \qquad (2.7)$$

where R and S are constants for each station. The frequency of longer runs, particularly of dry months, when calculated by this method was less than the observed frequency, indicating some persistence.

Averages for shorter periods

It is not customary to find that data are published in the form of weekly, daily or hourly averages. In fact even weekly totals are not often available and this is also the case for 10-day and 5-day totals, even though totals for these periods can be very useful in certain circumstances. There are also arguments in favour of averaging over natural periods, periods characterised by more or less uniform weather conditions. This approach would have caused considerable practical difficulties in the past, but with an increasing store of data on magnetic tape and the widespread use of computers, it is becoming possible to investigate whether the averaging over other periods would be advantageous.

One useful index of rainfall for short periods is the average number of days in the year when the catch exceeds a certain amount. This may be any figure one chooses, but the term 'rain day' is employed when 0.25 mm (0.01 in) or more is recorded while 1.0 mm (0.04 in) or more is needed for a 'wet day'. In the case of rain days *(Figure 2.15)*, even the driest parts of the country have a little less than half a year of rain days. At the other extreme, the wettest areas have on average less than 100 dry days. This large number of days with rain is of course one of the outstanding features of the British climate and it has been suggested[70] that the number of raindays is a better index of the wetness or 'raininess' of a month than the total rainfall. Use has been made of relations between certain threshold values (e.g. the average annual frequency of days with 10 mm or more) and mean annual rainfall.

Some of the best information on snow and snowfall is available in the form of number of days of snow falling or snow lying *(Figure 2.16)*. These maps are unusual by comparison with the maps of rainfall because they show gradients running from south-west to north-east. Manley has made a

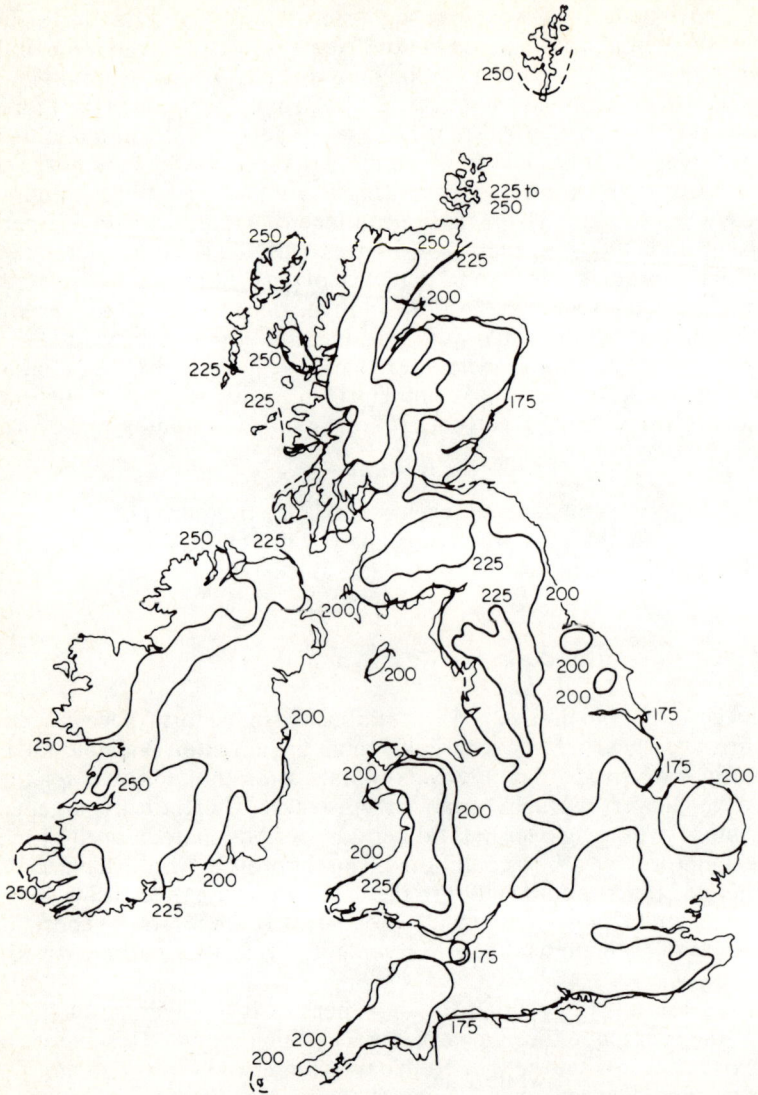

Figure 2.15   Average number of rain days per year, 1901–1930 (from the Climatological Atlas[48], by permission of the Controller, Her Majesty's Stationery Office)

number of studies of snow cover for particular areas[71] and the whole country[72,73], deriving some general relations between duration of snow cover and altitude *(Figure 2.17)*. More recently a map of median annual maximum snow depth *(Figure 2.18)* has been published[74].

One variant of the rainfall day approach is to employ the number of days in the year when more than a certain threshold value has been recorded, to calculate the probability of occurrence of a similar event for a day in the future. The same technique can be employed for individual sites, but should the probability density be required it is necessary to normalise the distribution of daily and weekly totals because of their skewed nature. One method is to take cube roots of the values, or alternatively to use special cube root probability paper *(Figure 2.19)*.

While persistence is not readily apparent from seasonal or yearly data, it is possible that this effect becomes more noticeable as the period shortens. In one example the frequency of runs of dry days and of rain days at Kew was analysed[75]. If N is the total number of days, the expected number of runs of

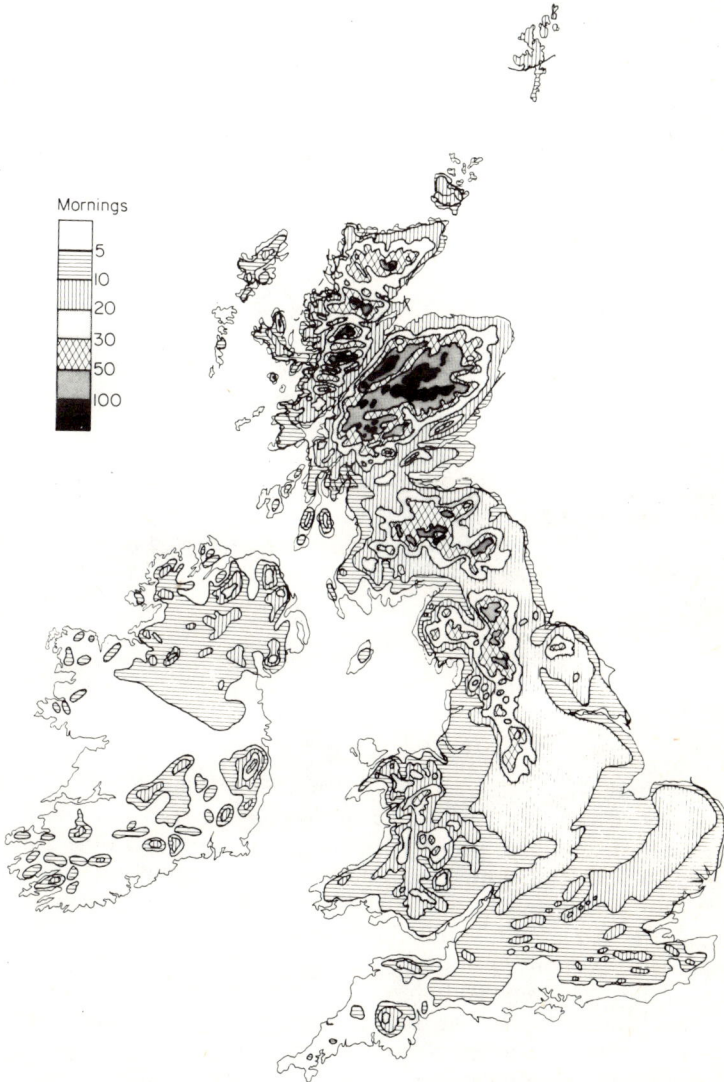

Figure 2.16 Average annual number of mornings with snow lying, 1912–1938 (from the Climatological Atlas[48], by permission of the Controller, Her Majesty's Stationery Office)

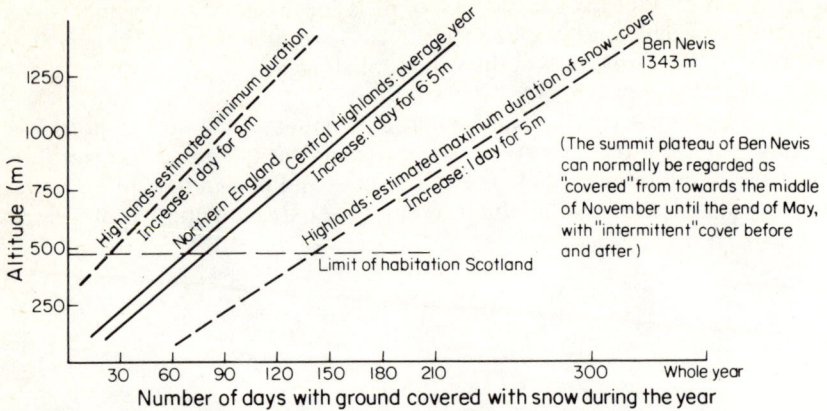

Figure 2.17　Generalised snow cover-altitude relationships (from Manley[73], courtesy Weather*)*

rain days are Nqp for a run of at least one day, Nqp^2 for a run of two days or more and so on, when there is no persistence. By comparison with the runs of wet days given by this theory, short runs are less frequent and long runs more frequent in the observed data, indicating persistence. There are several other examples of records being examined for sequences of wet and of dry days. A logarithmic series was employed in one case[76], but no evidence of persistence was found in either dry or wet spells.

As part of the study of the River Dee, a first-order Markov chain model was used to generate synthetic sequences of daily rainfalls[77]. Daily records from the Alwen Dam for 1925 to 1965 were employed in the model, the relationship between daily rainfalls in a wet spell being represented by:

$$P_n = c + bP_{n-1} + E_n \qquad (2.8)$$

where

$$\begin{aligned}
P_n &= \text{daily rainfall on day } n \text{ during a wet spell} \\
E_n &= \text{a random component} \\
c \text{ and } b &= \text{constants} \\
P_{n-1} &= \text{daily rainfall on day } n\text{-1}
\end{aligned}$$

For each month the range of observed rainfalls was divided into 29 classes and the amounts of rainfalls that occurred one day after days with class 1 rainfalls were listed. This was repeated for each class up to the 29th, days with zero rainfalls being omitted from these transition tables. Tables of the number of occurrences of dry spells of lengths from 1 to 25 days were also produced for each month, together with similar tables for wet spells.

To generate a month's record, the lengths of wet and dry spells were selected using a table of random numbers and the tables of spells for that month (it was assumed that the lengths of successive wet and dry spells were independent throughout the year). Then for each wet spell the transition tables were used to determine the actual rainfall amounts. A value for the first day's rain was selected at random from a table of amounts observed on the first day of a wet spell. This amount was employed to find the rainfall for the second day from the appropriate part of the transition tables, the process

Figure 2.18   Median annual maximum snow depths (mm), 1946–1964 (courtesy Institute of Hydrology)

being repeated to the end of the wet spell and started once more at the end of the intervening dry spell. The model generated fewer months of high and low rainfall than expected, but the mean annual rainfalls of four different 50-year sequences were within 5 per cent of the real mean.

Extremes

Exceptionally heavy falls of rain for durations of from one minute to one day and longer, the month and year with the greatest rainfall, and the longest period without measurable rain are some of the features characteristic of the extremes of rainfall.

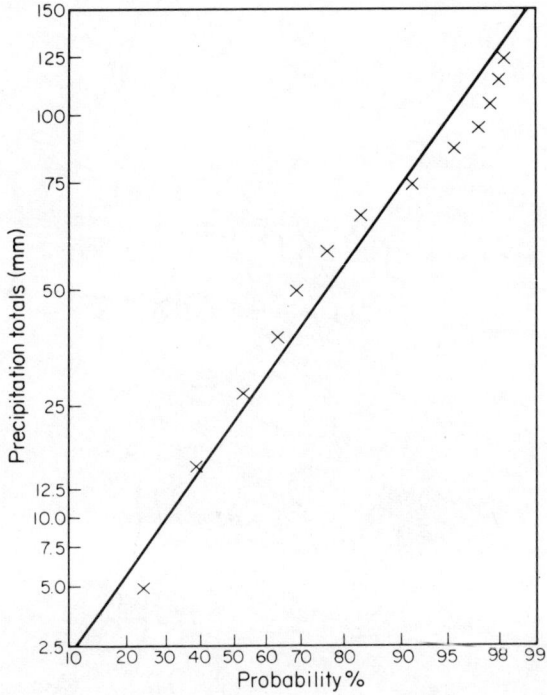

Figure 2.19   Cube root normal probability analysis of weekly totals (mm) for Clatteringshaws, Kirkcudbrightshire, 1948–1952

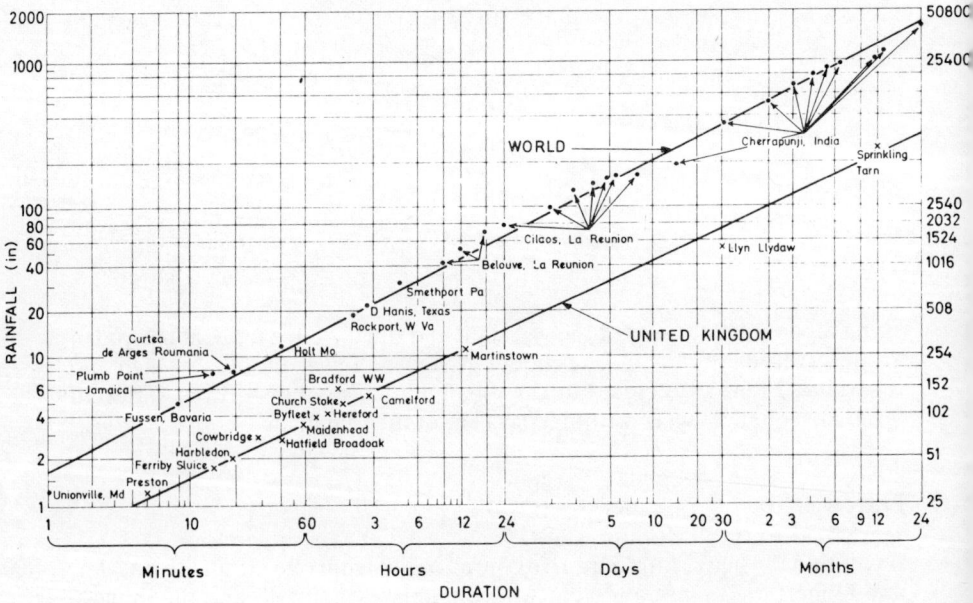

Figure 2.20   Magnitude–duration relationships for world and UK largest rainfalls (from Rodda[78], courtesy Institute of British Geographers)

Heavy rain

Because of the limitations of the rain recorder network, particularly in mountainous areas, observations of heavy rainfall are more readily available for the rainfall day than for any shorter duration. Of course, single storms can be split between two rainfall days, and there is also the difficulty that many totals now recorded as daily falls actually accumulated in less than that particular 24 hours. Knowledge of excessive rainfall amounts for periods longer than a day, for example five days or a week, is also limited, in this case simply because of the labour involved in distinguishing such falls from the remainder of the record, monthly and annual totals excepted.

The largest amounts ever recorded over a range of durations have been extracted from the lists of heavy falls to be found in *British Rainfall*. The relationship between magnitude and duration *(D)* for these falls was shown[78] to be:

$$R = D^{0.49} \qquad (2.9)$$

but a comparison of British and world maxima revealed that even the heaviest rains recorded in the UK were only about one quarter of the amounts registered elsewhere *(Figure 2.20)*.

Short duration rainfalls

Intense rainfalls for short duration have been subjected to a number of investigations[79,80]. Perhaps the best known work was undertaken by Bilham[81], who based his findings on records from 18 widespread rain recorders spanning the period 1925–1934. By averaging the results for the 18 gauges, he showed that the number of occurrences of a given rainfall amount in 10 years *(n)* could be found from:

$$n = 1.25\,t\,(r + 0.1)^{-3.55} \qquad (2.10)$$

where r = given rainfall (inches), and t = time (hours). Because most of Bilham's instruments were concentrated in the drier areas, some discrepancies arise between predicted and observed amounts at wet sites. Here the frequency of longer-duration falls is greater than estimated by the formula, but otherwise few faults have been found with Bilham's work[82], taking into account the records that have since accumulated. In addition, there has been an attempt[83] to make allowance for thunderstorm frequency in assessing the incidence of heavy rain. Other studies have been concerned with the distribution of storm rainfall in both space and time, particularly those undertaken by the Transport and Road Research Laboratory[84] where storm profiles were determined *(Figure 2.21)*.

Longer-duration rainfalls

There are few areas in the UK where a fall greater than 100 mm (4 in) in a rainfall day has not been recorded during the last 100 years[85] *(Figure 2.22)*. On the other hand, there were only six falls that attained 230 mm (9 in) or

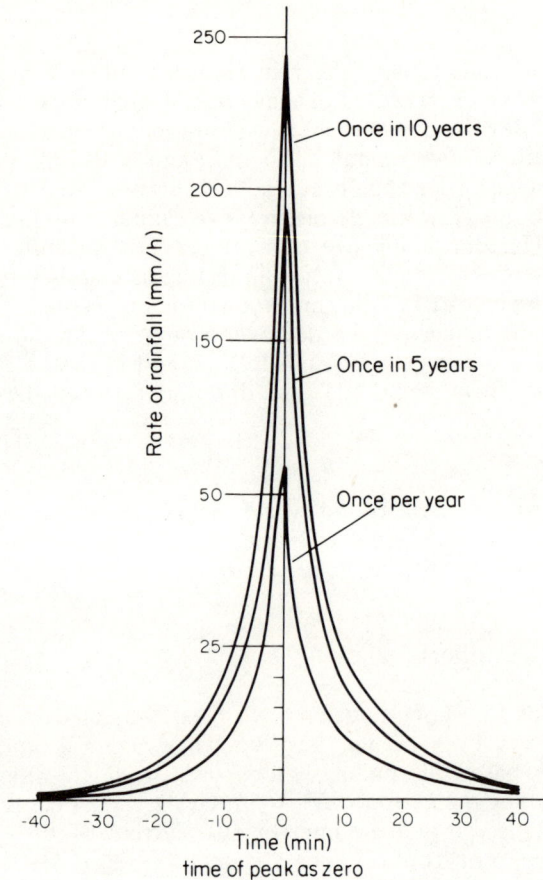

Figure 2.21   Examples of design storm profiles (from ref. 84, courtesy Transport and Road Research Laboratory)

more during this period; three of these falls were registered during the same storm and all the storms occurred in Somerset and Dorset. Why this part of the West Country should be especially susceptible to heavy rain is not completely clear, but it is worth noting that some of the heaviest rainfalls in France have been recorded just across the Channel in the Cotentin Peninsula. Accounts of heavy falls of rain and the flooding consequent on them abound in the literature, but apart from Bilham's threefold division into 'very rare', 'remarkable' and 'noteworthy' falls, not many attempts have been made to devise means for classifying intense rain.

One approach to this subject of the occurrence of heavy rain is through the 'extreme value' concept, introduced into hydrology by Gumbel[86]. Initially it was applied to floods, but other extreme events have been analysed in the same manner. A special probability paper was devised for use with the method by Powell[87], which linearises the distribution. There have been several modifications to the technique and Chow has proposed a general

WIDESPREAD FALLS

Areas with 150mm or more

Areas with 100-126mm

ISOLATED MEASUREMENTS
+ 150mm or more
▲ 127-149mm
· 100-126mm

KILOMETRES
0 20 40 60 80 100

LONGSTONE BARROW 229mm

CANNINGTON 239mm

BRUTON 243mm

MARTINSTOWN 280mm

Figure 2.22   Largest recorded daily rainfalls, 1863–1960 (from Rodda[78], courtesy Institute of British Geographers)

equation for hydrological frequency analysis[88]. Gumbel used the idea that the largest daily flow in a year was the upper extreme of the 365 daily flows, and that this value with other annual maxima formed part of an extreme-value series. The largest annual values give the annual maximum series and the smallest annual values the annual minimum series, which is used for drought flows. The time interval of one year is employed to reduce the chance of dependence existing between successive events—one of the main criticisms of the partial duration series (i.e. the series formed from the *n* largest flows in *n* years, regardless of when they occur). Gumbel found that the most suitable distribution for fitting to the annual maximum series was of the double exponential type[89], although others have suggested alternatives[90]. Annual maxima are fitted to the type 1 double exponential distribution:

$$1 - P_m(Y) = \exp\left[-\exp-\alpha(Y-u)\right] \qquad (2.11)$$

where $1 - P_m(Y)$ is the probability of an event not exceeding Y, and α and u are parameters of the distribution. Alternatively $P_m(Y)$ is the probability of Y being exceeded, and as P_m decreases so the period in years between occurrences in excess of Y increases. The average time in years between events in excess of Y, namely $T_m(Y)$, is known as the 'mean return period' or 'recurrence interval', or simply the 'frequency', where:

$$T_m(Y) = \frac{1 \text{ year}}{P_m(Y)} \qquad (2.12)$$

Some measure of the return period has to be computed from the limited number of annual maxima that are usually available in a series, and several formulae have been developed for this purpose. The one now in most common use is:

$$T_m = \frac{n+1}{m} \qquad (2.13)$$

where n = number of years of record and m = magnitude of the event, the highest being 1.

In most cases, errors in the basic data invalidate many of the more refined methods of determining these so-called 'plotting positions', and the method of fitting a straight line to the plotted points introduces further disparities. The example shown here *(Figure 2.23)* employs Gumbel's method of fitting, but a line drawn in by eye might be as acceptable in many cases. Alternatively, other ways of fitting the data can be employed[91] such as the method of moments or the method of maximum likelihood.

The Gumbel approach has been employed in an analysis of daily maxima for each year from 1911 to 1960 for 121 UK rainfall stations[92]. The relationship between magnitude and frequency was determined for each site and a series of maps constructed to show the distribution of rainfall for return periods of between 2 and 100 years *(Figure 2.24)*. Other maps have been drawn to show the distribution of return periods for specified daily falls. The dominance of highland Britain is again apparent, but certain of the excessive isolated falls do not conform to the pattern of the remainder. At these sites, the largest fall is between two and four times greater than the

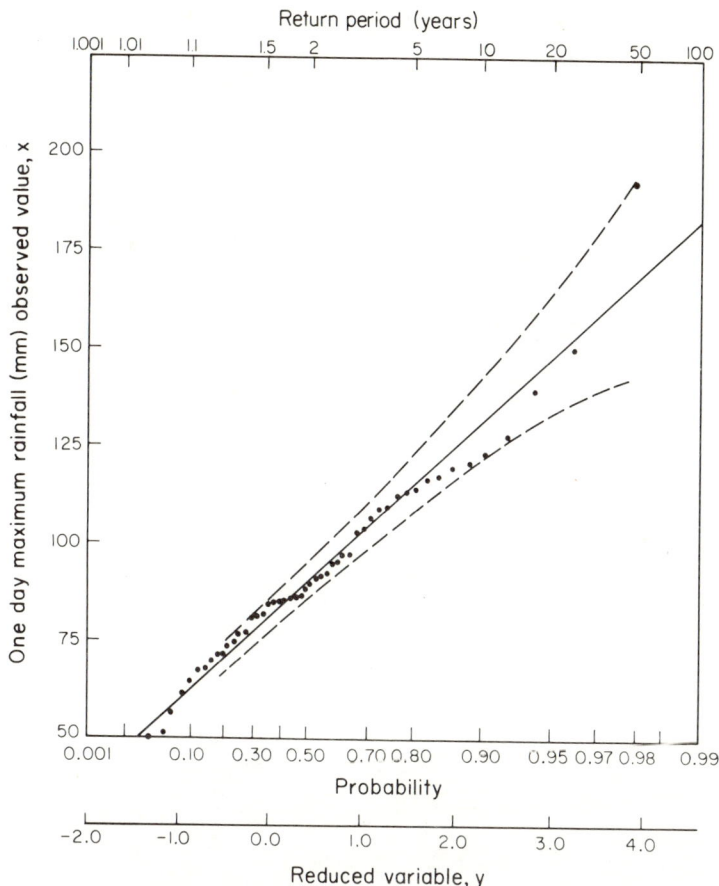

Figure 2.23   One-day rainfall magnitude–frequency relationship for Blaenau Ffestiniog (from Rodda[92], courtesy Elsevier Scientific Publishing Co.)

next largest and, according to Gumbel theory, they would have return periods in excess of one million years. On these grounds such events were excluded from the analysis; as a result it was found that a high correlation existed between the average annual rainfall at each site and the maximum daily fall for a particular return period. An alternative to the exclusion of these points would be to attempt another method of fitting, with the likely result that a much smaller return period would be obtained for them.

Probable maximum precipitation

Just how much rain can occur within a given time is a subject that has demanded attention for a long while. Initially estimates were based on increasing by some arbitrary factor the highest amount ever recorded. Then, particularly in the United States, several other methods were tried, such as

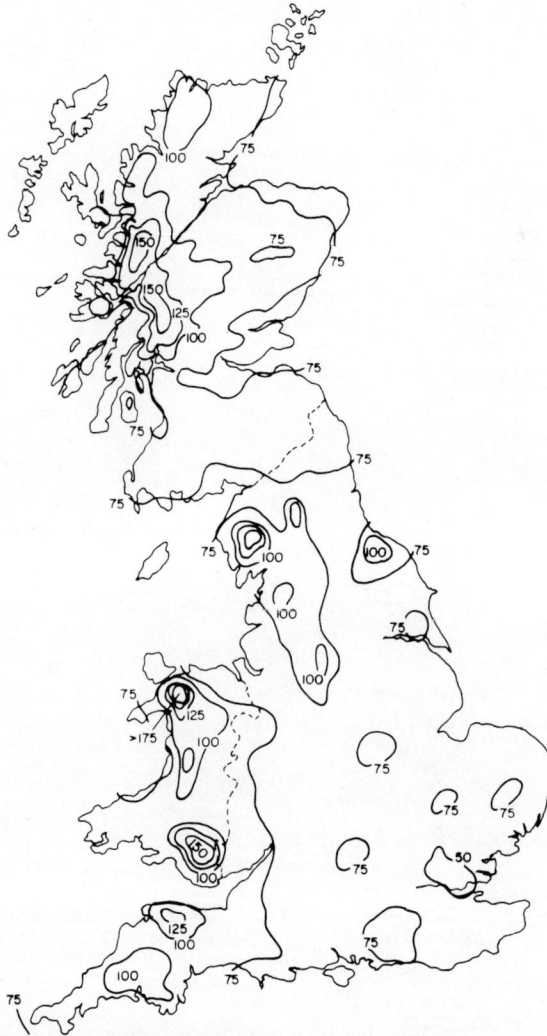

Figure 2.24   Highest rainfall (mm) in one day likely to be exceeded once in 100 years (from Rodda[92], courtesy Elsevier Scientific Publishing Co.)

storm transposition and attempts at physical maximisation. The latter approach embodies the probable maximum precipitation (PMP) concept. The PMP is the 'theoretical greatest depth of precipitation for a given duration that is physically possible over a particular drainage basin at a certain time of year'. The technique for estimating the PMP is well documented[93], but it requires three items to be maximised: the moisture content of the atmosphere above the basin, the rate of inflow of that moisture and the fraction of inflowing moisture that can be precipitated. Generalised charts exist for the United States showing the PMP for various durations. In the UK, the Meteorological Office has applied similar techniques in assessing the PMP for several areas overseas[94,95] and for the

Trent Basin[96]. There has been some criticism of both concept and method, however. One contention is that the term itself is misleading because of the statistical overtones resulting from use of the word 'probable'. Another is that no realistic limit can be placed on the amount of water that the atmosphere can release. For example, the PMP could always be increased by increasing the thickness of the air by one metre, by making the vertical rise of air greater, by increasing the condensation of moisture and so on.

An alternative approach to the estimation of the PMP has been developed by Hershfield[97] from a consideration of storms that have already occurred. The method employs Ven Te Chow's[88] general frequency formula:

$$X_T = \bar{X}_N + KS_N \tag{2.14}$$

where X_T = the rainfall for a return period of T years when a particular extreme value distribution is employed

$\bar{X}_N$ and S_N = the mean and standard deviation for a series of N annual maxima

K = the standardised variate using the Hershfield technique on the records from the 121 stations employed in the magnitude-frequency study[92]

Records for the stations employed in the magnitude frequency study were used to determine the maximum value for K for the UK. Several values in excess of $K = 15$ were found, the highest (26.7) being for Martinstown in Dorset. The rounded value $K = 27$ was then used to estimate the PMP for the other 120 stations using the mean and the standard deviation appropriate to each.

In common with most other rainfall maps of Britain, the PMP map for the rainfall day *(Figure 2.25)* repeats the contrast between the drier south-east and the wetter north-west. Point totals attain 750 mm (30 in) for parts of Snowdonia, but over most of the country including Northern Ireland the estimated PMP does not reach 400 mm (15.5 in).

It is very difficult to judge how realistic these values are, but they are of course the estimated upper limits to the likely samples of rainfalls. The method is very open to criticism, but it may be useful as a guide to PMP values.

Point to areal rainfall relations

So far this discussion of heavy rainfalls has been concerned with point amounts. How far the rainfall at a particular point is representative of that of the surrounding area has been the subject of a number of studies[98,99] including some concerned with heavy rainfall. In one case[100] means established from a fairly dense network were examined to see how they compared with the mean calculated from a few key gauges. In another[101] relationships were established between the rainfall at the centres of two 260 km² (100 mile²) areas and the areal rainfalls during prolonged rain. A small mean difference was found, suggesting that rates of rainfall from one central gauge were sufficient to assess the flood risk. Envelope curves have been

Figure 2.25   Probable maximum precipitation (mm) for the rainfall day (courtesy Institute of Hydrology)

constructed *(Figure 2.26)* to show the decay in intensity with increasing area, this example demonstrating the severity of the heavy rains in the west of Scotland on 16–17 December 1966[102].

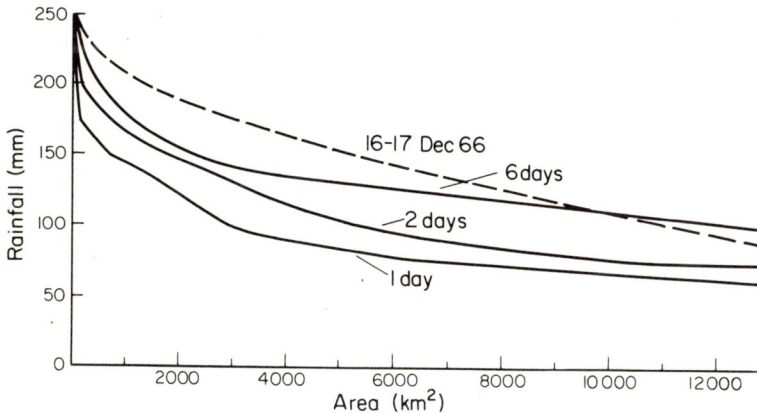

Figure 2.26   Area–intensity relationships for severe storms in the British Isles, 1890–1966 (from Reynolds[102], courtesy Weather)

Using the records from Cardington, Holland[103] proposed a formula for estimating areal rainfalls from:

$$\Gamma\left(\frac{\sqrt{(kA)}}{1-F}\right) = t \qquad (2.15)$$

where Γ = the gamma function
        F = reduction factor
        A = area (hectares)
        t = time of concentration (minutes)
        k = 0.009

Drought

That drought is not an extreme condition would be difficult to argue. On the other hand drought is usually prolonged, and as such it might be more appropriate for consideration in the context of monthly or seasonal totals. Indeed most of the studies already referred to that deal with wet and dry spells are important from the drought viewpoint.

Even with its relatively uniform climate, the UK has experienced some notable droughts. The years of 1864, 1887, 1921, 1944, 1947, 1949 and 1959 are particularly outstanding; drought straddled 1933 and 1934, while the three winters 1962–63 to 1964–65 were the three driest consecutive winters in much of south-east England since records commenced.

Exactly what constitutes a drought is something of a problem to decide. There are so many definitions[104] from which to choose, most employing some threshold value of rainfall as the sole criterion. A few, more comprehensive, use evaporation and rainfall together, and some attempt to assess the soil moisture deficit from a water balance.

The only definitions in current scientific use in the UK are those that have been adopted by the Meteorological Office. Thus a period of at least 15 consecutive days, to none of which is credited 0.25 mm (0.01 in) or more of rain, is defined as an absolute drought; while a partial drought is a period of at least 29 consecutive days, the mean daily rainfall of which does not exceed 0.25 mm[105]. On the basis of the first definition, it is very likely that each year some part of the country will experience an absolute drought. The longest drought recorded was in 1893 for 60 days from 17 March to 15 May, the drought affecting an area between Hastings and Lewes. For the year 1921 only 236 mm (9.29 in) were recorded at Margate—the lowest annual total ever registered in Britain. The definitions of drought, both absolute and partial, have long been recognised as rather unsatisfactory. Hence there have been attempts to discourage their use and provide some alternative simple method for classifying spells with little or no rainfall. The method put forward by Bleasdale[106] employs the rainfall recorded at a station in pentads, expressed first as a percentage of the annual average rainfall.

Another useful and similar index of drought, developed and used in Australia[107], is the Foley drought index. The method is to determine the median rainfall for each month and the annual median, then to calculate monthly departures of the observed falls from these medians, dividing each departure by the annual median. These values are then expressed cumulatively month by month, this index being shown *(Figure 2.27)* for the last 10 years of calculations made with Oxford rainfalls for the period 1815–1968[108]. The dryness of the three winters terminating in 1964–65 is readily apparent, but this and similar rainfall index approaches are subject to the criticism that factors other than lack of rainfall (such as antecedent soil

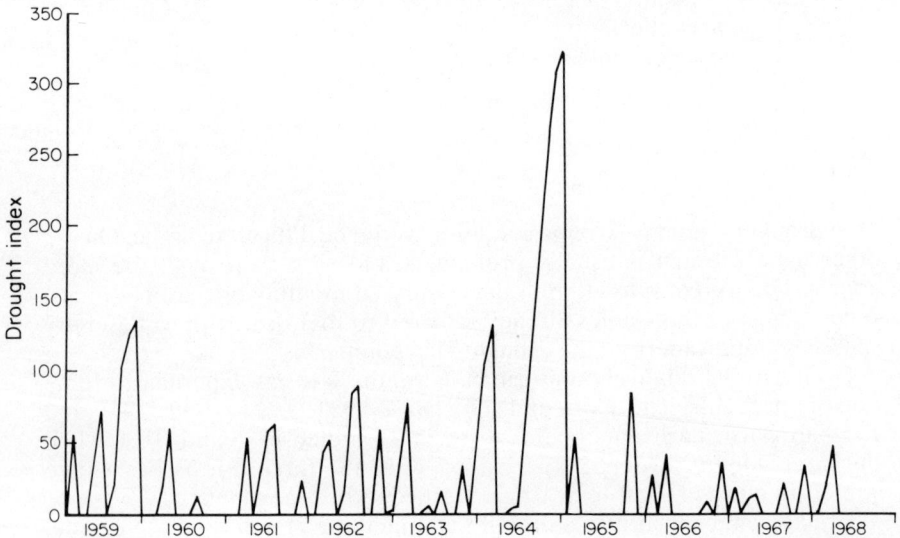

Figure 2.27   Foley drought index for Oxford, 1959–1968 (courtesy Institute of Hydrology)

moisture conditions) are important. Use of a water balance model overcomes most of these criticisms, but much more information is required and various assumptions have to be made. For example, in one study[109] estimates of the differences between the soil moisture content and field capacity were used as a drought index. Rainfall and potential transpiration values were used, along with an estimate of the soil water content at field capacity and assumptions about the effect of a reducing soil moisture content on the rate of transpiration. The results of these calculations are shown *(Table 2.7)* for Oxford and for part of the period used in the Foley index calculations.

Again the very dry character of the three winters is apparent, the exceptional nature of 1964–65 being demonstrated by a deficit continuing to December. The continuation of a deficit into the following year is, of course, extremely serious from the water supply point of view and for agriculture.

Reference has already been made to studies of sequences of wet and dry days and their frequencies. In one study[110] an analysis was made of the frequency of occurrence of spells of wet and dry weeks over the whole of England and Wales. A dry week was defined as one in which the mean rainfall was 50 per cent or less of the normal, and one with less than the average number of rain days. The number of occurrences of dry weeks in each month from 1948 to 1967 were tabulated and a relationship established between number of occurrences *(N)* and the length of dry spells *(S)* in weeks *(Figure 2.28)*.

As mentioned earlier, both geometric and logarithmic series have been applied to studies of sequences of wet and dry spells. In general the geometric series tend to underestimate the frequency of the longer dry

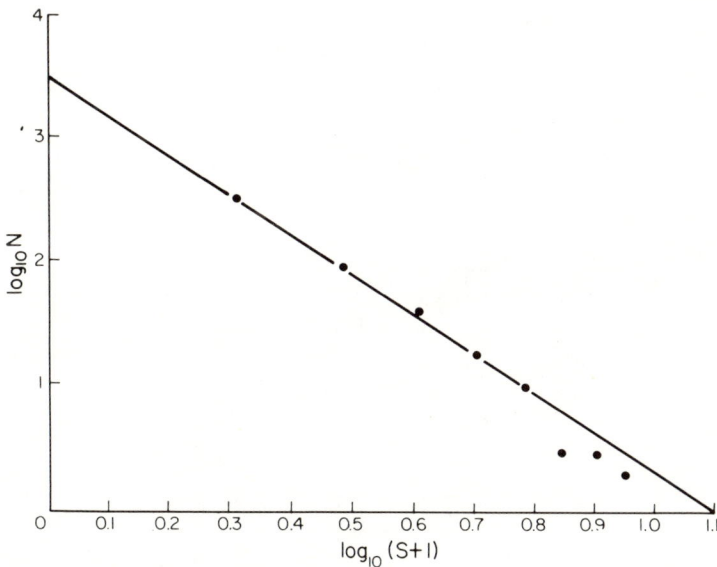

Figure 2.28   Relationship between the number (N) of occurrences of dry spells and their lengths (S, in weeks) (from Parker[110], courtesy Weather*)*

Table 2.7 INDEX OF DROUGHT FOR OXFORD PORTRAYED BY DEVIATIONS FROM FIELD CAPACITY DURING SEVERAL DROUGHT PERIODS (AFTER RODDA[109])

Year	Jan	Feb	Mar	Apr	May	June	July	Aug	Sept	Oct	Nov	Dec
1921	+59	-1	-12	-47	-98	-110	-114	-105	-105	-102	-57	-30
1922	+25	+43	+17	+21	-60	-102	-89	-17	-28	-41	-19	+41
1933	+38	+70	+9	-25	-57	-103	-107	-109	-103	-86	-64	-63
1934	-22	-33	-21	-14	-83	-105	-110	-106	-97	-78	-42	+97
1943	+90	+9	-17	-44	-78	-104	-109	-104	-103	-58	-30	-10
1944	+22	+11	-33	-72	-106	-105	-102	-105	-91	-51	+35	+33
1962	+87	-3	-6	-4	-44	-106	-105	-81	-17	-9	+43	+42
1963	+22	-3	+49	+6	-39	-85	-104	-67	-58	-29	+80	+13
1964	+9	+11	+74	+13	-37	-64	-105	-109	-107	-103	-84	-47
1965	+1	+1	+6	+1	-41	-71	—	—	—	—	—	—

Note. Positive values indicate runoff (mm) or percolation for months when soil was at field capacity. Negative values indicate soil moisture deficit (mm)

spells, because a constant probability *(P)* is assumed for a spell of *m* days continuing for another day. In other words, spells of dry days exhibit persistence. Geometric and logarithmic series were compared using records from 21 sites in south-west England. In this[111] and another study[112] the frequencies of spells of dry days predicted from the logarithmic series were found to match the observed frequencies closely. This is largely because *P* increasing as *m* increases is implicit in the logarithmic series.

Flood studies 1970–74

Volume Two of the Flood Studies Report[74] is concerned with the meteorological controls of floods, namely heavy rainfall and snow melt. The bulk of this volume is dealt with here rather than divided between the previous sections of this chapter. The aim of the meteorological studies is to present estimates and information for estimating: the magnitude–frequency –duration relationships for heavy rainfall, the distribution of rain storms in space and time, and details of snow cover and snow-melt rates. Examples are given to illustrate the methods developed for estimating relationships characteristic of these features of heavy rain, together with discussions of some of the severe storms of the past. The studies were based on analyses of records from more than 600 daily rainfall stations with an average of 60 years of record, supplemented by daily and monthly records from 6000 stations for the period 1961–1970. Hourly rainfalls for 150 stations equipped with rain recorders and having 20 years of records were also used, together with records from the Cardington and Winchcombe networks and the Jardi rate-of-rainfall recorders at Kew, Eskdalemuir, Aberdeen and Valencia. To facilitate analysis, most of these records were assembled on magnetic tape when they were not already in that form.

As a first step, regional analyses of point rainfalls were undertaken for England and Wales and also for Scotland and Northern Ireland. Using annual maximum 2-day falls of rain, extreme value analyses were carried out and the stations placed in seven classes according to the magnitude of the fall for a 5-year return period (M_5), the classes ranging from 40–50 mm to 200–300 mm. In the 40–50 mm class, there were 175 stations with an average length of record of 60 years; for each station, the ordered data were divided into the four quartiles and the four quartile means calculated. Then the four medians of the quartile means were obtained together with the medians of the highest falls H_1, the second, third and fourth highest falls H_2, H_3 and H_4. Next the 175 values of H_1 were subjected to a separate quartile analysis and the four quartile means determined, the second of these means being identical to the median for H_1. Taking these eleven values and those for M_5 and M_2, together with the appropriate values of their reduced variates (return periods), these points were plotted to show the relationship between magnitude and frequency *(Figure 2.29, line e)*. To extend the curve even further, the four highest values of H_1 were regarded as the upper members of a set of 10 500 (i.e. 175 × 60) annual maxima for plotting these four points, the appropriate values of the reduced variates were assumed to be 7.95, 8.27, 8.74 and 9.63, and the curve was extended accordingly. By repeating the exercise for each M_5 class and each chosen duration between 15 seconds and

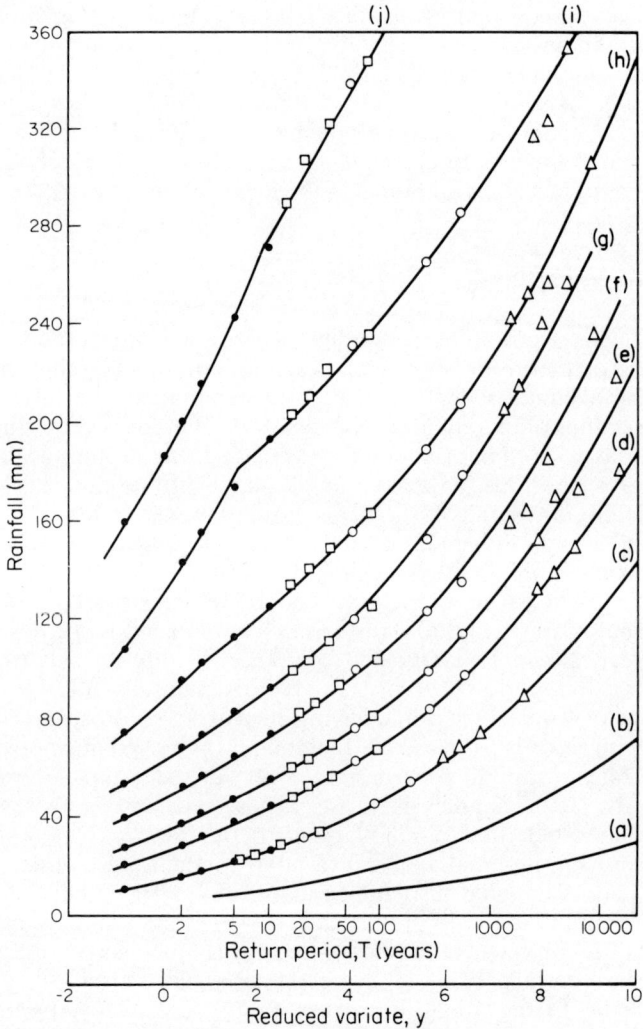

Figure 2.29   Rainfall 'growth' curves for England and Wales: (a) 5-min rainfall; (b) 15-min rainfall; (c) 2-hour rainfall; (d) 1-day rainfall for M_5 30–40 mm; (e) 2-day rainfall for M_5 40–50 mm; (f) 1-day rainfall for M_5 60–75 mm; (g) 4-day rainfall for M_5 75–100 mm; (h) 8-day rainfall for M_5 100–150 mm; (i) 25-day rainfall for M_5 150–200 mm; (j) 8-day rainfall for M_5 200–300 mm (courtesy Institute of Hydrology)

25 days, a set of magnitude–frequency curves was constructed, a selection of these curves being shown here *(Figure 2.29)*. In an examination of monthly and seasonal magnitude–frequency relationships, it was found that these could be expressed adequately as percentages of the annual values, e.g. of the M_5 and M_{100} values.

In order to display the countrywide variations in heavy rainfall, a series of maps was produced to show magnitude–frequency–duration relationships for annual maxima for various durations *(Figures 2.30, 2.31)* together with tables for seasonal monthly maxima. These maps were based on values of M_5 estimated by extreme value analyses for durations of 1, 2, 4, 8 and 25 days and for 1, 2 and 6 hours. Because not all the records used in these analyses were of the same length, a preliminary task was to standardise them. This was accomplished by determining the ratio p of the 2-day maximum fall for 1961 to that of the 2-day M_5 at the long-period stations, plotting values of p on a map of the UK and interpolating values of p for the short-period stations. The 2-day maximum fall in 1961 at these stations was R_2, and estimates of the long-period 2-day M_5 were made from R_2/p. This was repeated for each year from 1961 to 1970 and the median of these values was taken as the final 2-day M_5 estimate. A similar process was adopted to make estimates of the long-period 25-day M_5 for the short-period stations. One other problem was to convert rainfall days of rain

Figure 2.30 One-hour M_5 rainfalls (mm) (courtesy Institute of Hydrology)

Figure 2.31   Six-hour M_5 rainfalls (mm) (courtesy Institute of Hydrology)

into actual hours of rain, and from clock hours into minutes. A series of conversion factors was determined and these are shown in *Table 2.8*. A map of average annual rainfall for the period 1941–1970 was also constructed; it was found from this that relationships exist between average annual rainfalls and the magnitudes of maximum falls for particular durations and return periods, for falls of 24 hours and longer *(Table 2.9)*, for durations up to 120 minutes *(Table 2.10)*, and for durations from 60 minutes to 48 hours *(Table 2.11)*. Values for point rainfalls can be determined from these tables, or from the maps, or alternatively from a formula relating intensity *(I)* (i.e. rainfall–duration) to instantaneous intensity *(Io)*, duration *(D* hours), a continentality factor *(n)* and a factor *(B)* related to average annual rainfall as follows:

$$I = \frac{I_o}{(1+ BD)^n} \tag{2.16}$$

This formula applies to durations up to 2 days; typical values for the factors are $I_o = 150, n = 0.64, B = 25$ and $r = 0.26$, for an average annual rainfall of 1000–1400 mm.

To estimate probable maximum rainfalls, M_5 values of the dew-point were determined from maximum dew-points persisting for at least 6 hours at 60 stations. The amounts of precipitable water corresponding to a saturated column of air whose base temperature is this M_5 dewpoint were estimated and mapped. Then the 'efficiency' of major 2-hour and 24-hour storms was calculated from the ratio of the actual rainfall to the amount of precipitable water in the air column during the storm. For the 2-hour storm efficiency,

Table 2.8 FACTORS FOR CONVERTING CLOCK HOURS AND RAINFALL DAYS OF RAINFALL INTO MINUTES AND HOURS OF RAINFALL

Clock hours	1	2	6	
Factor	1.15	1.06	1.015	
Minutes of rain	60	120	360	

Rainfall days	1	2	4	8
Factor	1.11	1.06	1.03	1.015
Hours of rain	24	48	96	192

Table 2.9 VALUES OF M_5 AND M_{100} RAINFALLS FOR DURATIONS OF 24 HOURS TO 25 DAYS RELATED TO AVERAGE ANNUAL RAINFALL

AAR (hundreds of mm)	Approximate 2-day M_5 (mm)	Approximate M_5 and M_{100} values (mm) for durations					
		24 h	48 h	72 h	96 h	192 h	25 d
5– 6	44	40	47	51	55	71	114
		73	84	90	96	118	171
6– 8	50	44	53	59	65	84	144
		80	93	101	110	134	209
8–10	59	50	63	70	78	106	186
		89	107	116	126	161	262
10–14	71	58	75	85	97	134	253
		100	123	135	150	196	344
14–20	92	74	98	111	127	179	353
		121	152	168	188	254	462
20–28	124	98	131	151	174	248	485
		152	193	219	247	340	616
28–40	177	136	188	218	253	365	701
		199	265	303	344	478	876
40–	211	158	224	262	306	441	842
		228	309	356	410	564	1044

Precipitation

Table 2.10 RELATION BETWEEN $M_5(D)/M_5(60)$ AND ANNUAL AVERAGE RAINFALL, WHERE D IS DURATION UP TO 120 MINUTES AND 60 IS 60 MINUTES DURATION

AAR (hundreds of mm)	Percentage ratio $M_5(D)/M_5(60)$ D (minutes)						
	1	2	5	10	15	30	120
5– 6	12	21	38	54	64	83	120
6– 8	11	20	36	52	62	81	123
8–10	11	19	35	50	60	79	126
10–14	11	18	33	47	57	76	130
14–20	10	17	31	45	54	74	134
20–28	9	15	27	41	50	71	139
28–40	8	13	24	37	46	69	145
40–	7	12	23	35	45	67	149

Table 2.11 RELATION BETWEEN PERCENTAGE VALUES OF $M_5(D)/M_5(2\text{-DAY})$ AND r, WHERE r IS $M_5(60 \text{ MIN})/M_5(2\text{-DAY})$

r	Percentage values of $M_5(D)/M_5(2\text{-day})$						
	60 min	120 min	4 h	6 h	12 h	24 h	48 h
0.12	12	18	26	33	49	72	106
0.15	15	21	30	37	53	75	106
0.18	18	25	34	41	56	77	106
0.21	21	28	38	45	60	80	106
0.24	24	32	41	48	63	81	106
0.27	27	35	44	51	65	83	106
0.30	30	38	48	54	68	85	106
0.33	33	41	51	57	70	86	106
0.36	36	44	54	60	73	88	106
0.39	39	47	57	63	75	89	106
0.42	42	50	60	66	77	90	106
0.45	45	53	63	68	79	92	106

values approaching 3.86 were observed in all regions; this value was therefore taken as the countrywide probable maximum value. The corresponding values for 24 hours were 9.3 for summer storms and 12.2 for winter storms. The results of the calculations of probable maximum falls for durations from 1 minute to 25 days are shown in *Table 2.12*.

Many of the problems concerned with floods relate to catchment areas rather than point rainfalls. Hence a necessary part of these studies was to devise a method of converting point falls into areal rainfalls. In order to estimate the areal reduction factor (ARF) of point falls for various return periods, durations and sizes of area, investigations were made using a large number of storms that had occurred in different parts of England. For a given catchment and duration a list was made of rainfalls in the maximum areal event for each year; then the ratio of each station's maximum to the areal maximum was calculated and mapped. The catchment mean of these ratios was determined and the areal reduction factor was obtained from the mean of these values over a number of years. Since it was found that the ARF did not vary with return period, variations in the ARF could be related quite simply to area and duration *(Figure 2.32)*.

Table 2.12 PROBABLE MAXIMUM PRECIPITATION VALUES (mm) RELATED TO
AVERAGE ANNUAL RAINFALL FOR DURATIONS FROM 1 MINUTE TO 25 DAYS

AAR (hundreds of mm)	Minutes: 1	10	30	Hours: 1	2	Duration 24	48	72	96	Days: 8	25
5– 6						231	254	262	270	287	343
6– 8	7	40	72	92	111	246	267	276	283	301	379
8–10	7	37	67	90	109	261	278	287	293	333	439
10–14						275	290	302	320	369	529
14–20	9	53	96	122	156	289	323	343	358	430	646
20–28	8	46	64	106	135	323	365	390	421	521	805
28–40	11	64	117	150	189	369	442	484	526	661	1066
40–	9	52	93	118	158	400	491	542	594	750	1238

Note. For durations between 1 minute and 2 hours, the upper figures apply to England and Wales and the lower ones to Scotland and Northern Ireland

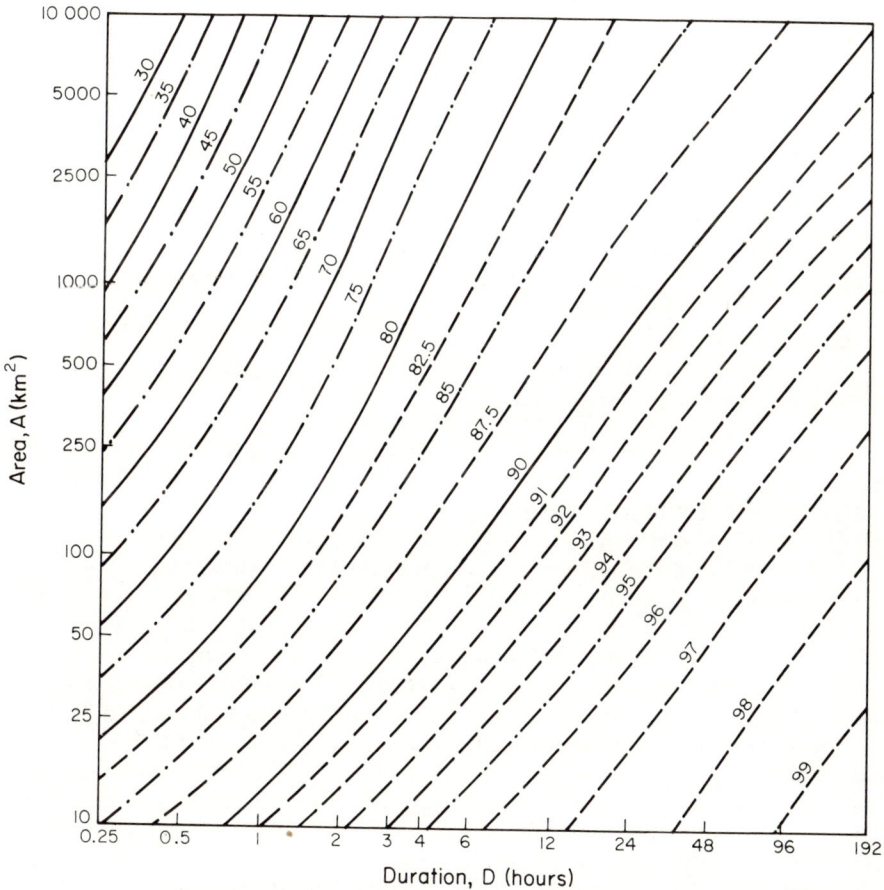

Figure 2.32   Areal reduction factor related to area and duration (courtesy Institute of Hydrology)

In order to examine the temporal distribution of rain during storms, various types of data were collected including hourly totals for storms of durations from 1 hour to 4 rainfall days, 2-minute totals from Cardington and Winchcombe, data from major flood-producing storms and for summer storms for an 8000 km² area in the Chilterns. To compare storm profiles each storm was centred on the shortest duration giving 50 per cent of the total rainfall, and the mean profile obtained for summer storms and winter storms separately. For each season, it was found that the mean profile of cumulative percentage of rainfall against cumulative percentage of duration was the same for all durations, the profile being symmetrical and the same for different return periods. Taking the 24-hour storms for each season, the median storm profile was determined, together with profiles for various percentages of the peak of the profile *(Figure 2.33)*, duration percentage intensity relations being shown as

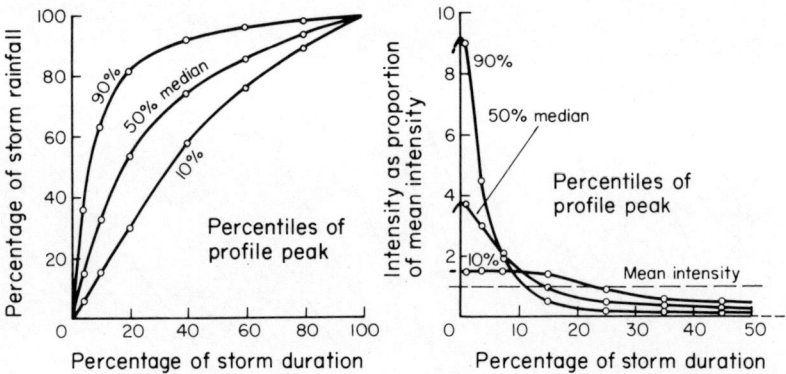

Figure 2.33   Summer storm profiles (courtesy Institute of Hydrology)

well as those for duration and percentage total. Little difference was found between the summer point profiles and the areal profiles constructed for the area in the Chilterns. Hence it was assumed that region-to-region differences and differences in rainfall type are incorporated in the seasonal profiles, and that these point profiles are adequate representations of the distribution of rainfall within storms.

Some of the country's most severe floods have resulted from the rapid melting of an extensive snow cover, but little work has been undertaken on this aspect of flood hydrology and few data exist that could be used as a basis for a study. The method adopted was first to scrutinise data on snow melt and snow density from abroad, to see if they were applicable to the UK. Then data on 9 a.m. snow depths, which had been collected since 1947 at some 100 stations, were assembled and analysed. The analysis concerned the annual maximum snow depths, and a relationship between return period and snow depth was established for the 'average' station *(Table 2.13)*. The value for the depth of snow that was equalled or exceeded on average once in two years was mapped *(Figure 2.18)*, but it was stressed that the map could be in error for mountainous areas because of lack of data.

Table 2.13 SNOW DEPTH–RETURN PERIOD RELATIONS FOR THE 'AVERAGE' STATION

Return period (years)	2	5	10	20	50	100
Snow depth (cm)	5.6	12.2	17.3	23.3	33.0	43.0

The density of lying snow and the snow water equivalent had been assessed at a number of Meteorological Office synoptic stations since the winter of 1964. These data, together with more collected in 1971–72 and data from other sources, were examined and a range of densities found from 0.05 to 0.42 g/cm^3. These values should be compared with the usually assumed 0.3 m of snow equals 25.4 mm of rain (a density of 0.083 g/cm^3).

The next step in the study was to establish the conditions giving rise to rapid snow melt. The return period of the 3-hour annual maximum temperature for days with snow lying was established *(Table 2.14)* and the

Table 2.14 ANNUAL MAXIMUM TEMPERATURE–RETURN PERIOD RELATIONS FOR THE 'AVERAGE' STATION WITH SNOW LYING

Return period (years)	2	5	10	25	50	100
Max. 3-hour temp. (°C)	4.2	5.4	6.2	7.2	7.9	8.6

relationship determined between the maximum air temperature and 24-hour changes in snow depth:

$$\text{Point snow melt} = 12.1\rho \text{ mm water d}^{-1}\,°C^{-1}$$

$$(2.17)$$

where ρ is the density of the lying snow.

For a typical density of 0.13 g/cm^3, this gives a point rate of melt of 1.56 mm water $d^{-1}\,°C^{-1}$, but for a more usual density of 0.30 it gives a melt rate of 3.6 mm. Densities have been found to reach 0.40; this figure combined with the estimated 100-year maximum temperature produces the rare snow melt rate of 4.2 mm/d, a value that might be typical of a lowland catchment. Such a melt rate might continue for two or three days with a depth of snow greater than 250 mm. The occurrence of rain during rapid snow melt such as this would obviously enhance the amount of runoff, but this topic was not covered in the report.

REFERENCES

1. Belasco, J. E., 'Characteristics of air masses over the British Isles', Meteorological Office *Geophysical Memoirs,* No. 87, HMSO (1952)
2. Shaw, E. M., 'An analysis of the origins of precipitation in northern England, 1956–1960', *Q. J. R. Met. Soc.,* **88,** 539–547 (1962)
3. Matthews, R. P., 'Variation of precipitation intensity with synoptic type over the Midlands', *Weather,* **27,** 63–72 (1972)
4. Mason, B. J. and Andrews, J. B., 'Drop-size distributions from various types of rain', *Q. J. R. Met. Soc.,* **86,** 346–353 (1960)

5. Best, A. C., 'The size distribution of raindrops', *Q. J. R. Met. Soc.,* **76,** 16–36 (1950)
6. Maidens, A. L., 'New Meteorological Office raingauges', *Met. Mag.,* **94,** 142–144 (1965)
7. Lowing, M. J. and Newson, M. D., 'Flood event data collation', *Water and Water Engng,* **77,** 91–95 (1973)
8. Meteorological Office, *Observers' Handbook,* 3rd edn, HMSO, 139 (1969)
9. Rodda, J. C., 'A note on the operation of rain-recorders during cold weather', *Met. Mag.,* **92,** 335–338 (1963)
10. Goodison, C. E. and Bird, L. G., 'Telephone interrogation of rain gauges'. *Met. Mag.,* **94,** 144–147 (1965)
11. Reynolds, G., 'Automatic rain gauges in North Scotland,' *Weather,* **23,** 88–93 (1968)
12. Lacy, R. E., 'Some measurements of snow density', *Weather,* **19,** 353–356 (1964)
13. Harrold, T. W., 'The measurement of rainfall using radar', *Weather,* **21,** 247–258 (1966)
14. Propert-Jones, J. R., 'The radar equation in meteorology', *Q. J. R. Met. Soc.,* **88,** 485–495 (1962)
15. Harrold, T. W., Bussell, R. and Grinsted, W. A., 'The Dee weather radar project', *Proc. Symp. on distribution of precipitation in mountainous areas, WMO No. 326,* **2,** 47–61 (1972)
16. Harrold, T. W., English, E. J. and Nicholass, C. A., 'The Dee weather radar project: the measurement of area precipitation using radar', *Weather,* **28,** 332–338 (1973)
17. Harrold, T. W., English, E. J. and Nicholass, C. A., 'The accuracy of radar-derived rainfall measurements in hilly terrain', *Q. J. R. Met. Soc.,* **100,** 331–350 (1974)
18. Kessler, E. and Wilk, K. E., *Radar measurement of precipitation for hydrological purposes,* Report No. 5, WMO Reports on IHD Projects (1968)
19. Harrold, T. W., *Estimation of rainfall using radar—a critical review,* Meteorological Office Scientific Paper No. 21 (1965)
20. Rodda, J. C., 'The systematic error in rainfall measurement', *J. Instn Water Engrs,* **21,** 173–177 (1967)
21. Robinson, A. C. and Rodda, J. C., 'Rain, wind and the aerodynamic characteristics of rain-gauges', *Met. Mag.,* **98,** 113–120 (1969)
22. Green, M. J. and Helliwell, P. R., 'The effect of wind on the rainfall catch', *Proc. Symp. on distribution of precipitation in mountainous areas, WMO 326,* **2,** 27–46 (1972)
23. Rodda, J. C., 'The rainfall measurement problem', *Proc. IASH General Assembly, Bern 1967,* Pub. No. 78, 215–231 (1968)
24. Gold, E., 'The splashing of rain', *Met. Mag.,* **66,** 153–158 (1931)
25. Ashmore, S. E., 'The splashing of rain', *Q. J. R. Met. Soc.,* **60,** 517–522 (1934)
26. Nipher, F. E., 'On determining the true rainfall in elevated gauges', *Am. Assoc. for the Advancement of Science,* **27,** 103–108 (1878)
27. Alter, J. C., 'Shielded storage precipitation gauges', *Monthly Weather Review,* **65,** 262–265 (1937)
28. Huddleston, F., 'A summary of seven years experiments with rain gauge shields in exposed positions, 1926–1932, at Hutton John, Penrith', *British Rainfall 1933,* **73,** 274–293 (1934)
29. Bleasdale, A., 'The measurement of rainfall', *Weather,* **14,** 12–18 (1959)
30. Winter, E. J. and Stanhill, G., 'Rainfall measurements at ground level', *Weather,* **14,** 367 (1959)
31. Green, M. J., 'Some factors affecting the catch of rain gauges', *Met. Mag.,* **99,** 10–20 (1970)
32. Green, M. J., 'Effects of exposure on the catch of rain gauges', *Proc. IASH Symp. on results of research in representative and experimental basins,* Wellington, IASH Pub. no. 97, Vol. II, 399–412 (1972)
33. Crawford, S. G., 'A recording gravimetric raingauge—towards an absolute reference instrument', *Met. Mag.,* **101,** 368–374 (1972)
34. Lacy, R. E. and Shellard, H. C., 'An index of driving rain', *Met. Mag.,* **91,** 177–184 (1962)
35. Rodda, J. C., Langbein, W. B., Kovzel, A. G., Dawdy, D. R. and Szesztay, K., *Hydrological network design—needs, problems and approaches,* Report No. 12, WMO Reports on IHD Projects (1969)
36. Glasspoole, J., 'Communication on the systematic error in rainfall measurement', *J. Instn Water Engrs,* **21,** 624–630 (1967)
37. Law, F., 'Measurement of rainfall, interception and evaporation losses in a plantation of sitka spruce trees', *Proc. IASH General Assembly, Toronto 1957,* Pub. No. 45, 397–411 (1958)

38. Shaw, E. M., 'The Devon River Authority raingauge network', *Weather,* **21,** 291–297 (1966)
39. Hutchinson, P., 'Estimation of rainfall in sparsely gauged areas', *Bull. IASH,* **14,** 101–119 (1969)
40. Holland, D. J., 'The Cardington rainfall experiment', *Met. Mag.,* **96,** 193–202 (1967)
41. Sutcliffe, J. V., 'The assessment of random errors in areal rainfall estimation', *Bull. IASH,* **11,** 35–42 (1966)
42. Stephenson, P. M., 'Objective assessment of adequate numbers of rain gauges for estimating areal rainfall depths', *Proc. IASH General Assembly, Bern 1967,* Pub. No. 78, 252–264 (1968)
43. Herbst, P. H. and Shaw, E. M., 'Determining raingauge densities in England from limited data to give a required precision for monthly areal rainfall estimates', *J. Instn Water Engrs,* **23,** 218–230 (1969)
44. Clarke, R. T. and Edwards, K. A., 'The application of the analysis of a variance to mean areal rainfall estimation', *J. Hydrology,* **15,** 97–112 (1972)
45. Langbein, W. B., 'Hydrologic data networks and methods of extrapolating or extending available hydrologic data', in *Hydrologic networks and methods,* WMO Flood Control Series No. 15, 13–41 (1960)
46. Bleasdale, A., 'Raingauge network development and design with special reference to the United Kingdom', *Proc. WMO/IASH Symp. on design of hydrological networks,* Quebec, IASH Pub. No. 67, 146–157 (1965)
47. Bleasdale, A. and Rowsell, H., *Average annual rainfall 1916–1950,* Director-General, Ordnance Survey, Chessington (1967)
48. *Climatological atlas of the British Isles,* HMSO (1953)
49. Wales-Smith, B. G., 'Monthly and annual totals of rainfall representative of Kew, Surrey, from 1697 to 1970', *Met. Mag.,* **100,** 345–362 (1971)
50. Manley, G., 'Manchester rainfall since 1765', *Mem. and Proc. Manchester Lit. and Phil. Soc.* **114,** 1–20 (1972)
51. Nicholas, F. J. and Glasspoole, J., 'General monthly rainfall over England and Wales, 1727–1931', *British Rainfall 1931,* **72,** 299–306 (1932)
52. Barrett, E. C., 'Regional variations in rainfall trends in North England, 1900–59', *Trans. Inst. Brit. Geographers,* **38,** 41–58 (1964)
53. Poulter, R. M., 'Rainfall and configuration', *Q. J. R. Met. Soc.,* **62,** 49–79 (1936)
54. Rodda, J. C., 'An objective method for the assessment of areal rainfall amounts', *Weather,* **17,** 54–59 (1962)
55. Bleasdale, A. and Chan, Y. K., 'Geographic influences on the distribution of precipitation', *Proc. Symp. on distribution of precipitation in mountainous areas,* WMO 326, **2,** 322–333 (1972)
56. Glasspoole, J., 'The fluctuations of annual rainfall', *British Rainfall 1921,* **61,** 288–300 (1922)
57. Gregory, S., 'Annual rainfall probability maps of the British Isles', *Q. J. R. Met. Soc.,* **83,** 543–549 (1957)
58. Senior, M. R., 'Changes in the variability of annual rainfall over Britain', *Weather,* **24,** 354–359 (1969)
59. Bleasdale, A., Boulton, A. G., Ineson, J. and Law, F., 'Study and assessment of water resources', *Proc. Symp. on conservation of water resources in the UK,* Instn Civil Engrs, 121–136 (1963)
60. Davis, N. E., 'Classified central England temperatures and England and Wales rainfall', *Met. Mag.,* **101,** 205–217 (1972)
61. Crowe, P. R., 'A new approach to the study of the seasonal incidence of British rainfall', *Q. J. R. Met. Soc.,* **66,** 285–316 (1940)
62. Holland, D. A., 'The prediction of monthly rainfall as exemplified by data from south-east England', *J. Agric. Sci.,* **58,** 327–342 (1962)
63. *Meteorological Office Annual Report 1967,* HMSO, 90 (1968)
64. Glasspoole, J., 'The reliability of rainfall over the British Isles', *J. Instn Water Engrs,* **35,** 174–199 (1930)
65. Bilham, E. G., 'Notes on sequences of dry and wet months in England and Wales', *Q. J. R. Met. Soc.,* **60,** 514–516 (1934)
66. Murray, R., 'Sequences in monthly rainfall over England and Wales', *Met. Mag.,* **96,** 129–135 (1967), and 'Sequences in monthly rainfall over Scotland', *Met. Mag.,* **97,** 181–183 (1968)

67. Stephenson, P. M., 'Seasonal rainfall sequences over England and Wales', *Met. Mag.*, **96**, 335–342 (1967)
68. Reynolds, G., 'Rainfall at Bidston, 1867–1951', *Q. J. R. Met. Soc.*, **79**, 137–149 (1953)
69. Beer, A., Drummond, A. J. and Furth, R., 'Sequences of wet and dry months and the theory of probability', *Q. J. R. Met. Soc.*, **72**, 74–86 (1946)
70. Murray, R. and Miles, M. K., 'Simple measure of the raininess of a month', *Met. Mag.*, **94**, 1–7 (1965)
71. Manley, G., 'On the frequency of snowfall in metropolitan England', *Q. J. R. Met. Soc.*, **84**, 70–72 (1958)
72. Manley, G., 'Snow cover in the British Isles', *Met. Mag.*, **76**, 28–36 (1947)
73. Manley, G., 'Snowfall in Britain over the past 300 years', *Weather*, **24**, 428–437 (1969)
74. Natural Environment Research Council, *Flood Studies Report* (1975)
75. Brooks, C. E. P. and Caruthers, N., *Handbook of statistical methods in meteorology*, HMSO (1953)
76. Williams, C. B., 'Sequences of wet and dry days fitted to a logarithmic series', *Q. J. R. Met. Soc.*, **76**, 91–96 (1950)
77. Sherriff, J. D. F., *Synthetic rainfall sequences*, TP72, Water Research Association (1970)
78. Rodda, J. C., 'Rainfall excesses in the United Kingdom', *Trans. Inst. Brit. Geographers*, **49**, 49–60 (1970)
79. MacLean, D. J., 'Rainstorm data', *Surveyor*, **104**, 36 and 56 (1945)
80. Norris, W. H., 'Sewer Design and the frequency of heavy rain', *Proc. Inst. Municipal Engrs*, **75**, 350–363 (1948)
81. Bilham, E. G., 'Classification of heavy falls in short periods', *British Rainfall 1935*, **75**, 262–280 (1936)
82. Holland, D. J., 'Rain intensity frequency relationships in Britain', *Meteorological Office Hydrological Memo No. 33* (1964)
83. Collinge, V. K., 'The frequency of heavy rains in the British Isles', *Civil Engng and Public Wks Rev.*, **56**, 341–344 and 497–500 (1961)
84. Road Research Laboratory, 'A guide for engineers to the design of storm sewer systems', *Road Research Note No. 35* (1963)
85. Bleasdale, A., 'The distribution of exceptionally heavy daily falls of rain in the United Kingdom, 1863–1960', *J. Instn Water Engrs*, **17**, 45–55 (1963) and **24**, 181–189 (1970)
86. Gumbel, E. J., 'Probability interpretation of the observed return period of floods', *Trans. Am. Geophys. Union*, **22**, 836–849 (1941)
87. Powell, R. W., 'A simple method of investigating flood frequency', *Civil Engr*, **13**, 105–107 (1943)
88. Chow, V. T., 'A general formula for hydrologic frequency analysis', *Trans. Am. Geophys. Union*, **31**, 231–237 (1951)
89. Gumbel, E. J., 'Statistical theory of floods and droughts', *J. Instn Water Engrs*, **12**, 157–184 (1958)
90. Jenkinson, A. F., 'The frequency distribution of the annual maximum or minimum values of meteorological elements', *Q. J. R. Met. Soc.*, **81**, 158–171 (1955)
91. Lowry, M. D. and Nash, J. E., 'A comparison of methods of fitting the double exponential distribution', *J. Hydrology*, **10**, 259–275 (1970)
92. Rodda, J. C., 'A countrywide study of intense rainfall for the United Kingdom', *J. Hydrology*, **5**, 58–69 (1967)
93. Myers, V. A., 'The estimation of the extreme precipitation as the basis for design floods, résumé of practice in the United States', *Proc. WMO/UNESCO/IASH Int. Symp. on floods and their computation*, **1**, 84–104 (1969)
94. Tucker, G. B., 'Some meteorological factors affecting dam design and construction', *Weather*, **15**, 3–113 (1960)
95. Singleton, F. and Helliwell, N. C., 'The calculation of rainfall from a hurricane', *Proc. WMO/UNESCO/IASH Int. Symp. on floods and their computation*, **1**, 450–461 (1969)
96. Meteorological Office, *Annual Report 1967*, 11 (1968)
97. Hershfield, D. M., 'Estimating the probable maximum precipitation', *J. Hydraulics Div., Proc. Am. Soc. Civil Engrs*, **87**, 99–116 (1961)
98. Watkins, L. H., 'Variations between measurements of rainfall made with a grid of gauges', *Met. Mag.*, **84**, 350–354 (1955)
99. Rigg, J. B., 'A statistical study of the distribution of rainfall over small and medium sized areas', *Weather*, **15**, 377 (1960)

100. Collinge, V. K. and Jamieson, D. G., 'The spatial distribution of storm rainfall', *J. Hydrology*, **6**, 45–57 (1968)
101. Burns, F., 'The relationship between point and areal rainfall in prolonged heavy rain', *Met. Mag.*, **93**, 308–312 (1964)
102. Reynolds, G., 'Heavy rains in the West of Scotland, 16–17 December, 1966', *Weather*, **22**, 224–228 (1967)
103. Holland, D. J., *Appendix to hydrological memoranda No. 33*, Meteorological Office, Met.O.8 (1968)
104. Subrahmanyam, V. P., 'Incidence and spread of continental drought', *Report No. 2., WMO Reports on IHD Projects* (1967)
105. *British Rainfall 1887*, **27**, 15–22 (1888)
106. *British Rainfall 1961*, **101**, 26–37 (1967)
107. Foley, J. C., 'Droughts in Australia', *Bulletin No. 43*, Bureau of Meteorology (1957)
108. Hounam, C. E., private communication
109. Rodda, J. C., 'A drought study in south-east England', *Water and Water Engng*, **69**, 316–321 (1965)
110. Parker, A. E., 'Dry and wet weeks and their relationship with the 500 millibar flow', *Weather*, **24**, 146–151 (1969)
111. Lawrence, E. N., 'Application of mathematical series to the frequency of weather spells', *Met. Mag.*, **83**, 195–200 (1954)
112. Chatfield, C., 'Wet and dry spells', *Weather*, **21**, 308–310 (1966)
113. Edwards, K. A. and Rodda, J. C., 'A preliminary study of the water balance of a small clay catchment', *Proc. Symp. on results of research in representative and experimental basins*, Wellington, IASH Pub. no. 97, 187–199 (1972)

Chapter 3

Evaporation

Instruments and instrument techniques

Evaporation is the process by which water is converted from a liquid to a vapour by the application of energy. It is a process governed by a number of interrelated controls and one that has been considered almost impossible to measure directly[1]. To add to these difficulties, various workers have adopted different names for the sum total of the water 'lost' to the atmosphere from the land surface. Perhaps Symons's[2] remark of 1867 is as applicable today as it was then: 'evaporation is the most desperate branch of the desperate science of meteorology'.

For a given set of conditions, evaporation (E) will take place from a water surface (at a temperature T) in proportion to the difference between the saturation vapour pressure at the water surface (e_w) and the vapour pressure of the air above (e_a). Molecules will continue to leave the water surface until e_a and e_w are equal. Condensation occurs when the net movement of molecules is reversed. Energy is required for evaporation; at 20°C some 585 calories of heat are required to evaporate one gram of water. The source of this energy can be solar radiation acting directly, the atmosphere's sensible heat, and the heat stored in the body of water itself. The rate of evaporation from a body of water is also controlled by wind speed, for the movement of air carries away water molecules as they leave the water surface. Other factors of lesser importance are atmospheric pressure, the quality of the water, and the size, shape and depth of the water body.

Evaporation can take place from a wet soil surface, from snow and ice surfaces and from vegetation wetted by rain. In addition, vegetation conveys water from the soil to the atmosphere by evaporation mainly through stomata in the leaves—a process known as transpiration. Some transpiration occurs through the leaf cuticle itself, but most occurs from the internal leaf surface to diffuse in a gaseous state through the stomata[3]. As a leaf transpires it develops a suction that has the effect of drawing water through the plant, a movement initiated by absorption of water from the soil through the root hairs. A state of equilibrium is assumed to exist within the plant between absorption from the soil, use of water within the plant and transpiration into the atmosphere. A plant may transpire several times its own weight of water on a hot summer's day. This water brings a supply of inorganic material into the plant and it helps to regulate the temperature of the leaves exposed to bright sunshine. As the soil dries it releases water less readily until finally its moisture content is so small, and the force retaining the water in the soil so large, that the plant can no longer absorb any more water.

This intake limit for plants is known as the *permanent wilting point;* soil water is no longer available, the plant cannot maintain its turgor and it wilts. At nearly the opposite end of the scale of available moisture is the state known as *field capacity.* Then the soil is holding all the water it can against the force of gravity, a state usually reached some two to three days after the soil has been saturated by heavy rain. At field capacity, the force exerted by the plant to absorb water through its root hairs is at a minimum. How the rate of transpiration depends on the soil moisture state has not been clearly established: some workers maintain that water is equally available for transpiration between field capacity and wilting point, but others consider that, at states less than field capacity, the transpiration rate is a function of the proportion of water remaining in the soil. Whether this is a smooth relationship or a stepped one is another matter for argument.

At field capacity, the plant has an abundant supply of moisture available in the soil, so its rate of transpiration is governed only by the climatic conditions pertaining. This rate is known as the *potential evapotranspiration rate* (Thornthwaite)[4] or the *potential transpiration rate* (Penman)[5]. It is considered to be largely independent of the kind of crop, but the crop should be short, about the same colour green as grass, and should completely cover the ground. When the soil moisture becomes limited, transpiration occurs at the actual rather than the potential rate, but it is not known at exactly what stage in the soil drying process this occurs. Because of these complications of the terminology, Penman terms are used in this text wherever possible: 'evaporation' generally, 'potential transpiration' where a crop has an unlimited water supply, and 'actual transpiration' where the water supply is limited.

Direct measurement of evaporation

Some methods of measurement require very complex instrumentation, others little more than a bucket. However, the cost and complexity of the apparatus involved is no guide to the reality of the measurement it produces.

Evaporation pans and tanks

The basis of evaporation determination by pan or tank is the establishment of a water balance. This water balance assumes the form:

$$E_o = P - \Delta W \qquad (3.1)$$

where P = precipitation
   ΔW = changes in water level
   E_o = evaporation from the pan or tank

There are various types of pan and tank in use. The British tank[6] is 610 mm (2 ft) deep and has sides 1.83 m (6 ft) long *(Figure 3.1a).* It is installed

(a)

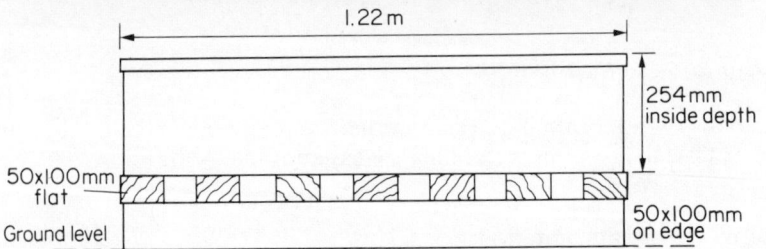

(b)

Figure 3.1   (a) UK evaporation tank; (b) US class A pan (courtesy Institute of Hydrology)

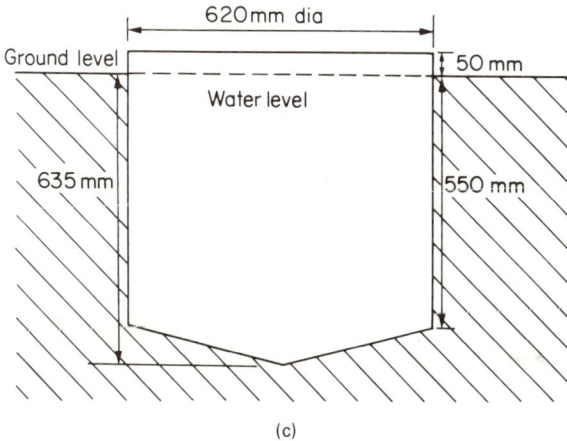

Figure 3.1   (c) USSR GGI–3000 pan (courtesy Institute of Hydrology)

with its rim protruding 76 mm (3 in) above ground level and filled so that the water surface is approximately level with the ground. Changes in water level are measured with a micrometer hook gauge usually fixed to one corner of the tank and surrounded by a stilling well. The water level is allowed to deviate by about 30 mm (1.2 in) in either direction before water is added, or alternatively removed.

The US Weather Bureau class A pan and the GGI–3000 pan of the Hydrometeorological Service of the USSR are in use at a number of sites in the UK. The former is a cylinder 1.22 m (4 ft) in diameter and 254 mm (10 in) in depth made of galvanised iron and supported by a framework of timber that stands on the ground *(Figure 3.1b)*. This framework allows air to circulate under the pan. In contrast, the GGI–3000 is sunk into the ground with its rim just protruding above the surface *(Figure 3.1c)*. It is cylindrical in shape with a surface area of 0.3 m², a depth of 600 mm and a cone-shaped bottom. In both instruments the day-to-day evaporation is determined by measuring the amount of water that has to be added, or taken out, to maintain a constant water level (178 mm depth for the class A pan); a point gauge set in a stilling well is used as a reference. Certain auxiliary equipment is employed with most pans and tanks, including an anemometer for determining wind speed over the pan, a rain gauge (which should be set at the same height as the pan) and maximum and minimum thermometers for water-temperature measurement. Wire mesh screens have been used in some countries for covering the water surface to stop interference by animals and birds, but this practice is usually avoided in the UK.

Atmometers

An alternative to a pan or tank is the atmometer. There is a wide range of shapes and sizes, but most consist of some porous material simulating the evaporating surface with a reservoir of distilled water behind it. The Piche atmometer is widely used in ecological studies and has also been favoured for hydrological purposes[7].

Lysimeters

Lysimeters are employed for assessing the evaporation from bare soil, or the transpiration from vegetation such as grass and other crops—even trees. Most lysimeters consist of a container that is installed so that its rim is level with the ground. Some lysimeters have been filled with an undisturbed block of soil, but this is a difficult task. It is simpler to try to replace the soil in the correct sequence and to compact it by some means. Another possibility is to build the lysimeter around the soil block, but this is very difficult unless the bottom is naturally water-tight.

The simplest form of lysimeter is based on a 44 gallon (200 litre) oil drum[8]. A drum is filled with soil and sunk into the ground almost to its rim *(Figure 3.2)*. A drain in the bottom of the drum conducts the percolate to a nearby container standing at the bottom of a shallow pit[9]. The soil in the drum is covered with turf; each day the turf is irrigated to maintain the soil at field capacity and the percolate is measured so that:

$$E_T = P + I - D \tag{3.2}$$

where E_T = potential transpiration amount
          I = amount of irrigation
          D = amount of drainage

Ideally, during dry weather the area around the lysimeter should also be irrigated so that an 'oasis' effect is not generated. An alternative to irrigating a lysimeter is to monitor the soil moisture changes that take place within it, in order to assess the actual transpiration. Another use for these gauges is to obtain a measure of the percolation through the soil into the bedrock at a particular site. A number of percolating gauges are in use at different sites, possibly the most widely known being the one thousandth of an acre gauges at Rothamsted Experimental Station.

The method normally used for obtaining measurements of actual transpiration is to weigh the soil and its container. Weighing lysimeters of

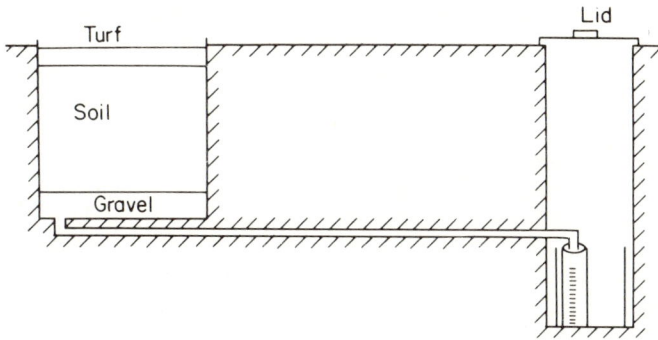

Figure 3.2   Simple lysimeter (not to scale) (courtesy Institute of Hydrology)

considerable size and cost have been installed in other countries: some operate on a beam balance, some on load cells and some use various hydraulic devices. Several types of hydraulic lysimeter have been built in the UK, the simplest employing a water-filled motor-scooter inner tube[10]. The inner tube supports the soil container and changes in weight are indicated by a manometer. A 1.5 tonne lysimeter[11] was installed at Rothamsted in the late 1950s, and this produced a record of weight changes registered through a beam balance.

A large forest lysimeter was the basis for the studies of the effects of forests undertaken at Stocks Reservoir[12,13]. Here a concrete wall was constructed to surround about 0.04 hectare (0.1 acre) of sitka spruce in a mature stand of 4.5 hectares (11 acres). The wall was founded on an impervious clay layer that was considered to seal the bottom, evaporation being deduced from measurements of rainfall on and drainage out of the lysimeter. This assumption of no leakage has also been employed to estimate evaporation in studies of a number of catchment areas. For example, the water balance of the Stour[14] has been examined and that of other basins such as the Ystwyth[15] and the Nidd[16]; usually long period balances have been effected to overcome the problem of lack of measurements of storage change. More recent studies have incorporated soil moisture measurements and assessments of soil heat storage[17] in order to refine the precision of the study.

Errors of measurement

Probably the biggest problem arising in the use of any instrument to directly measure evaporation and transpiration is that of representativeness. Pans sited on rafts in lakes are less likely to be criticised in this respect than those used on land, but the question arises: what do these measurements mean? How does the pan- or tank-measured evaporation relate to the loss from an imaginary lake or reservoir at that site? Can the results from a weighing lysimeter planted with wheat be extrapolated to the wheat fields surrounding it?

Pans and tanks suffer from some common disadvantages; foreign matter, growth of algae, interference by bird and animals, undetected leaks, splash in and splash out are factors that can distort records. These problems tend to be greater in sunken than in raised instruments, but sunken pans have the advantages of disturbing the air flow less and of possessing thermal regimes that more closely resemble those of lakes. Lysimeters suffer from some of these disadvantages and a number of additional ones, which have to be minimised in an attempt to make the lysimeter's thermal and moisture regimes match those of its surroundings as closely as possible. For example, a lysimeter with uncompacted soil has infiltration characteristics that differ from those of its surroundings: water may drain down the inside walls and at the bottom the soil will become saturated. To counteract this last tendency some lysimeters have been fitted with a means of exerting a suction—of, say, several atmospheres—on the bottom layers of soil.

Of course the siting of a lysimeter is particularly important. The centre of a large area of uniform vegetation is ideal, so that the air passing over the lysimeter has travelled across a large fetch. In this case an oasis effect should be avoided, but it is not clear exactly how large the fetch should be.

Instruments for indirect methods employing empirical or semi-empirical formulae

Indirect methods of assessing evaporation invariably depend upon measurements of other factors made at climatological stations. At the simplest type of station, observations are made once a day at 0900 GMT, but at some sites additional observations are made of certain instruments at 1500, 1800 or 2100 hours[18]. The instruments most commonly installed in a climatological station are: thermometers for observing the maximum and minimum temperatures, wet and dry bulb thermometers, a grass minimum thermometer, earth and soil thermometers at various depths, a sunshine recorder, a wind vane, a cup counter anemometer, a standard raingauge and a rain recorder. Most stations possess a hair hygrometer and a bi-metallic thermograph, and these record temperature and humidity on charts. In addition to these instruments some stations are equipped with anemographs, radiometers of different types, pans, tanks and lysimeters.

The ideal site for a climatological station is on level ground away from obstacles where an enclosure of about 9 m by 9 m (30 ft by 30 ft) can be erected.

Temperature, humidity and wind

The maximum and minimum thermometers are housed in a Stevenson Screen, designed to allow a relatively free passage of air and keep out direct sunlight. The screen also contains the wet and dry bulb thermometers, and usually the hygrometer and bi-metallic thermograph as well. Relative humidity, dew-point and vapour pressure can be determined from the wet and dry bulb reading using tables or a 'humidity' slide rule. It is assumed that these thermometers are well ventilated, but another source of error is the assumption that a single reading at 0900 provides a good estimate of the mean daily humidity. Speed or force, gustiness and direction are the parameters normally employed to describe wind. Most observations of direction are made by eye by using a wind vane, while force is estimated from the Beaufort scale. Mean wind speed increases with height above ground in a logarithmic fashion; most meteorological measurements are made at the 'standard' height of 10 m, but for climatological and hydrological purposes 2 m is the standard height. *Table 3.1* shows how speeds at different heights compare. Speed can be measured by an anemometer, the majority being of the three-cup type revolving horizontally. In the simplest the drive through a central shaft is geared to a counter, which is usually observed daily for run of wind. Others, the contact or generator types, record remotely, including the special lightweight models that have been developed[19]. The Dines pressure-tube anemograph records wind speed and direction on a chart.

Table 3.1 VARIATION OF WIND SPEED WITH HEIGHT ABOVE GROUND (AFTER McCULLOCH[63])

Height (m)	3	4	5	10	15	20
Ratio $\dfrac{\text{speed at 2 m}}{\text{speed at given height}}$	0.94	0.92	0.89	0.78	0.72	0.68

Radiation

Most solar radiation is contained within the range of wavelengths between 0.3 and 4 μm (1 μm or micron $=10^{-6}$ m); this includes the direct and the diffuse radiation that has been scattered and reflected by the atmosphere. Some of the solar radiation incident to the ground surface is reflected, the reflection coefficient (albedo) varying with the type of surface from about 0.05 for water to about 0.90 for fresh snow. Terrestrial radiation has a much wider range of longer wavelengths (4–100 μm) than solar radiation. The units of measurement are usually calories, milliwatts or langleys.

The Campbell-Stokes sunshine recorder *(Figure 3.3a)* is one of the simplest radiation instruments. It records duration of bright sunshine by focusing the sun's rays to scorch a line on a special card. The length of the burn gives an estimate of the total solar radiation (R_c) from:

$$R_c = R_A (a + bn/N) \tag{3.3}$$

where R_A = total solar radiation that would have been recorded at the site in the absence of an atmosphere
n = duration of bright sunshine (hours)
N = maximum possible duration of sunshine (hours)
a and b = local constants ranging from 0.08 to 0.22 and 0.69 to 1.04 in Britain[20]

Values for a and b are obtained by measuring R_c with a more precise radiometer sited near the sunshine recorder and by substituting for R_c in equation 3.3, which is usually known as the Ångström equation.

Another simple instrument is the Gunn-Bellani distillometer[21] *(Figure 3.3b);* water contained in a glass bulb exposed to radiation at ground level is

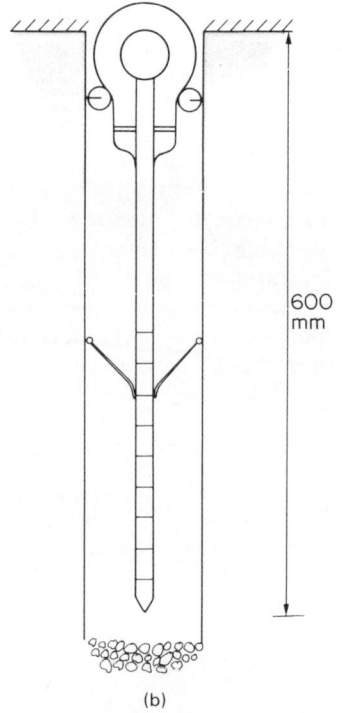

Figure 3.3   Radiometers: (a) Campbell–Stokes sunshine recorder; (b) Gunn–Bellani distillo-meter; (c) bi-metallic actinograph; (d) Kipp and Zonen solarimeter (courtesy Institute of Hydrology)

(e)

(f)

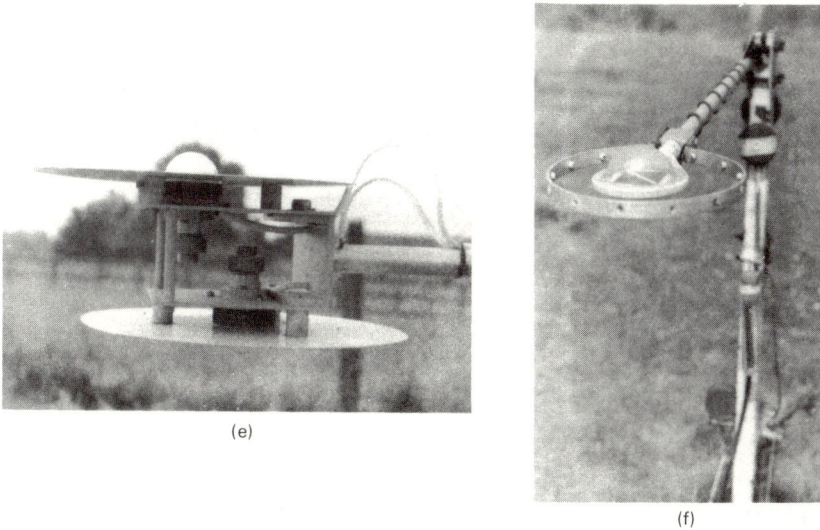

Figure 3.3   (e) albedo meter, (f) Funk net radiometer (courtesy Institute of Hydrology)

evaporated and condenses in the cool stem of the device's glass envelope buried some 600 mm below ground level. The bi-metallic actinograph[22] *(Figure 3.3c)*, another simple instrument, has a central blackened strip that senses the radiation and two shielded compensating strips alongside. The movement of the bi-metallic strip is transmitted mechanically to a pen that records on a daily chart driven by clockwork. Both these devices require calibrating against a more precise radiometer, the volume of the distillate or the expansion of the strip being proportional to the solar radiation.

These more precise solar radiation measurements are usually made with Moll-Gorcynski solarimeters manufactured by Kipp and Zonen *(Figure 3.3d)*. The output from one of these thermopiles can be recorded on a potentiometric recorder, by an electrolytic integrator[23] or by means of an integrating motor coupled to a printout mechanism or a paper-tape punch. One small light solarimeter[24] has been designed for field use to operate with a galvanometer; mains power is not required but observations of the galvanometer have to be made. The albedo of any surface can be measured using two of these solarimeters, one pointing up and the other down *(Figure 3.3e)*. Normally the two solarimeters are attached to a bar about 1 m long mounted horizontally on a photographic tripod, but at sites where solar radiation is already being recorded it may be sufficient to record the output from a second solarimeter inverted over one or more representative surfaces. Another method is to mount a pair of solarimeters on an aircraft[25] in order to obtain more representative measurements over a wider range of surfaces than by a ground survey.

Measurements of the net flux of solar and terrestrial radiation have been made with radiometers based on a design described by MacDowall[26]. In this instrument, a thermopile is housed between the upper and lower surfaces of a flat plate about 70 mm square and 3 mm thick. The plate, which is painted black, is exposed horizontally about 1 m above the surface in question and is ventilated artificially by an electric blower to keep its immediate surroundings as constant as possible. In the Funk[27] net radiometer *(Figure 3.3f),* which has largely replaced the MacDowall instrument, the plate, which is much smaller, is covered by a polythene dome kept inflated by a supply of nitrogen piped from a cylinder.

Measurements of soil heat flux can be made by means of heat flux plates installed in the soil. Alternatively, it can be determined from soil-temperature profile changes and measurements of soil heat capacity. For a review of instruments for heat flux and radiation measurement, readers are advised to consult Monteith's *Survey of Instruments for Micrometeorology*[28].

Automatic weather stations

Several different types of self-contained station have been designed and operated that can replace the conventional climatological station. The automatic weather station (AWS) is most usually battery powered, built to operate at isolated sites and to withstand the extremes of weather that may be experienced there. Earlier types of AWS tended to use chart recording, but the advantages of recording on tape soon became apparent. Progression to tape recording (magnetic tape in most instances but paper tape in some) led to the concept[29] of a system that could handle the data automatically from start to finish. The sensor, the transducer, the logger and the computer for quality control, storage and analysis of the data are the essential parts in this system.

Features common to all types of AWS are a mast, on which the sensors are supported by short booms, and a logger box. In some types the logger box is fixed to the mast and in others it is separate from it *(Figure 3.4).* Exhaustive tests of the different types of AWS have been carried out[30,31] in the laboratory and in the field; sometimes pairs of stations have been operated together[32]. Their hourly and daily records have been compared statistically to show whether or not there were significant differences between stations. Printouts on, say, an hourly basis *(Table 3.2)* in themselves provide very interesting material, with a detail unmatched by the conventional station. In addition to this type of listing, three main items of information are usually required[33]. These are a daily summary of the climatic variables, rainfall totals and intensities, and an estimate of potential evaporation.

Networks

Like the raingauge network, the national evaporation network has not been planned scientifically and relatively few studies have been made of its adequacy. Evaporation pans and tanks are in use at a number of sites, but the only true network of instruments for the direct measurement of the water 'lost' to the atmosphere is the one consisting of thirty or so simple lysimeters that has been developed by F. H. W. Green. Acquisition of records from this network commenced in the mid-1950s and the data are published annually in *British Rainfall*[34].

Figure 3.4   Automatic weather station (courtesy Institute of Hydrology)

Table 3.2   TYPICAL PRINTOUT FROM AN AUTOMATIC WEATHER STATION (EISTEDDFFA GURIG, PLYNLIMON, WALES) (COURTESY INSTITUTE OF HYDROLOGY)

```
09 00   20 /11 /1971
```

	SOLAR RAD C/SQC	NET RAD C/SQC	DEPRN DEG.C	TEMP DEG.C	WIND RUN KLMS	WIND DIR DEGS	RAIN MMS	N	NE	E	SF	S	Sw	W	Nw
09	001.8	+000.6	00.00	-02.0	012.2	169	00.00	004	000	000	000	000	005	002	001
10	004.1	+001.6	00.00	-01.7	017.9	213	00.00	000	000	000	000	001	011	000	000
11	001.9	+000.8	00.02	-01.5	017.2	213	00.00	000	000	000	000	002	010	000	000
12	001.1	+000.4	00.01	-01.5	022.9	204	00.00	000	000	000	000	007	005	000	000
13	002.2	+000.6	00.04	-01.5	031.6	202	00.00	000	000	000	000	008	004	000	000
14	001.3	+000.4	00.00	-01.6	028.7	209	00.00	000	000	000	000	005	007	000	000
15	000.5	+000.0	00.00	-01.6	037.4	199	00.00	000	000	000	000	010	002	000	000
16	000.0	+000.0	00.00	-01.6	031.9	204	00.00	000	000	000	000	008	004	000	000
17	000.0	+000.0	00.05	-01.3	032.9	212	00.00	000	000	000	000	003	009	000	000
18	000.0	+000.0	00.08	-00.1	037.2	224	08.50	000	000	000	000	000	012	000	000
19	000.0	+000.2	00.04	+01.6	031.8	222	13.00	000	000	000	000	001	011	000	000
20	000.0	+000.4	00.07	+04.3	024.5	232	15.50	000	000	000	000	000	011	001	000
21	000.0	+000.0	00.18	+07.8	036.0	251	14.50	000	000	000	000	000	004	008	000
22	000.0	-000.2	00.00	+08.3	034.5	247	08.50	000	000	000	000	000	007	005	000
23	000.0	-000.4	00.00	+08.4	033.8	251	05.00	000	000	000	000	000	005	007	000
00	000.0	-000.4	00.00	+08.3	031.9	237	07.50	000	000	000	000	000	010	002	000
01	000.0	-000.5	00.00	+08.2	035.1	241	02.00	000	000	000	000	000	008	004	000
02	000.0	-000.6	00.00	+08.1	036.4	241	05.00	000	000	000	000	000	008	004	000
03	000.0	-000.6	00.00	+07.9	036.4	244	02.50	000	000	000	000	000	008	004	000
04	000.0	-000.6	00.00	+07.6	037.6	243	01.00	000	000	000	000	000	009	003	000
05	000.0	-000.6	00.00	+07.4	037.6	248	03.00	000	000	000	000	000	008	004	000
06	000.0	-000.6	03.00	+06.6	036.5	259	02.00	000	000	000	000	000	003	009	000
07	000.0	-000.6	00.00	+05.8	036.6	250	04.50	000	000	000	000	000	006	006	000
08	000.8	-000.2	00.01	+04.8	030.1	301	02.50	000	000	000	000	000	001	004	007

The network of climatological and agrometeorological stations *(Table 3.3),* which supplies the records for use in indirect methods of evaporation estimation, reached a total of over 500 stations by the end of 1971. Only a small proportion of these stations is equipped with radiometers of one sort or another. Similarly, there is only a small number of sites, probably not more than thirty, where automatic weather stations are used operationally. Many of these are in representative and experimental basins where network design has usually been on the basis of the distribution of altitude within a basin[35]. There are indications, however, that within basins of diverse topography potential tra.ispiration may vary considerably. For example, Cole and Green[36] in their study of the Derwent used measurements of windspeed, temperature and radiation at a small number of sites to calculate weekly potential transpiration amounts for a grid of points covering the 118 km² catchment at 1 km intervals. For the higher areas, the effects of wind and radiation combined to give values of E_T that were 1.3 times greater than the basin mean, while for the low sheltered areas values were only 0.8 times the mean. A similar variability has been found in the results from AWS in the Plynlimon catchments[37] *(Figure 3.5);* although there are differences between pairs of stations at the same site, the more significant differences are between the lower sheltered site at Tanllwyth and the other more exposed stations.

Table 3.3 STATIONS REPORTING CLIMATOLOGICAL INFORMATION (INCLUDING THOSE NOT IN THE METEOROLOGICAL OFFICE ORGANISATION) ON 31 DECEMBER 1971

Region of UK	Station					Autographic record		
	Observatories	*Synoptic*	*Agrometeorological*	*Climatological*	*Rainfall**	*Sunshine*	*Rainfall*	*Wind*
Scotland, north	1	9	0	30	332	28	10	15
Scotland, east	0	10	8	63	596	53	19	11
Scotland, west	1	13	1	54	545	29	20	14
England, north-east	0	10	4	25	515	27	16	10
England, east	0	11	14	20	571	29	18	11
England, Midlands	0	13	17	46	1311	60	32	15
England, south-east (including London)	1	19	18	43	870	63	33	18
England, south-west	0	13	8	33	562	33	10	6
England, north-west	0	5	3	24	507	24	12	14
Wales, north	0	2	3	16	291	11	4	2
Wales, south	0	7	9	15	383	22	4	4
Isle of Man	0	2	0	1	19	3	1	2
Scilly and Channel Isles	0	3	0	3	21	7	0	2
Northern Ireland	0	7	7	49	308	25	28	9
Total	3	124	92	422	6831	414	207	133

*Includes stations in earlier columns

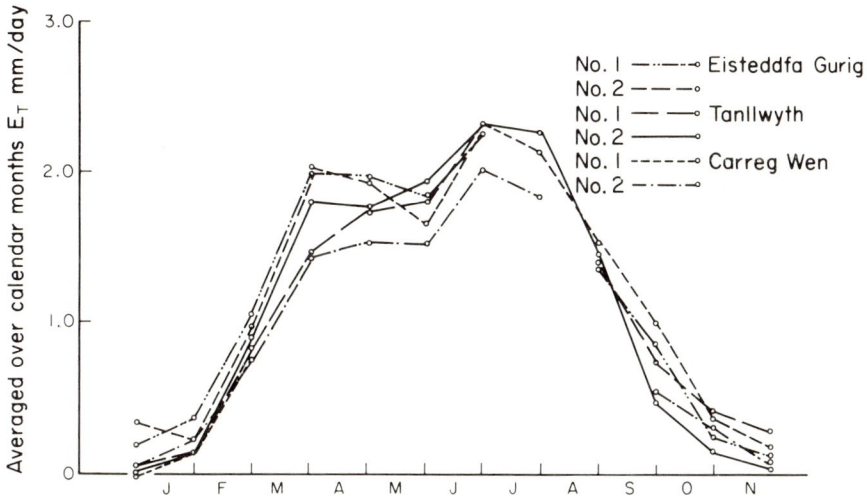

Figure 3.5   Penman estimate of potential transpiration using net radiation in the energy term; data taken from three pairs of automatic weather stations, Plynlimon, 1972 (from ref. 37, courtesy Institute of Hydrology)

Indirect approaches

Methods abound for the indirect determination of the amount of water 'lost' to the atmosphere from the land surface. Some methods are based on physical theory, some on very broad empiricisms and others on a mixture of the two. To be strict, most of the methods that claim a basis in theory fall into one or other of two categories, the energy budget approach and the aerodynamic (vapour flow) approach.

The energy budget approach

As in the water balance method, this approach employs an equation with evaporation as the unknown; the equation is solved in this case for the amount of energy required for evaporation. All other sources and sinks are accounted for. With any type of natural surface a balance is set up between income and expenditure of energy. Penman[38] described this heat balance *(H)* as:

$$\text{Income} \quad H = R_c (1 - r) - R_B \tag{3.4}$$

$$\text{Expenditure} \quad H = E + K + (S + C + M) \tag{3.5}$$

where
R_c = incoming short-wave radiation reaching the surface
r = albedo (reflection coefficient of the surface)
R_B = outgoing long-wave radiation from the surface
E = energy used in evaporation (latent heat)
K = energy used in heating the air (sensible heat)
S = heat transfer to the surface (sensible heat)

C = heat conducted in or out of the surface (sensible heat)

M = energy used in melting snow (latent heat)

Terms for the energy used in photosynthesis, the net energy stored in a crop and any lateral heat transfer are not often considered, and for most practical purposes (excluding assessing lake evaporation) S, C and M are neglected, S less frequently than the others.

The measurement of R_c, and also its estimation through use of the Ångström equation, has already been discussed. Values for r have been determined for a number of natural surfaces *(Table 3.4)*, variations

Table 3.4 ALBEDO VALUES FOR DIFFERENT SURFACES

r	*Surface*
0.80–0.90	new snow
0.60–0.80	old snow
0.40–0.60	melting snow
0.40–0.50	ice
0.20–0.28	lawn, pasture, green cereals
0.16–0.24	rough pasture and stubble
0.16–0.24	bracken
0.15–0.20	coniferous forest
0.15–0.17	houses
0.12–0.19	deciduous forest
0.12–0.16	heather
0.17–0.18	gorse
0.05–0.10	wet peat
0.05–0.10	water

occurring with sun angle and weather conditions *(Figure 3.6)*. A map of summer albedo values[39] has also been constructed for the whole of England and Wales. Values of R_B can be calculated from an equation developed by Brunt[40]:

$$R_B = \sigma T_a{}^4 \ (0.56 - 0.09 \sqrt{e_d}) \ (1 - 0.09m) \tag{3.6}$$

where $\sigma T_a{}^4$ = theoretical black-body radiation at $T_a(\mathrm{K})$

e_d = actual vapour pressure of the air (mm of Hg) at height h (usually the screen height)

m = fraction of sky covered by cloud (tenths)

For practical purposes equation 3.5 reduces to:

$$H = E + K \tag{3.7}$$

Because K cannot be measured, division of the remaining energy between that used for evaporation and that for heating the air has to be based on an assumption and the measurement of other factors. This assumption is that the transport of heat and the transfer of vapour by eddy diffusion are similar and then the ratio of K/E, the Bowen[41] ratio (β), is given by:

$$K/E = \beta = \frac{\gamma(T_s - T_a)}{(e_s - e_d)} \tag{3.8}$$

where γ = the psychrometric constant
T_s = mean surface temperature
T_a = mean air temperature
e_s = saturation vapour pressure at the evaporating surface at temperature T_s

Thus

$$E = \frac{H}{(1 + \beta)} \qquad (3.9)$$

A typical value of β for low vegetation with an adequate supply of soil moisture is 0.2; this is fortunate because, when the value of β lies between -0.5 and 1.0, the relative error in the evaporation estimate is smaller than the Bowen ratio itself[42]. On the other hand, β values from 1 to 4 were found to apply on a dry sunny day over the Scots and Corsican pine in Thetford Forest[43].

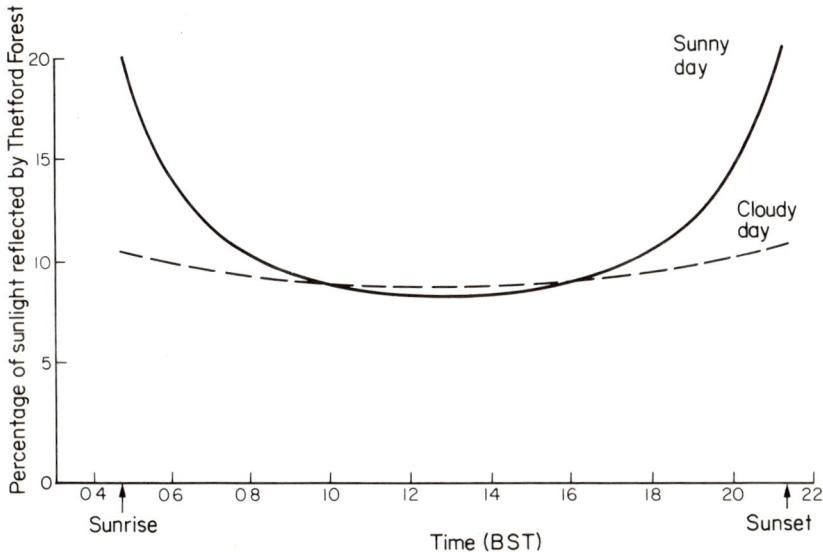

Figure 3.6 Variation of albedo of pine forest during typical sunny and cloudy days in June 1970 (courtesy Institute of Hydrology)

Estimation of the evaporation from the energy budget has been the technique adopted by a number of workers. For some purposes E (mm) has been obtained from:

$$E = H/L \qquad (3.10)$$

where $L = 58.5$ cal/cm². Values of E estimated in this manner are obviously most reliable when H is measured directly by net flux radiometer, a further refinement being measurements of the soil heat flux (S). Where only R_c is measured, Davies[44] showed that the net daytime radiation (R_n) could be estimated from R_c in units of g cal cm^{-2} day^{-1} by:

$$R_n = 0.62\,R_c - 24 \qquad\qquad (3.11)$$

He demonstrated that for some West African sites the water equivalent of R_n gave an excellent agreement with the potential evapotranspiration measured by simple lysimeters. When the Bowen ratio is employed, it requires measurements of the vertical profiles of temperature and humidity for its derivation. Average values of the gradients are established from series of measurements, in one case a special computer being constructed[45] to analyse the results. In another case[46] evaporation calculated by use of the Bowen ratio was employed to determine the surface resistance of several crops.

In the very detailed study undertaken in Thetford Forest *(Figure 3.7)*, temperature and humidity profiles were obtained from wet and dry bulb aspirated quartz-crystal thermometers mounted at six heights above and three within the canopy[47]. Two sets of instruments were used mounted on booms facing in opposite directions, those on the upwind side of the mast being used for analysis. Once a minute the outputs from the thermometers were fed into a computer-controlled data acquisition system. The computer

Figure 3.7   Instrument towers in Thetford Forest (courtesy Institute of Hydrology)

simplified the data, calculated averages and output the results on punched tape.

The aerodynamic approach

Evaporation depends on two requirements being met: sufficient energy to provide the latent heat of vaporisation, and a means for removing the vapour produced. Aerodynamic (vapour flow) methods concern the second of these matters. Some approaches are relatively simple and empirical, others involve an appreciation of the turbulent transport of vapour by the process of eddy diffusion. The former usually need simple instruments, the latter have more complex requirements.

Most empirical formulae have the form of John Dalton's original equation:

$$E = a (1 + bu) (e_s - e_d) \tag{3.12}$$

where a and b = constants, u = wind speed measured at height h.

Expressions of this type have been widely used for estimating open water evaporation; for example, in the study of Lake Hefner in the United States, Penman[48] adjusted the Rohwer version of this formula so that it could be used with wind speed measurements at the normal 2 m height:

$$E = 0.40 (1 + 0.17u_{2m}) (e_s - e_d) \tag{3.13}$$

Profile techniques are concerned with the turbulent transfer of water vapour at a small distance (z) above the ground. One of the three principal vertical fluxes, the water vapour flux (g cm^{-2} s^{-1}), is given by:

$$E = -\rho K_w \frac{\partial q}{\partial z} \tag{3.14}$$

where ρ = air density
K_w = eddy transfer coefficient for water vapour
q = specific humidity

Within a metre or two of the ground surface, it is assumed that storage and sideways effects are negligible, and there is evidence of the identity of the eddy diffusivities for vapour and momentum. Assuming this part of the atmosphere to be in an adiabatic state and horizontally uniform conditions to apply, Thornthwaite and Holzman[49] derived the following relationship from equation 3.14:

$$E = \frac{\rho k^2 (q_1 - q_2) (u_2 - u_1)}{[\log_e (z_2/z_1)]^2} \tag{3.15}$$

where u_1 and u_2 = wind speeds at two levels z_1 and z_2
q_1 and q_2 = specific humidities
k = von Karman's constant (approx 0.4)

Pasquill[50] showed that equation 3.15 could be rewritten as:

$$E = Bu_2 (e_1 - e_2) \tag{3.16}$$

and

$$B = \frac{k^2 M[1 - (u_1/u_2)]}{RT[\log (z_2/z_1)]^2}$$ (3.17)

where e_1 and e_2 = vapour pressures at the two levels
               M = molecular weight of water
               R = universal gas constant
               T = air temperature (K)

He applied this formula to observations from flat pasture land, where profiles of temperature, humidity and windspeed observations were made at 3-minute intervals over 15-minute periods, the sensors being at levels up to 200 cm. Evaporation amounts were determined from hourly and daily means of these observations for five selected days in May, the former ranging from 1.4 to 3.6 mm/day and the latter from 1.0 to 2.7 mm/day. Later, Rider[51] employed a modified form of equation 3.15, accounting for a zero plane displacement (*d*) to assess hourly evaporation rates from an oatfield. In a second study of several surfaces[52] equation 3.15 was used in the form:

$$E = \frac{3.46 \times 10^{-5} (e_1 - e_2) (u_2 - u_1)}{T_A [\log_e (z_2 - d)/(z_1 - d)]^2}$$ (3.18)

where e = vapour pressure (mb)
               u = wind speed (cm/s)
     z and d = heights (cm)
             T_A = the absolute temperature
               E = evaporation (g cm^{-2} s^{-1})

Measurements of humidity were made at two heights and wind speed at five heights over various crops and bare soil, with three sets of apparatus. Instrumental errors, the difficulties of correctly assessing the height of the zero plane displacement and the limited periods when equation 3.15 applied were some of the problems encountered. Nevertheless daily water losses were determined, and from the larger amounts that applied to the peas and sprouts it was concluded that the concept of equal potential transpiration for different crops should be treated with caution.

Subsequently this profile approach has been modified by various workers but, as Oliver[53] points out, its practical application is governed by four conditions. These are: the site should be surrounded by a sufficiently large, uniform fetch; the lowest sensors should be high enough to avoid any irregular influences of the vegetation and errors in the assessment of the zero plane displacement; the conditions should be steady; the instrumentation should be good enough to measure temperatures to 0.01°C and wind speeds to better than 1 per cent.

Eddy flux measurement

For the sake of completeness, this approach to the assessment of evaporation is mentioned here, although few studies involving the direct measurement of the eddy flux have been undertaken in the UK. Following initial work in Australia with the evapotron, the fluxatron has been tested in a number of different environments, most recently over the surface of the sea[54,55].

Heat flow measurements of sap flux

At the University of Oxford, a study was made of the technique of injecting a pulse of heat into the sap flow of a tree and detecting it farther up the stem to give a measure of the sap flux and hence the water consumption of the tree[56]. By enclosing the tree in a plastic bag and measuring the moisture content of the air passing into and out of the bag, the transpiration from the tree can be determined and the heat pulse method calibrated *(Figure 3.8)*.

Figure 3.8   Plastic bag used for determination of transpiration, Bagley Wood, near Oxford (courtesy Dr. L. Leyton)

Empirical and semi-empirical methods

Because of the instrumental and other problems involved in applying the approaches based on physical theory, much effort has been devoted to devising methods for estimating evaporation from readily available records of climate. Probably the most widely used method, in global terms, is the one devised by Thornthwaite, but there are others that have also been employed extensively. Penman, Turc, Ferguson, Blaney and Criddle, McIlroy and Budyko are but a few of the originators of these methods. Their efforts and those of others have been compared, modified and the results applied to assessing evaporation in many and differing circumstances with varying degrees of success. Some methods are now employed on a routine basis; others are simply hydrological museum pieces.

Penman

The Penman method and its several modifications must be regarded as one of the most widely known and respected systems[57]. This is a combination method, the energy budget and aerodynamic approaches being combined to eliminate the seldom measured temperature of the evaporating surface, a measurement the two methods require when used separately. In its original form (1948) the Penman method[48] was essentially a two-stage process involving the calculation of the evaporation from a hypothetical open water surface *(E_o)*, then the conversion of E_o to the evaporation from turf *(E_T)* from:

$$E_T = fE_o \qquad (3.19)$$

Results from the Rothamsted lysimeters and an evaporation tank showed that average values of f were: 0.6 (November, December, January, February), 0.7 (March, April, September, October) and 0.8 (May, June, July, August). Combining the energy budget and aerodynamic approaches to estimate E_o gives:

$$E_o = \frac{\Delta H + \gamma E_a}{\Delta + \gamma} \qquad (3.20)$$

where Δ = slope of the saturation vapour pressure curve against temperature at T_a in mm Hg per °F
γ = 0.27 mm Hg per °F at sea level

$$H = (1 - r)R_A (0.18 + 0.55n/N)$$
$$- \sigma T_a^4 (0.56 - 0.09\sqrt{e_d}) (0.10 + 0.90n/N) \qquad (3.21)$$

$$E_a = 0.35 (1 + 9.8 \times 10^{-3} u_2) (e_a - e_d) \qquad (3.22)$$

Following his original publication, Penman[58] modified the wind term in the light of the Lake Hefner experience so that it became:

$$E_a = 0.35 (0.5 + 10^{-2} u_2) (e_a - e_d) \qquad (3.23)$$

where e_a = saturation vapour pressure at air temperature T°F.

Further work on albedo measurements[59] showed that the two-stage process was unnecessary[60] and that for turf and similar short green vegetation with $r = 0.25$, the heat budget could be determined from:

$$H = 0.75R_A (a + bn/N) -$$
$$\sigma T_a^4 (0.56 - 0.09\sqrt{e_d}) (0.10 + 0.90n/N) \qquad (3.24)$$

To account for the extra roughness of a crop compared to open water the aerodynamic term became:

$$E_a = 0.35 (1 + 10^{-2}u_2) (e_a - e_d) \qquad (3.25)$$

resulting in

$$E_T = \frac{\Delta H + E_a\gamma}{\Delta + \gamma} \qquad (3.26)$$

Later, Monteith[61] introduced factors concerned with the nature of the crop, namely, its aerodynamic resistance (r_a) and surface (stomatal) resistance (r_s) to estimate the latent heat flux (λE) from an unsaturated surface:

$$\lambda E = \frac{\Delta H + \rho c\,[e_s(T) - e_d]/r_a}{\Delta + \gamma\,[1 + (r_s/r_a)]} \qquad (3.27)$$

with c being the specific heat of the air, $e_s(T) = e_a$, i.e. the saturation vapour pressure at the air temperature T. This development has the aim of supplementing the largely meteorological approach with the physiological concepts of the control of the passage of water vapour from the leaf to the atmosphere; one control is the size and distribution of stomata, the other the aerodynamic properties of the plant surface. The main problem with this variant of the Penman approach is the difficulty of obtaining adequate measurements of the crop factors. Several studies have been made in which values of r_s and r_a have been determined[62] and one set of results (*Table 3.5*) shows considerable variations from month to month and surface to surface.

Table 3.5 VALUES OF r_a AND r_s (sec/cm) AND EVAPORATION (mm/day) FOR VARIOUS SURFACES IN SOUTHERN ENGLAND (AFTER SZEICZ *et al.*[62])

		April	*May*	*June*	*July*	*Aug*	*Sept*
Open	r_a	1.0	1.11	1.25	1.25	1.25	1.25
water	r_s	0	0	0	0	0	0
	E	1.64	2.77	3.42	3.30	3.10	2.46
Pine	r_a	0.03	0.03	0.03	0.03	0.03	0.03
forest	r_s	1.14	1.06	0.98	1.25	1.50	1.20
	E	1.00	1.90	2.70	2.00	1.34	1.10
Potatoes	r_a	0.65	0.63	0.61	0.45	0.33	0.38
	r_s	1.30	1.26	1.13	0.45	0.60	1.10
	E	0.85	1.52	2.05	2.67	2.20	1.08
Lucerne	r_a	0.54	0.44	0.60	0.43	0.33	0.52
	r_s	0.70	0.56	0.64	0.35	0.30	0.60
	E	1.20	2.36	2.46	2.90	2.72	1.40
Penman estimate	E_T	1.45	2.60	3.15	2.78	2.12	1.29

Making Penman estimates

A number of workers have devised ways of simplifying the calculation of potential evaporation and potential transpiration. There are nomograms, tables and computer programs available, as well as worked examples of the method. For instance, McCulloch[63] simplified the procedure so that values for 10-day periods could be determined from the sum of three products, the first two representing the heat budget and the third the aerodynamic term. The appropriate values for calculating the products are obtained from tables with the altitude and latitude of the station being taken into account.

Table 3.6   INFORMATION NEEDED TO MAKE PENMAN ESTIMATES (AFTER YOUNG[64])

	Latitude	Time of Year	Radiation			Air temp.	Vapour pressure	Wind speed
			Solar	Net	Sunshine			
No radiation	Yes	Yes	No	No	Yes	Yes	Yes	Yes
Solar radiation	Yes	Yes	Yes	No	Yes	Yes	Yes	Yes
Net radiation	No	No	No	Yes	No	Yes	Yes	Yes

In one of the several studies describing computer programs[64], the information needed for making Penman estimates is set out *(Table 3.6)*. This program was designed so that all the four variables dependent upon basic data (R_A, N, e_a and Δ) are calculated directly as follows:

$$R_A = \frac{1440R}{59\pi}\ (h \sin L \sin D + \sin h \cos L \cos D)\ \text{mm/day} \qquad (3.28)$$

where L = latitude
       D = sun's mean daily declination
       R = 1.94 cal cm^{-2} min^{-1}
       h = $\cos^{-1} \tan L \tan D$ (radians)

$$N = \frac{24h}{\pi}\ \text{hours} \qquad (3.29)$$

$$e_a = \exp\left[47.226 - \frac{6463}{273 + T_a} - 3.927 \log_e (273 + T_a)\right] \text{mm Hg} \ (3.30)$$

$$\Delta = \frac{e_a}{(273 + T_a)}\left[\frac{6463}{(273 + T_a)} - 3.927\right] \qquad (3.31)$$

Automatic weather station data are usually processed so as to produce Penman estimates, the results from three pairs of weather stations being shown in *Figure 3.5*.

Worked examples of the calculations for sites in Southern England and tables of average potential transpiration values for most of the country have appeared in two publications[65,66], some changes being made in the later one to take the advances into account. E_T is calculated from:

$$E_T = \frac{H\Delta/\gamma + E_a}{\Delta/\gamma + 1} \qquad (3.32)$$

where H = $(1 - r)R_c - R_B$ (3.33)
       r = 0.25 for a green crop such as grass

$R_c = R_A f(n/N)$, where $f(n/N)$ takes the following forms:
$0.135 + 0.68 n/N$ for smoky areas with $n/N \leqslant 0.40$
$0.16 + 0.62 \ n/N$ for latitudes south of $54\frac{1}{2}°N$
$0.155 + 0.69 \ n/N$ for latitudes $54\frac{1}{2}°N$ to $56°N$ (with 0.01 being added for every degree north of $56°N$ to a maximum of 0.04 at $60°N$)

$$R_B = \sigma T_a^4 (0.47 - 0.075 \ \sqrt{e_d}) (0.17 + 0.83 \ n/N) \qquad (3.34)$$
$$E_a = 0.35 (1 + 10^{-2} u_2) (e_a - e_d) \qquad (3.35)$$

Thornthwaite

The great advantage (and weakness) of the Thornthwaite (1948) formula is that it only requires measurements of mean temperature. Hence it can be applied in areas where the most meagre climatic records exist—a more important factor in the US than in the UK. For a 30-day month, the basic equation is:

$$E_T = 1.6 \ (10 T_a/I)^a \qquad (3.36)$$

where T_a = mean air temperature (°C)
I = heat index, being the sum of 12 one-month heat indices (i)
$i = (T_a/5)^{1.514}$
a = a constant depending on latitude

The use of nomograms and tables that were included in the original paper reduces the labour involved in this method, which has been applied a number of times in the UK.

Crowe

This method[67] employs daily maximum temperature (T °F) with tabular values of day length and a weighting factor for the rainfall of the month when potential evapotranspiration is at a minimum in:

$$E_T = 0.16 \ (T_m - 43) \qquad (3.37)$$

Turc

For assessment of evaporation[68] on a yearly basis, measurements of precipitation (P mm) and mean air temperature (T °C) are required and are used in:

$$E_T = \frac{P}{\sqrt{[0.9 + (P/L)^2]}} \qquad (3.38)$$

where $L = 300 + 25T + 0.05T^3$ (mm).

Averages, comparisons and actual transpiration

There is a relative abundance of stations with long-period averages of
rainfall, but few sites with lengthy observations of evaporation. One reason
for this is that evaporation seems to be regarded as a conservative
phenomenon, with little variation in either space or time from year to year
and month to month, but this is a view that needs justification, particularly
for actual transpiration.

One site with lengthy records of evaporation is Camden Square where
observations of a tank were maintained for 70 years, starting in 1885.
Evaporation has been estimated on a monthly basis for several long-
established stations. For example, the 1948 Penman method was used to
determine potential transpiration values for Radcliffe Observatory, Oxford,
for the years 1881–1966 *(Figure 3.9)*, the annual mean being 480 mm.

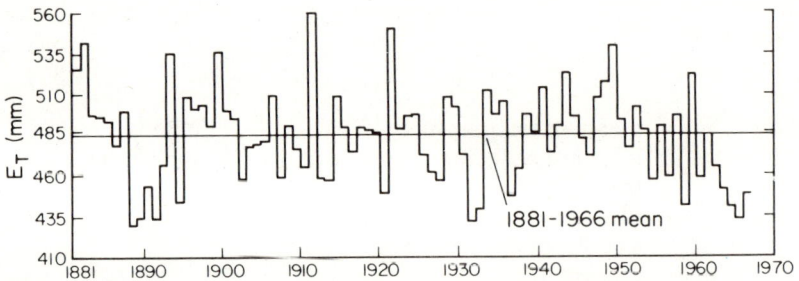

*Figure 3.9   Penman values of potential transpiration for Oxford, 1881–1966 (courtesy Institute
of Hydrology)*

Year-to-year variations are considerable, from a minimum of some 430 mm
to a maximum of 560 mm. A similar study was undertaken with the records
for Edgbaston Observatory for the years 1900–1968[69]. Annual values were
calculated by several different methods and 13-year moving means plotted
(Figure 3.10). A significant trough occurs centred on 1930, which correlates
well with the pattern for Oxford.

Average and annual amounts of potential transpiration

Averages for six stations for the period 1916–1950 are given in *British
Rainfall 1959* and *1960*, but *British Rainfall 1961* contains several
evaporation maps[70], one showing average annual totals at 19 sites. A
number of maps has also been constructed[71] from the records of the network
of simple lysimeters to demonstrate the variations of potential transpiration.
The map included here *(Figure 3.11)* for 1967 contains the familiar pattern
of contrast, the south and east against the north and west; but gaps in the
lysimeter network may be responsible for some of the features. A similar
pattern is apparent in the map of average annual values calculated by the
Thornthwaite method[72], although these values are some 125–250 mm
greater. The map of average summer potential transpiration obtained from

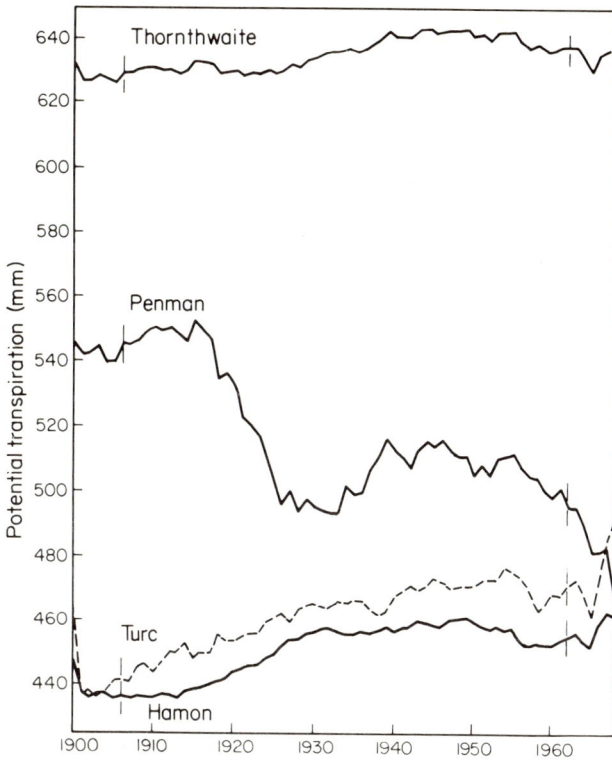

Figure 3.10 Comparison of 13-year moving averages of annual potential transpiration estimates for Edgbaston observatory (from Takhar and Rudge[69], courtesy Elsevier Scientific Publishing Co.)

Penman calculations[65] also shows the same gradient from south-east to north-west, but with smaller totals. This map suggests that some coastal zones might experience higher potential rates, and this feature is continued in the publication *Potential Transpiration*[66] where separate maps are shown for coastal areas. The map included here *(Figure 3.12)* shows the average annual potential transpiration adjusted to the mean county height. Some 540 mm is attained around the Thames Estuary, but north of Scotland averages reach only 380 mm a year. The altitude correction was made from an examination of station records: this suggested that, in summer, totals decrease on average by some 17 mm per 100 m in England and Wales and by over 20 mm for Scotland and Northern Ireland. Comparable figures for the winter were 13 mm and 8.5 mm respectively.

These publications are directed towards the assessment of irrigation need, employing the information on evaporation, together with the concept of the 'root constant'. This, as it was originally conceived[5], was the quantity of readily available water held within the depth of rooting. More recently, it has been defined as 'the maximum soil moisture deficit that can be built up without checking transpiration'. It is considered essentially a plant characteristic, but one that can be modified by the soil and the age of the

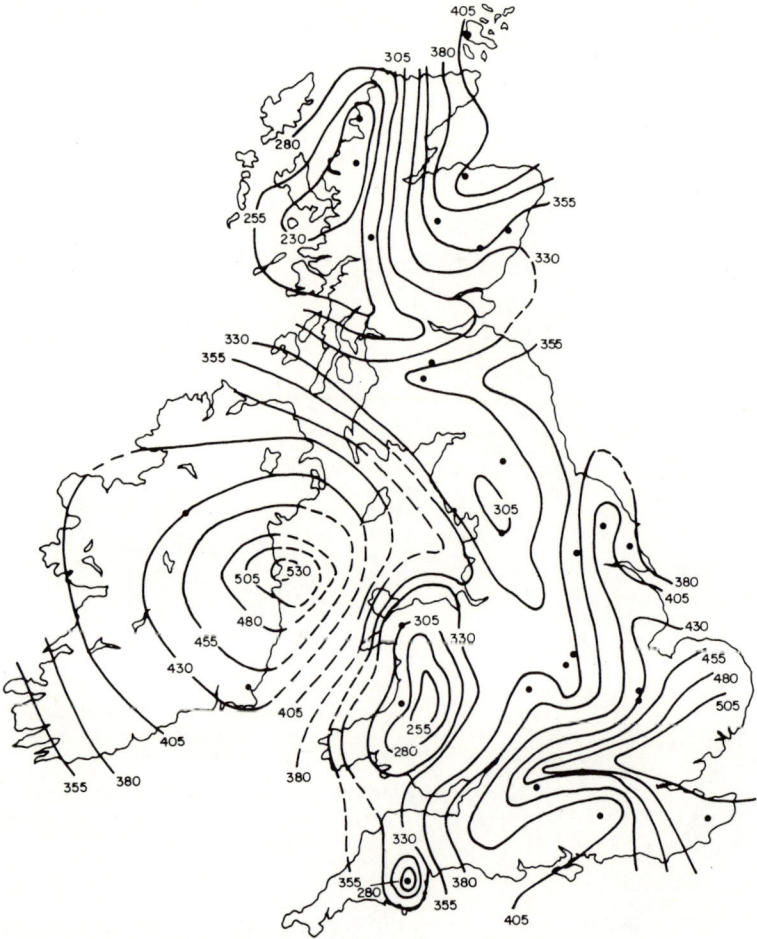

Figure 3.11   Potential transpiration (mm), April–September 1967, from lysimeter observations (from Green[71], courtesy Elsevier Scientific Publishing Co.)

plant. For grass, the root constant is supposed to be about 70–130 mm, a sufficiently wide margin of deficit to circumvent the lack of knowledge of when plant growth is checked during soil moisture depletion. The map of frequency of irrigation need *(Figure 3.13)* shows the number of years out of ten when the total rainfall for the summer six months falls below the amount required to produce a soil moisture deficit of 100 mm or less. The pattern is similar to those exhibited by the maps of evaporation, but with perhaps a more marked gradient from south-east to north-west. As one of the steps in obtaining a synthetic monthly record of runoff from ungauged British catchments, Barton[73] produced a synthetic monthly potential transpiration matrix. In this matrix *(Table 3.7)*, which it is claimed applies to the whole of Britain, the monthly means are expressed as a percentage of the annual mean. Thus only the appropriate annual mean is needed to establish the matrix for any locality.

Figure 3.12 Average annual potential transpiration (mm), 1930–1949 (from Smith[66], reproduced by permission of the Controller, Her Majesty's Stationery Office)

Table 3.7 MATRIX OF PARAMETERS FOR THE GENERATION OF MONTHLY POTENTIAL TRANSPIRATION AMOUNTS (AFTER BARTON[73])

	Mean	C. Var.	C. Skw	C. Reg.	C. Cor.
October	5.1	0.8	−0.1	0.0	0.0
November	2.2	0.5	0.9	0.0	0.0
December	1.0	0.5	0.5	0.0	0.0
January	0.9	0.4	−0.2	0.6	0.3
February	1.9	0.8	0.7	0.9	0.5
March	4.9	1.5	0.2	0.2	0.2
April	10.0	1.6	0.3	0.1	0.1
May	15.0	1.7	−0.2	0.4	0.3
June	17.7	2.7	0.1	0.0	0.0
July	17.3	2.4	0.4	0.4	0.5
August	14.4	2.1	0.1	0.4	0.6
September	9.6	1.4	0.7	0.2	0.2

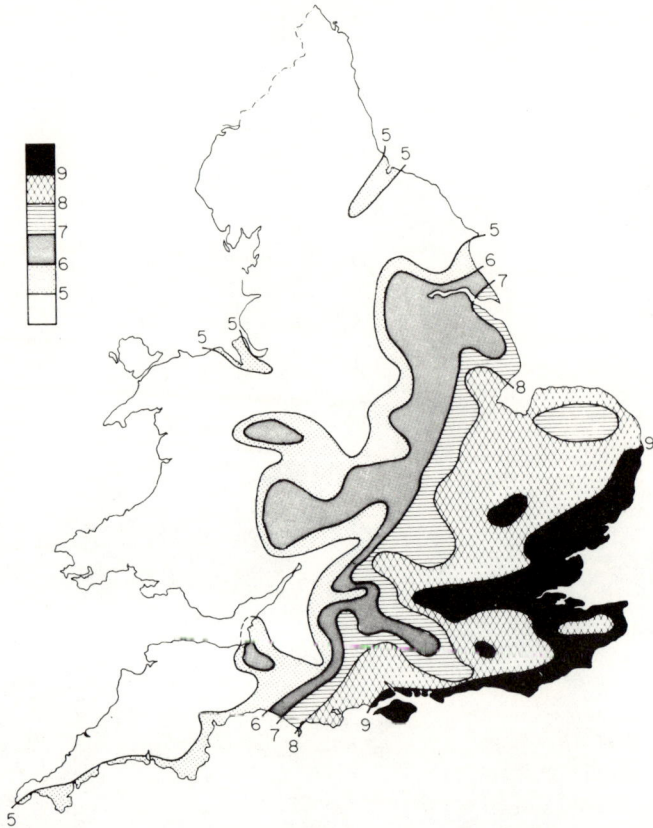

Figure 3.13   General frequency of irrigation need (years in ten), assuming a root constant of 75 mm (from Pearl[65], reproduced by permission of the Controller, Her Majesty's Stationery Office)

Comparisons

Penman has commented that his method has been used rather than tested. Indeed this seems true in the broader sense, for although there have been an appreciable number of comparisons of the various methods, few have really been tested adequately. For example, Ward[74] employed a pair of the simple lysimeters to measure potential evaporation and compared these measurements with Thornthwaite- and Penman-calculated amounts. It was found that Thornthwaite values were generally higher than those from the lysimeters and Penman values lower, but of course both methods have their basis in lysimeter measurements. The same type of study was made by Stanhill[75] using weighed lysimeters for comparison with Thornthwaite and Penman estimates, with similar results.

Penman himself has attempted to test estimates of potential evaporation by setting them against independent measurements of 'loss' from catchments obtained from the difference between rainfall and runoff. For instance, in the study of the Stour, he matched annual rainfall-runoff differences with the estimates of evaporation, taking into account the changes in storage within the catchment. To refine the estimate, the catchment was divided into three: the riparian area, those parts of the catchment remote from the river, and the intermediate areas. It was assumed that within the riparian area there was never any check to transpiration, but that root constants of 200 mm and 130 mm applied to the intermediate and remote areas. Six estimates of mean basin evaporation were made, each employing different proportions of the three areas; the one for 25 per cent riparian, 25 per cent intermediate and 50 per cent remote turned out to be exactly the same as the mean rainfall-runoff difference, namely 505 mm. Penman recognised that there may be errors in the data, errors of observation and errors in the method itself. It was considered that these might in some way compensate for each other, but that finally an estimate is likely to be within 5 per cent of the true value.

The same type of study has been undertaken for the Thames[76] and several other catchments[77,78], for the British Isles[1] and for part of Europe[79], with, in general, good agreement between estimates and measurements of loss. It has been suggested[80] that for catchments where evaporation estimates and losses do not agree within ±10 per cent, this may indicate leaks, differences between the surface and groundwater watersheds and errors of measurement, as well as errors in basic assumptions.

In another trial of the Penman method[81], the evaporation from one of a pair of reservoirs at Kempton Park was employed as a datum. During this period there was no inflow into the reservoir and no outflow. Daily measurements of rainfall and water level were made and the evaporation amounts determined from them were compared with Penman estimates, and with observations from a pan and a tank. The pan was installed only after the study had been under way for three years, climatological records from Kew were used in the Penman estimate, and there were some problems with water level and rainfall measurements and with possible leaks from the reservoir. Nevertheless, it was found that annual and monthly Penman totals agreed well with reservoir evaporation, although a seasonal bias was apparent in the calculated values—winter ones tending to be lower and summer ones higher than those measured (*Table 3.8*). Using measurements of water temperature to assess the heat stored in the reservoir, evaporation totals were 'corrected' from a deep-water to a shallow-water rate, with a consequent improvement in the agreement with the other totals. The pan and tank factors were found to be 0.7 and 1.1 respectively, and it was concluded that these devices and the Penman method were likely to produce monthly totals that could differ by as much (or as little) as ±13 mm from reservoir evaporation.

Assessments of the heat storage term, in this case for soil, were included in the water balance study of a small clay catchment by Edwards and Rodda[17]. Measurements of rainfall, runoff and soil moisture storage were employed to assess the losses from the basin, and these were compared with Penman potential transpiration amounts. As might be anticipated, there were differences between the two sets of values month by month from 1965 to 1968. Some of these differences represented the margin between actual and potential transpiration; the Penman figures exceeded the values determined from the water balance in the summer while the reverse was true during the

Table 3.8 SUMMARY OF RESULTS FROM EVAPORATION STUDY AT
KEMPTON PARK RESERVOIR (mm) (AFTER LAPWORTH[81])

	Reservoir evaporation	Penman	Tank	Pan (average for 4 years)	Reservoir evaporation corrected for heat storage
January	15	8	5	15	13
February	18	15	13	25	20
March	28	41	30	46	41
April	48	69	56	84	69
May	76	104	89	132	94
June	94	124	109	173	117
July	107	117	102	165	112
August	94	94	84	132	89
September	74	61	58	99	58
October	58	28	33	53	36
November	33	8	15	23	13
December	18	3	8	13	8
Year	663	672	612	960	670

winter. A major part of the rest of these differences was accounted for by heat stored in the soil *(Figure 3.14)*, these quantities being estimated from soil temperature records in conjunction with soil moisture records.

Further studies of basin water balances have involved forested catchments; the differences caused by evaporation of the precipitation intercepted by the canopy are dealt with in Chapter 8.

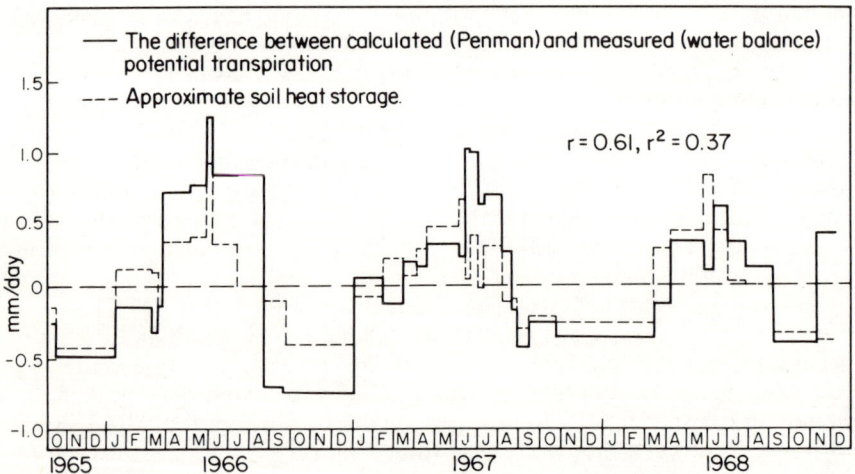

Figure 3.14   Soil heat storage (from Edwards and Rodda[17], courtesy Hydrological Sciences Bulletin*)*

Actual transpiration

Many of the methods and techniques already discussed are concerned with assessing the transfer of water to the atmosphere when the only control is the atmosphere's evaporative demand. Under natural conditions, however, unless rainfall is frequent and copious, the soil is not always wet enough for water to be readily available to plants, so transpiration is likely to be occurring at a rate other than potential, for some of the time at least. As mentioned earlier, there are differing opinions on the point when potential and actual transpiration rates depart from one another and by how much. One school of thought considers that these rates are the same throughout the whole range of depletion to wilting point. Another maintains that actual and potential rates depart immediately the soil dries below field capacity and that thereafter the difference between them increases.

Depending upon the nature of the soil and its depth, the nature of the vegetation and its rooting depth, the true position is probably some way between these extremes. It seems likely that the transpiration rate and growth are related so that the potential rate pertains until growth is restricted by shortage of water in the leaves. Milthorpe[82] in summarising results from a number of studies concluded that the weight of evidence was for a progressive decline in the rate of transpiration as the soil water deficit increased. He suggested that the point at which the two rates depart varies considerably and that the root growth and potential transpiration amounts were its most important controls. Stanhill[75] also gave evidence for a steady decline in the ratio of actual to potential values from field capacity to wilting point. On the other hand, Penman's[5] concept of a root constant seems attractive for its simplicity. Transpiration from a crop proceeds at the potential rate until the 'root reservoir' has been depleted. From that point about 25 mm more water can be withdrawn from the soil at close to the potential rate, before a change occurs and transpiration proceeds at an actual rate that is about one twelfth of the potential *(Figure 3.15)*. The amount of water within the root range is usually considered as 75–125 mm for grass. By taking soil samples and carrying out tests in the laboratory, soil moisture depletion curves can be constructed and the actual amount of water held within the soil can be determined. Use of a value for the water

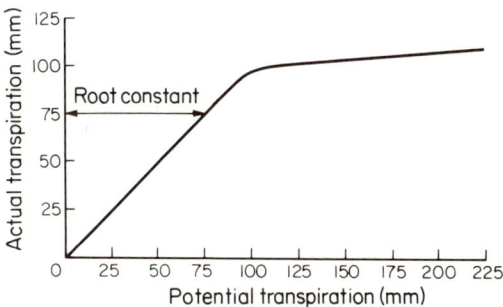

Figure 3.15 Relationship between potential and actual transpiration, based on the root-constant concept

Table 3.9 MONTHLY WATER BALANCE AT OXFORD FOR 1921 (in)

	Precipitation, P	Potential transpiration, PE	Change in soil moisture, ΔS	Runoff or percolation, R	Soil moisture in 2 ft of profile $\Sigma_0^2 S$	Actual transpiration, AE	$\Sigma_0^2 S + P = P'$	$P' - 1.5 = N$	$(AE - N) \times 0.1 = M$
January	2.47	0.14	2.33	2.33	5.50	0.14	—	—	—
February	0.56	0.59	−0.03	—	5.47	0.59	—	—	—
March	0.92	1.38	−0.46	—	5.01	1.38	—	—	—
April	0.91	2.28	−1.37	—	3.64	2.28	—	—	—
May	1.50	3.50	−2.00	—	1.64	3.50	1.98	0.48	0.33
June	0.34	3.78	−3.44	—	1.17	0.81	1.45	—	—
July	0.28	4.30	−4.02	—	1.02	0.43	2.70	1.20	0.15
August	1.68	2.74	−1.06	—	1.35	1.35	3.05	1.55	0.04
September	1.70	1.92	−0.22	—	1.36	1.69	2.63	1.13	0.01
October	1.27	1.22	0.05	—	1.49	1.14	3.26	—	—
November	2.03	0.26	1.77	—	3.26	0.26	4.31	—	—
December	1.28	0.23	1.05	—	4.31	0.23	—	—	—

held within the top 250–500 mm of the soil profile between certain moisture tensions can add some reality to the Penman concept, but its scientific merit remains in doubt. However, computing the month-by-month or week-by-week values of actual transpiration requires only simple arithmetic calculations, and because of this the method has been employed a number of times.

For a study of drought[83] in which records of rainfall and potential transpiration were used to make a water balance, tests on representative soil samples showed that in the top 500 mm of the profile there were 140 mm of available water; it was assumed that 100 mm of this could be removed at the potential rate and then a rate of one tenth potential would apply. When the soil moisture state reached 140 mm and rainfall occurred that exceeded the potential transpiration amount, runoff and percolation were assumed to take place. When the reverse happened, the potential demand was made up from the soil moisture store unless 100 mm had been removed already. From that point actual transpiration took place at one tenth the potential rate *(Table 3.9)*. For the period in question, a renowned drought year (rainfall for Oxford 380 mm), although the potential transpiration total calculated by the Penman (1962) method[60] reached 565 mm, the actual transpiration was estimated to be only 350 mm. This implies that for a dry year the amount of actual transpiration may be considerably less than during one with average or wetter than average conditions; some experimental evidence exists to support this view. In a study of transpiration of grass cut for hay[84], lysimeter records showed that the energy used in evaporation was, on average, only 0.8 of the net heat available. In other words transpiration was occurring at 0.8 of the potential rate, but during a dry spell this figure dropped to 0.3.

Of course, one of the most important points to decide is the value for the root constant. Various values have been adopted *(Table 3.10)* ranging from 50 mm for grass to 200 mm for woodland. Actual transpiration amounts have been calculated for the different land-use units in the area in question, together with an estimate of the amount of infiltration. For example, in the study of the Lincolnshire Limestone[85], actual transpiration amounts for nine different areas were estimated to be between 50 mm and 80 mm less than potential values. Similar differences are given in the London Basin study[86] *(Table 3.11),* where it was considered that, for the Chalk as a whole, actual transpiration is about 450 mm a year, some 50 mm less than potential. A 75 mm root constant[87] was used for a study of clay catchments in Kent and one of 25mm for an investigation of the Havant and Bedhampton Springs[88]. For his study of chalk catchments in Hampshire, Headworth[89] used six different root constants between 12.5 and 75 mm and concluded that, in terms of the

Table 3.10 ROOT CONSTANTS

Temporary grass	50 mm
Permanent grass	75 mm
Cereals	140 mm
Potatoes	65 mm
Beet	115 mm
Woodland	200 mm

Table 3.11 RAINFALL, TRANSPIRATION AND INFILTRATION AMOUNTS FOR THE CHALK OF THE LONDON BASIN (mm) (FROM REF. 86)

Unit	Sub-unit	Rainfall	Potential trans-piration	Actual trans-piration	Infil-tration
I Wey*	IA	780	523	485	295
	IB	780	523	485	295
II Mole*	IIA	760	508	465	295
	IIB	833	508	465	368
III Wandle*		762	503	457	305
IV Ravensbourne*		808	523	472	336
V Medway*		838	498	460	378
VI Darent*	VIA	683	511	450	233
	VIB	803	498	452	351
VII Burnham*		676	554	498	178
VIII Colne	Misbourne*	749	516	478	271
	Chess*	775	508	472	303
	Gade*	744	500	457	287
	Ver*	711	506	457	254
	Upper Colne†	663	544	467	196
	South Colne*	678	533	483	195
XI Lee	Upper Lee†	663	521	457	206
	Mimram†	660	521	457	203
	Beane†	658	523	462	196
	Rib†	653	518	455	198
	Ash†	645	526	462	183
	Stort†	643	521	447	196
X Thames*		752	506	452	300
XI Great Ouse†		635	526	465	170
XIII Grays*		533	561	483	50

XII Greenwich unit not included because it is densely built-up
*Figures relate to surface catchments
†Figures relate to surface catchments. Groundwater units differ significantly in size. Evaporation data are probably similar for both cases, rainfall is not necessarily so

measured runoff, a value of 25 mm gave the most meaningful estimates of actual transpiration for short-rooted crops. Changing the root constant has also been investigated by the Meteorological Office[86]; the results show that actual transpiration is some 10 and 5 per cent less for 25 and 50 mm root constants than for one of 75 mm. Grindley[91] mapped average annual actual transpiration for the area of one river authority. Starting with a potential transpiration map and a map of average annual rainfall, he calculated the actual transpiration for each point on a 10 km grid, using information about the land use around each grid point to estimate the root constant. From the resulting map *(Figure 3.16)*, the areal estimate of actual transpiration was found to be 480 mm while the lumped value was 488 mm.

From the frequency of irrigation need map *(Figure 3.13)*, an estimate of the incidence of actual transpiration can be made for a particular catchment. This map is based on a root constant of 75 mm and if a smaller value were to be adopted, either for the whole map or part of it, there could be some changes in the pattern. To give a broad indication of the distribution of actual transpiration amounts for the average year, the potential transpiration rainfall difference was first calculated. The deficit that resulted was first met from the 75 mm root constant and then by transpiration at one tenth the

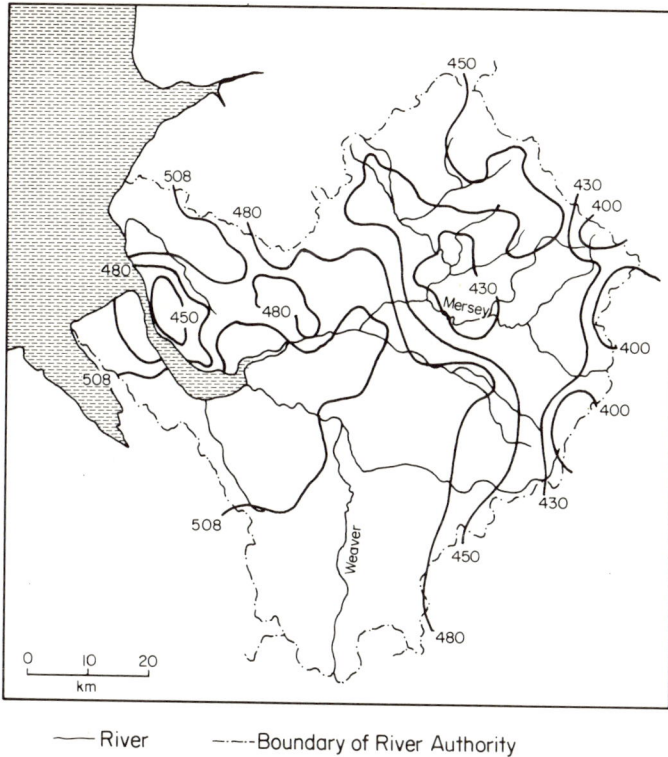

— River   —·—·Boundary of River Authority

Figure 3.16.   Average annual actual transpiration (mm) for the area of the Mersey and Weaver River Authority (from Grindley[90], courtesy Hydrological Sciences Bulletin)

Table 3.12   POTENTIAL TRANSPIRATION AMOUNTS (mm) CORRECTED FOR THE SOIL HEAT FLUX

	1966		1967	
	E_T	E_{TC}	E_T	E_{TC}
January	4.4	5.9	3.5	5.3
February	13.7	9.6	10.8	6.6
March	27.9	23.2	34.5	29.9
April	42.6	33.9	46.7	39.2
May	86.2	74.4	73.0	59.5
June	100.7	82.5	93.5	81.5
July	89.3	83.7	100.4	92.7
August	72.9	68.5	71.4	71.4
September	42.5	47.1	34.8	42.3
October	14.5	22.8	20.1	27.7
November	6.8	18.9	2.8	8.8
December	3.6	11.5	3.0	4.5
Year	505.1	482.0	494.5	469.4

E_{TC} = potential transpiration corrected for soil heat flux

Figure 3.17   Guide to actual transpiration totals (mm) in the average year (Note: for most years potential rates apply outside the area for which actual amounts are shown)

potential rate. The results are shown in terms of county values *(Figure 3.17)*, but only for those areas experiencing actual transpiration. Apart from the Isle of Wight, the greatest difference from potential occurs in Cambridgeshire, the value being 100 mm. During a very dry year it is probable that these amounts and their distribution across the country would alter radically. Take as an example the situation in a 1-in-10 dry year: rainfall totals would be approximately 75 per cent of normal and potential transpiration amounts could be about 5 per cent greater than average. Actual transpiration would be taking place over a much larger part of the country, including the west, while in the south and east actual amounts could be 100 mm below the potential rate.

It must be remembered that these estimates of actual transpiration are based upon a concept that has not been confirmed scientifically, an assumption that the land use is entirely grass and an estimate of potential transpiration that may not be completely realistic. There appears to be a considerable need for research to establish the reliability of this and other methods. Such research is being undertaken to a considerable extent, through studies of the evaporation process and through studies of representative and experimental basins. For example, in the study referred to earlier concerned with a small clay catchment[17], potential transpiration amounts were corrected for soil heat storage. The results for 1966 and 1967 *(Table 3.12)* show how summer rates are decreased and winter ones increased by taking the soil heat flux into account. Use of measured values of the flux would improve these estimates appreciably and further refine the estimation of actual transpiration.

REFERENCES

1. Penman, H. L., 'Evaporation over the British Isles', *Q. J. R. Met. Soc.,* **76,** 372–383 (1950)
2. Symons, G. J., 'Evaporators and evaporation', *British Rainfall 1867,* **8,** 9–10 (1868)
3. Rutter, A. J., *Transpiration,* Oxford Biology Readers (1972)
4. Thornthwaite, C. W., 'An approach toward a rational classification of climate', *Geographical Review,* **38,** 85–94 (1948)
5. Penman, H. L., 'The dependence of transpiration on weather and soil conditions', *J. Soil Sci.,* **1,** 74–89 (1949)
6. Bilham, E. G., *The climate of the British Isles,* Macmillan & Co. (1938)
7. Stanhill, G., 'Use of the Piche evaporimeter in the calculation of evaporation', *Q. J. R. Met. Soc.,* **88,** 80–82 (1962)
8. Garnier, B. J., 'A simple apparatus for measuring potential evapotranspiration', *Nature,* **190** No. 4320, 286–287 (1952)
9. Garnier, B. J. and Lewis, W. V., 'Potential evapotranspiration: an appeal for its measurement', *Weather,* **9,** 243–245 (1954)
10. Winter, E. J., 'A new type of lysimeter', *J. Horticultural Science,* **38,** 160–168 (1963)
11. Morris, L. G., 'A recording weighing machine for the measurement of evapotranspiration and dewfall', *J. Agric. Engng Research,* **4,** 161–173 (1959)
12. Law, F., 'The effect of afforestation on the yield of water catchment areas', *J. Brit. Waterworks Assoc.,* **10,** 489–494 (1956)
13. Law, F., 'Measurement of rainfall, interception and evaporation losses in a plantation of sitka spruce trees', *Proc. IASH General Assembly, Toronto 1957,* Pub. No. 45, 397–411 (1958)
14. Penman, H. L., 'The water balance of the Stour catchment area', *J. Instn Water Engrs,* **4,** 457–469 (1950)
15. Howe, G. M. and Rodda, J. C., 'An investigation of the hydrological cycle in the catchment area of the R. Ystwyth during 1958', *Water and Water Engng,* **64,** 10–16 (1960)
16. Smith, K., 'Rainfall, runoff and water balance in the Nidd Valley, Yorkshire', *Water and Water Engng,* **69,** 5–10 (1965)
17. Edwards, K. A. and Rodda, J. C., 'A preliminary study of the water balance of a small clay catchment', *Proc. IASH Symp. on results of research in representative and experimental basins,* Wellington, IASH Pub. no. 97, 187–199 (1972)
18. Meteorological Office, *Observers' Handbook,* 3rd edn, HMSO (1969)
19. Jones, J. I. P., *Physics Exhibition Handbook,* Institute of Physics, London (1971)
20. Day, G. J., 'Distribution of total solar radiation on a horizontal surface over the British Isles and adjacent areas', *Met. Mag.,* **90,** 269–284 (1961)
21. Pereira, H. C., 'Practical field instruments for estimation of radiation and evaporation', *Q. J. R. Met. Soc.,* **85,** 253–260 (1959)
22. Jacobs, L., 'Radiation recording in the Meteorological Office', *Met. Mag.,* **90,** 284–289 (1961)
23. Trickett, E. S., Moulsey, L. J. and Edwards, R. I., 'Measurements of solar artificial radiation with particular reference to agriculture and horticulture', *J. Agric. Engng Research,* **2,** 86–110 (1957)

24. Monteith, J. L. and Szeicz, G., 'Simple devices for radiation measurement and integration', *Arch. Met. Geophys. und Bioklim.,* **B II,** 491–500 (1962)
25. Barry, R. G. and Chambers, R., 'Albedo variations in southern Hampshire and Dorset', *Weather,* **22,** 60–65 (1967)
26. MacDowall, J., 'Total radiation flux meter', *Met. Mag.,* **84,** 65 (1955)
27. Funk, J. P., 'Improved polythene-shielded net radiometer', *J. Sci. Instruments,* **36,** 267–270 (1959)
28. Monteith, J. L., *Survey of instruments for micrometeorology,* International Biological Programme Handbook No. 22 (1972)
29. Strangeways, I. C. and McCulloch, J. S. G., 'A low-priced automatic hydrometeorological station', *Bull. IASH,* **10,** 57–62 (1965)
30. McGuigan, P. J., 'Evaluation of Plessey automatic climatological equipment', Paper No. 11, *Water Resources Board Symp. on data retrieval and processing* (1969)
31. Whitaker, R. J., 'Field evaluation of Normalair–Garrett climatological station', Paper No. 10, *Water Resources Board Symp. on data retrieval and processing* (1969)
32. Strangeways, I. C., 'Automatic weather stations for network operation', *Weather,* **27,** 403–408 (1972)
33. Lavis, J. J. and Gleave, J. E., 'Recording and using climate data', *Water and Water Engng,* **76,** 443–447 (1972)
34. Green, F. H. W., 'Potential evaporation measurements', *British Rainfall 1958,* **98,** 10–14 (1962)
35. McCulloch, J. S. G., 'Hydrological networks for the measurement of evaporation and soil moisture', *Proc. Symp. on design of hydrological networks,* IASH Pub. No. 68, 579–587 (1965)
36. Cole, J. A. and Green, M. J., 'Measurements of net radiation over vegetation and of other factors affecting transpiration losses in water catchments', *Proc. IASH General Assembly, Berkeley,* Pub. No. 62, 190–202 (1963)
37. Institute of Hydrology, *Research 1972–73,* Natural Environment Research Council (1973)
38. Penman, H. L., 'Evaporation: an introductory survey', *Netherlands J. Agric. Sci.,* **4,** 9–29 (1956)
39. Barry, R. G. and Chambers, R. E., 'A preliminary map of summer albedo values over England and Wales', *Q. J. R. Met. Soc.,* **92,** 543–548 (1966)
40. Brunt, D., *Physical and dynamical meteorology,* Cambridge U.P. (1939)
41. Bowen, I. S., 'The ratio of heat losses by conduction and by evaporation from any water surface', *Phys. Rev.,* **27,** 779–787 (1926)
42. World Meteorological Organisation, 'Measurement and estimation of evaporation and evapotranspiration', *WMO Tech. Note No. 83,* 121 (1966)
43. Institute of Hydrology, *Research 1971–72,* Natural Environment Research Council (1972)
44. Davies, J. A., 'A note on the relationship between net radiation and solar radiation', *Q. J. R. Met. Soc.,* **93,** 109–115 (1967)
45. House, G. J., Rider, N. E. and Tugwell, C. P., 'A surface energy balance computer', *Q. J. R. Met. Soc.,* **86,** 215–231 (1960)
46. Szeicz, G. and Long, I. F., 'Surface resistance of crop canopies', *Water Resources Research,* **5,** 622–633 (1969)
47. Stewart, J. B. and Thom, A. S., 'Energy budgets in a pine forest', *Q. J. R. Met. Soc.,* **99,** 154–170 (1973)
48. Penman, H. L., 'Natural evaporation from open water, bare soil and grass', *Proc. Royal Soc.,* **A 193,** 120–146 (1948)
49. Thornthwaite, C. W. and Holzman, B., 'Measurement of evaporation from land and water surfaces', *US Dept. Agric. Tech. Bull. No. 817* (1942)
50. Pasquill, F., 'Some estimates of the amount and diurnal variation of evaporation from a clay-land pasture in fair spring weather', *Q. J. R. Met. Soc.,* **75,** 249–256 (1949)
51. Rider, N. E., 'Evaporation from an oat field', *Q. J. R. Met. Soc.,* **80,** 198–211 (1954)
52. Rider, N. E., 'Water losses from various land surfaces', *Q. J. R. Met. Soc.,* **83,** 181–193 (1957)
53. Oliver, H. R., *Evaporation,* unpublished MSc dissertation, Department of Geophysics, University of Reading, 115 (1969)
54. Hicks, B. B. and Dyer, A. J., 'Measurements of eddy fluxes over the sea from an off-shore oil rig', *Q. J. R. Met. Soc.,* **96,** 523–528 (1970)
55. Thompson, N., 'Turbulence measurements over the sea by tethered balloons', *Q. J. R. Met. Soc.,* **98,** 745–762 (1972)

56. Leyton, L., Reynolds, E. R. C. and Thompson, F. B., 'Forest hydrology research in the United Kingdom', *Proc. Int. Symp. on forest hydrology* (ed. W. Sopper and H. Lull), Pergamon Press, 95–98 (1967)
57. Thornthwaite, C. W. and Hare, F. K., 'The loss of water to the air', in *Agricultural Meteorology*, Am. Met. Soc. Monograph No. 28, 163–180 (1965)
58. Penman, H. L., 'Estimating evaporation', *Trans. Am. Geophys. Union*, **37**, 43–50 (1956)
59. Monteith, J. L., 'The reflection of short-wave radiation by vegetation', *Q. J. R. Met. Soc.*, **85**, 386–392 (1959)
60. Penman, H. L., 'Woburn irrigation, 1951–1959', *J. Agric. Sci.*, **58**, 343–379 (1962)
61. Monteith, J. L., 'Evaporation and environment', *Symp. Soc. Exp. Biol.*, **19**, 205–234 (1965)
62. Szeicz, G., Endrodi, G. and Tajchman, S., 'Aerodynamic and surface factors in evaporation', *Water Resources Research*, **5**, 380–394 (1969)
63. McCulloch, J. S. G., 'Tables for the rapid computation of the Penman estimate of evaporation', *EAAFRO Report* (1963)
64. Young, C. P., 'A computer programme for the calculation of mean rates of evaporation using Penman's formula', *Met. Mag.*, **92**, 84–89 (1963)
65. Pearl, R. T. (ed.), 'The calculation of irrigation need', *Minist. Agric. and Fisheries Tech. Bull. No. 4* (1954)
66. Smith, L. P., 'Potential transpiration', *Minist. Agric. Fish. and Food Tech. Bull. No. 16* (1967)
67. Crowe, P. R., 'Some further thoughts on evapotranspiration: a new estimate', *Geographical Studies,* **4**, 56–75 (1957)
68. Turc, L., 'Le bilan d'eaux des sols: relations entre les précipitations, l'évaporation et l'écoulement', *Ann. Agron.*, **5**, 491–596 (1954)
69. Takhar, H. S. and Rudge, A. J., 'Evaporation studies in standard catchments', *J. Hydrology*, **11**, 329–362 (1970)
70. Holland, D. J., 'Evaporation', *British Rainfall 1961*, pt III, HMSO, 5–22 (1967)
71. Green, F. H. W., 'Some isopleth maps based on lysimeter observations in the British Isles in 1965, 1966 and 1967', *J. Hydrology,* **10**, 127–140 (1970)
72. Howe, G. M., 'The moisture balance in England and Wales based on the concept of potential evapo-transpiration', *Weather*, **11**, 74–82 (1956)
73. Barton, B. M. J., 'Synthetic monthly runoff for ungauged British catchments', *J. Instn Water Engrs*, **27**, 149–162 (1973)
74. Ward, R. C., 'Observations of potential evapotranspiration (PE) on the Thames flood plain, 1959–1960', *J. Hydrology*, **1**, 183–194 (1963)
75. Stanhill, G., 'The accuracy of meteorological estimates of evapo-transpiration', *J. Instn Water Engrs*, **12**, 377–383 (1958)
76. Penman, H. L., 'Components in the water balance of a catchment area', *Q. J. R. Met. Soc.*, **81**, 280–283 (1955)
77. Pegg, R. K., 'Evapotranspiration and the water balance', in *The role of water in agriculture* (ed. J. A. Taylor), UCW Aberystwyth Memo No. 12 (1970)
78. Ward, R. C. and Pegg, R. K., 'Evapotranspiration from a small clay catchment', *J. Hydrology*, **15**, 149–166 (1972)
79. Penman, H. L., 'Evaporation over parts of Europe', *Proc. IASH General Assembly, Rome,* Pub. No. 38, 168–176 (1954)
80. Prus-Chacinski, T. M., Discussion of 'Some aspects of the hydrology of the Thames basin' (F. M. Andrews), *Proc. Instn Civil Engrs*, **24**, 247–287 (1963)
81. Lapworth, C. F., 'Evaporation from a reservoir near London', *J. Instn Water Engrs*, **19**, 163–181 (1965)
82. Milthorpe, F. L., 'The income and loss of water in arid and semi-arid areas', in *Plant-water relationships in arid and semi-arid conditions,* UNESCO Reviews of Research, Arid Zone Research, XV (1960)
83. Rodda, J. C., 'A drought study in south-east England', *Water and Water Engng*, **69**, 316–321 (1965)
84. Monteith, J. L., 'The photosynthesis and transpiration of crops', *Experimental Agriculture Review,* **2**, 1–14 (1966)
85. Downing, R. A. and Williams, B. P. J., 'The groundwater hydrology of the Lincolnshire Limestone', *Water Resources Board Publication No. 9* (1969)
86. *The hydrogeology of the London Basin,* Water Resources Board, Reading (1972)
87. Kent River Authority, *Water Resources Act 1963, Section 14, First Periodic Survey* (1967)

88. Day, J. B. W., 'Infiltration into a groundwater catchment and the derivation of evaporation', *Geological Survey of Great Britain Research Report No. 2* (1964)
89. Headworth, H. G., 'The selection of root constants for the calculation of actual evaporation and infiltration for chalk catchments', *J. Instn Water Engrs,* **24,** 431–446 (1970)
90. Grindley, J., 'Estimation and mapping of evaporation', *Proc. Symp. on the world water balance,* IASH Pub. No. 92, **1,** 200–213 (1970)

Chapter 4

The Unsaturated Zone

Introduction

This chapter is concerned with the movement and storage of water above the water table. In the soil and rock matrices forming this zone water can exist in all three phases; as a liquid it is a weak solution of salts and organic substances. Water only partly fills the space between the soil and rock particles, the remainder being taken up by water vapour and air. Water in the unsaturated zone is often divided into three classes[1] distinguished by the increasing force with which it is held by the matrix: *gravitational water, capillary water,* and *hygroscopic moisture.* There is also the water that is chemically combined with the soil minerals, but this is ignored for hydrological purposes.

Where the soil cover is well developed it acts as a significant reservoir, releasing water for plant growth and storing rainfall to modify the pattern of drainage from the land surface. In fact irrigation and field drainage, designed to prevent the soil from becoming too dry on the one hand and too wet on the other, are two agricultural practices in which the hydrologist has an important part to play.

Infiltration is the process by which rain enters the soil due to the combination of gravity and capillary forces. Infiltration raises the moisture content of the surface layers, and, when no more water can be accepted by the soil, *surface ponding* commences followed by *overland flow* to the nearest channel. There is no clear distinction in the literature between infiltration and *percolation,* but here the former term will be confined to entry of the water at the air–soil interface and the latter to downward movement of water through the profile to the saturated zone. The low vertical permeability of some horizons, the presence of fissures, the influence of capillary forces and the slope of the land may cause *interflow.* This is the lateral movement of water through the unsaturated zone, but particularly through the soil, that terminates in natural channels and drains.

The most marked upward movement of soil water takes place within the *capillary fringe* immediately above the water table, and within the *root zone* extending down from the soil surface. Evaporation from bare wet soil, which causes drying, usually results in upward movement of soil-water by capillary action.

Some of the basic concepts of soil–plant–water relationships were

125

introduced in Chapter 3, together with several of the terms employed in soil physics. Probably the soil water characteristics of greatest importance to the hydrologist are:

1. the maximum water content of a soil;
2. the soil water content at field capacity;
3. the soil water content at wilting point;
4. the fall in infiltration rate as the soil wets up from (2) to (1).

For most purposes the water content of a sample of soil is determined by weighing, drying the sample in an oven at 105°C and reweighing. The water content may then be expressed as the dry weight fraction *(M_d)*, which is rather simpler to obtain than the moisture volume fraction (MVF), the more useful parameter[2].

$$M_d = \frac{\text{weight of water expelled at } 105°C}{\text{oven-dry weight of soil}}$$

$$MVF = \frac{\text{volume of water expelled at } 105°C}{\text{volume of soil before drying}}$$

Another parameter that is used fairly frequently is the dry bulk density *(D_d)*:

$$D_d = \frac{\text{oven-dry weight of soil}}{\text{volume of soil before drying}}$$

Soil types and classifications

A soil is normally described in terms of its horizons down to the parent material from which it has been derived. Usually three horizons can be recognised in most soils:

Horizon A   top soil; zone of cultivation, root growth and organic decay
Horizon B   sub-soil; weathered parent material, low in organic matter
Horizon C   fragmented parent material

Although soil mapping of the country is far from complete[3], a considerable number of different soil series has been identified and named and also grouped in soil associations. There have been a number of attempts at further groupings into the major soil types for several different purposes; one classification by the Agricultural Advisory Council produced eleven types for England and Wales[4]. *Table 4.1* shows two separate groupings, one by Russell[5] and the other for Wales by Taylor[6].

In practice hybrid combinations of these different soils are fairly common, but from a hydrological point of view the criteria used as a basis for these classifications are not necessarily the most appropriate. Although the soil surveyor will normally have taken a soil's drainage characteristics into account before typing it, he may not place as great a weight on these as on its other properties, such as colour and texture. To get around this point a number of classifications of soils specific to hydrology have been devised in recent years, based largely on inferred drainage values. For example in two studies[7,8] in which Painter was involved the Musgrave system[9], developed for American conditions, was used to translate the normal soil maps produced by the Soil Survey of England and Wales into ones showing minimum infiltration rates *(Figure 4.1)*. The minimum infiltration rate is

Table 4.1 SOIL CLASSIFICATIONS

After Russell[5]

1. Sandy soils
2. Loams
3. Clay soils
4. Chalk and limestone soils
5. Peat, moor and fenland soils
6. High-lying soils

After Taylor[6]

1.	Brown earths (or brown forest soils)	
2.	Podsols	(a) developed podsols
		(b) peaty podsols
		(c) crypto-podsols
		(d) truncated podsols
3.	Gley soils	(a) non-calcareous
		(b) calcareous
		(c) podsolised
		(d) peaty podsolised
4.	Soils derived from calcareous parent material	
5.	Organic soils	(a) hill peats
		(b) lowland peats
6.	Dune soils	
7.	Alluvial soils (undifferentiated)	
8.	Skeletal or mountain soils	
9.	Rock-dominant areas	

defined as the limiting value occurring when all the available storage has been taken up in those horizons above the one possessing the lowest permeability: *Table 4.2* gives the assumed infiltration rates for the different soil types and associations.

For the UK Flood Study[10] the Soil Survey was asked to interpret the normal soil classification so that a generalised map could be produced showing infiltration characteristics in relation to flood runoff. A fivefold classification based on the winter rain acceptance—broadly the infiltration potential—was devised. Where a soil is highly permeable, the water table is at a considerable depth and the ground slope is gentle, the acceptance is Class 1; low permeability, a high water table and a steep slope place a soil in Class 5 *(Table 4.3)*.

The soils of many of the catchments included in the representative and experimental basin programme have been subjected to detailed examination by the organisations involved in these studies and by the Soil Survey. Specially prepared soil maps exist for some of these basins, while the soil-water characteristics of numbers of sites have been established experimentally. Probably the most extensive study is that of the Dee Basin, made by the Soil Survey[11] so that soil data could be included in the flow simulation model that was being developed by the Water Resources Board. Duplicate undisturbed core samples were taken of the various horizons in each of 40 profiles chosen to represent the soil groups and landscape units of the basin in proportion to their frequency of occurrence. *Table 4.4* shows the

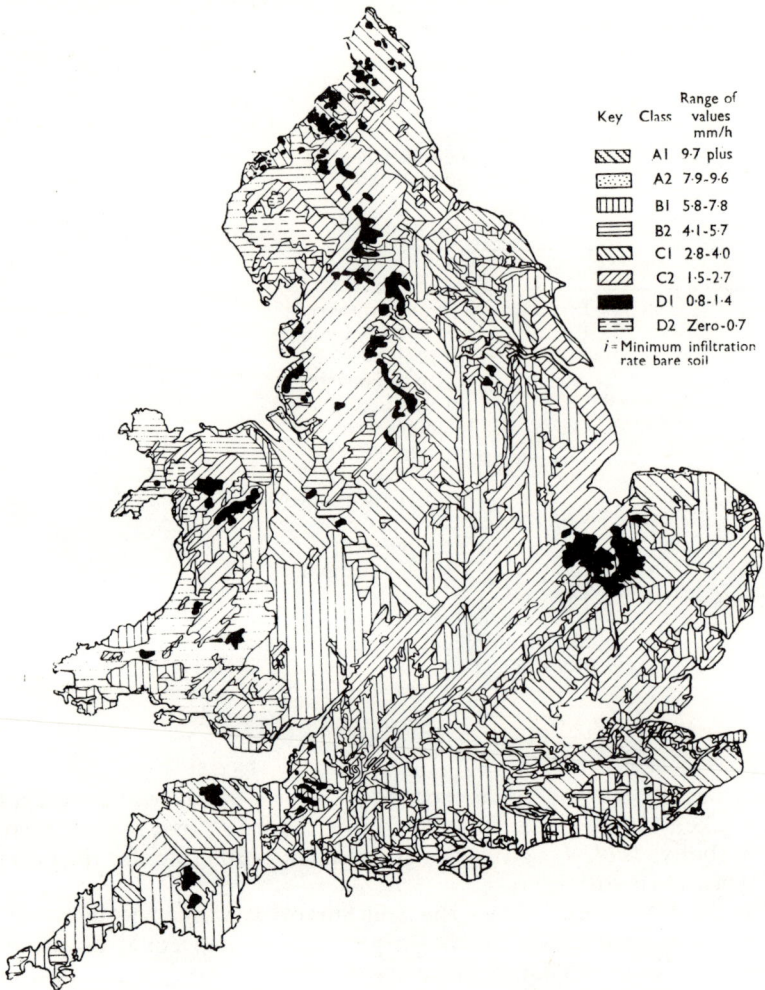

Figure 4.1 Hydrological map of soils in England and Wales (no soils classified as A1) (from Painter[8], courtesy Institution of Civil Engineers)

results of the laboratory tests, which are examined in greater detail later in this chapter.

There have been very few hydrological studies of a particular soil association or a particular soil series similar to the hydrogeological studies of particular aquifers. One exception is peat. Peat has been the subject of studies to assess its extent[12], its rate of erosion[13] and the effects of draining it in different ways[14]. For example Conway and Millar[15] measured the runoff from four small peat-covered catchments at Moor House in the Pennines. They found that flood peaks were both earlier and higher on the drained and eroded catchments and that the flow from these basins ceased on some occasions, compared with the catchments in a natural state.

Table 4.2 INFILTRATION RATES ASSUMED FOR SOIL
ENGLAND AND WALES (AFTER PAINTER[8])

Type	Group	Soil associations	Infiltration rate (mm/h)
Mainly sands and gravels	A2	Regosols, deep, well-drained, non-calcareous sandy soils.	7.9 to 9.6
Sandy loams to sandy clay loams	B1	Brown forest soil, well drained, non-calcareous medium-textured soils; calcareous soils, well drained medium-fine textured soils over chalk or limestone, often very shallow in lowland areas.	5.8 to 7.8
	B2	Acid brown soils on well drained slopes; podsolic soils of lowlands, deep, imperfectly drained, non-calcareous, sandy or coarse-textured soils, often with a moderately high water table.	4.1 to 5.7
Clay loams to silty clay loams or with impeding layer	C1	Brown forest soils with gleying, imperfectly drained non-calcareous medium textured soils; grey-brown podsolic soils, imperfectly drained, medium-fine textured soils, usually non-calcareous but occasionally having free calcium carbonate at approximately 1 m, commonly with a textural B-horizon; podsolic soils of the mountains, poorly drained non-calcareous, medium-textured soils; warp soils, alluvial soils of variable texture and drainage often with a moderately high water table.	2.8 to 4.0
	C2	Gley soils of the lowlands, either poorly drained, coarse-medium textured soils with a high water table, or poorly drained non-calcareous medium-fine textured soils with almost impermeable sub-strata; warp soils with gleying (as C1) but with high water table; calcareous soils, often very shallow, in mountain limestone areas.	1.5 to 2.7
Clay soils with panning or high water table or impervious materials	D1	Hill and basin peat, usually saturated.	0.8 to 1.4
	D2	Lithosols and shallow podsol soils, shallow and skeletal soils or hard rock dominant areas.	0 to 0.4

Note. Regosols — dry sandy soils
Podsols — a layered soil, often of acidic decomposing plant material, over a white sand leached of its iron and manganese over a pan-layer brown with these materials and precipitated alumina
Gleys — mottled soils due to changes in iron compounds, following water table movements
Warp soils — silt soils
Lithosols — stony materials, as in recent moraines, shingle banks and scree

Table 4.3 WINTER RAIN ACCEPTANCE INDICES: 1 = VERY HIGH; 2 = HIGH; 3 = MODERATE; 4 = LOW; 5 = VERY LOW (AFTER UK FLOOD STUDIES REPORT[10])

Drainage group	Depth to impermeable layer (mm)	Slope classes								
		0–2°			2–8°			>8°		
		Permeability rates above impermeable layers								
		Rapid	Medium	Slow	Rapid	Medium	Slow	Rapid	Medium	Slow
Rarely waterlogged within 600 mm at any time	>800		1	—	1		—	1	2	3
	400–800	2		—	3	2	—	3	—	4
	<400	3		—	—		—	—	—	—
Commonly waterlogged within 600 mm during winter	>800		3	—	3		—	3	—	—
	400–800			—			4	—	—	—
	<400			—			—	—	—	—
Commonly waterlogged within 600 mm winter and summer	>800		5			5		—	—	—
	400–800							—	—	—
	<400							—	—	—

Note. Upland peat and peaty soils are in Class 5; urban areas are unclassified

Table 4.4 SOIL MOISTURE CHARACTERISTICS FOR SOILS IN THE DEE BASIN
(AFTER RUDEFORTH AND THOMASSON[11])

Soil group	Soil	Dominant soil series	No. of profiles	TPS (mm)	AWC (mm)	G (mm)	S (mm)
1	Peat, peaty gley, peat over rock	Caron, Ynys	8	418	225	36	261
2	Peaty gleyed podsol	Hiraethog	2	146	74	5	79
3	Gley soil	Cegin	3	199	106	32	138
4	Gleyed brown earth	Sanan	6	304	139	79	218
5	Brown earth (mull)	Denbigh	17	305	103	125	228
6	Brown earth (mor)	Manod	4	337	119	130	249

Note. TPS = total pore space
AWC = available water capacity from 0.1 to 15 atmospheres tension
G = gravitational water pore space (G = TPS — 0.1 atmosphere water content pore space)
S = AWC + G

TPS, AWC, G and S are expressed in mm for the upper 600 mm of soil in Groups 4, 5 and 6, and above impermeable layers at 510 mm in Group 1, 248 mm in Group 2 and 395 mm in Group 3.

Infiltration

Infiltration rates

The maximum rate that a soil, in a given condition, absorbs rainfall is referred to as its *infiltration capacity*. If the infiltration rate is less than the infiltration capacity there is no surface runoff. For any given rain period the infiltration rate declines with time and the rate of decline also reduces with time, tending towards a minimum value that represents the equilibrium infiltration rate of the soil profile.

The infiltration rate obtaining at a particular time depends upon a number of factors including:

1. Physical nature of the soil, for example pore size distribution, soil structure and proportion of colloidal material.
2. Soil moisture content—dry soils absorb water more readily than moist ones.
3. Permeability of the soil profile.
4. Rainfall intensity—intense rain can compact the soil surface, especially bare soils, and reduce the ability of the soil surface to accept rainfall.
5. Vegetation cover and whether it is cultivated—infiltration rates are generally higher where the surface supports vegetation.

Infiltration measurement

There are few direct measurements of infiltration rates for soils in the UK, and knowledge of limiting infiltration rates during specific storms is sparse. *Figure 4.1* and *Table 4.2* give infiltration rates for different soil types in England and Wales. Hills[16] investigated the interaction of soil characteristics and land management in the Bristol area. Infiltration capacity was measured using 100 mm diameter steel cylinders driven 50 mm into the soil and fed from a calibrated feeder bottle maintaining a head of 50 mm over the soil. Allowance was made for the lateral spread of water from the instrument by a series of graphical correction factors obtained from laboratory experiments[17]. Infiltration capacity was measured at 5-minute intervals for the first 15 minutes and then at 15-minute intervals for the remainder of the test, which lasted for between 30 and 120 minutes. The mean capacity for each run was computed and averaged from 20 samples. The results of the experiments are summarised in *Table 4.5*. Of the various soil types examined Nibley and Haselor are more freely draining soils, while Evesham, Spetchley and Charlton Bank are classed as poorly drained. The Worcester Group generally tends to be poorly drained, but at the sites studied it had a light texture and was more closely related to the freely draining group. *Table 4.5* indicates a wide range of infiltration capacities for individual soil types. As would be expected land management was a significant factor and values were considerably higher in summer. Hills[16] concluded that most soils in the UK, with their characteristic vegetation cover, are capable of storing the majority of British rainfalls. The surfaces where rain is most likely to exceed infiltration capacities are bare or lightly vegetated compacted ground when rain is falling at moderate intensities for long periods in winter.

The Field Drainage Experimental Unit[18] has employed a ring infiltrometer with a neutron probe to investigate the efficiency of different drainage

Table 4.5 INFILTROMETER RESULTS FOR SOILS IN THE BRISTOL AREA (mm/h) (AFTER HILLS[16])

Soil series	Site's National Grid reference	Site characteristics	Infiltration capacity	
			Winter	Summer
Worcester (heavy)	ST 534695	Orchard, bare ground (chemical), compaction by vehicle	0.5	9
	ST 534696	Part bare (biotic), compaction by vehicle	—	29
	ST 534697	Woodland	50	—
	ST 534696	Untouched control plot	27	—
(light)	ST 538697	Cultivated blackcurrant bushes, bare ground (chemical), compaction by vehicle	—	2
	ST 538697	Cultivated as above but not compacted	5	—
	ST 539697	Orchard	47	—
	ST 538697	Cultivated (disturbed soil)	63	—
Nibley	ST 651813	Vegetated (compaction by cattle)	23	11
	ST 651813	Light pasture	26	115
Haselor	ST 538693	Heavy pasture	9	—
Evesham	ST 539694	Heavy pasture (compaction by cattle)	0	164
	ST 540693	Cultivated	53	—
Spetchley	ST 543694	Partly bare (compaction by vehicle)	8	142
	ST 543694	Orchard	86	—
	ST 534694	Cultivated	25	—
Charlton Bank	ST 754882	Vegetated (compaction by cattle)	23	—
	ST 754882	Heavy pasture (compacted)	—	55
	ST 742876	Woodland	191	383
	ST 754882	Light pasture	68	366

techniques at a number of sites. Fenwick[19] discussed the relative merits of tube infiltrometers and irrigated infiltration plots. The advantage of the former was that individual horizons could be examined, but infiltration plots needed less replication to provide adequate data.

Manufacturers in the United Kingdom specialising in the production of spray irrigation equipment have investigated the maximum rates at which soils can accept water droplets. It is known that large droplets cause 'capping' as the crumb structure at the surface is broken down. Soil porosity decreases and infiltration rates decay more rapidly, especially if the soil is heavy. Thus infiltration rates can never be considered constant given the wide variation in natural droplet sizes between, say, thunderstorms and light drizzle.

The inability to be precise is shown in Shockley's figures for soil intake rates[20] *(Table 4.6)*. These figures are for bare ground and would be higher with a cover of vegetation. As they were prepared with irrigation in view, they are naturally higher than would apply with saturated ground[21]. The magnitude of the figures for light soils exceed all but the rarest of British rain intensities.

Table 4.6 SOIL INTAKE RATES (AFTER SHOCKLEY[20])

Soil type	Intake rate (mm/h)
Coarse-textured sands, fine sands and loamy sands	25–13
Moderately coarse-textured sandy loams and fine sand loams	19–9
Medium-textured very fine sandy loams, loams, sandy clay loams, silt loams	10–6
Moderately fine-textured clay loams and silt clay loams	8–5
Fine-textured sandy clays, silty clay and clay	4–1

During the course of artificial recharge studies, the Water Resources Board examined the feasibility of recharging the Bunter Sandstone by irrigation techniques at a site near Mansfield[22]. The aquifer is a fine to medium grained sandstone that is uncemented near the surface. Ten lines of 35 mm diameter aluminium tubes were suspended 1 m above ground level and fitted with mist nozzles that delivered water at a nozzle pressure of up to 40 m head. The maximum irrigation rate achieved was about 50 mm/hour for each of three separate 1-hour periods in a day. The vegetated soil surface received no special preparation.

Seasonal and annual infiltration and percolation

As the different factors affecting infiltration vary throughout the year, infiltration rates are also subject to seasonal variations. However, the main factors controlling annual percolation to the groundwater zone are evaporation and transpiration, and most percolation occurs in the winter.

Where permeable deposits, such as sands, gravels and limestones, occur above the water table, the annual percolation reaching the water table is equivalent to the difference between rainfall and actual evaporation. Recharge of groundwater storage is theoretically possible when soil moisture deficits have been replenished and the storage available in the soil profile has been filled. However, percolation to the water table may commence when soil moisture deficits still exist, no doubt because water can move rapidly through the larger interconnected pores or through small channels formed by roots.

In eastern and southern England, where summer rainfall is normally less than evaporation, limited natural recharge of groundwater storage during the summer is not unusual. In the north and west, deep percolation is more general in the summer where geological conditions are suitable.

The map in Chapter 8 showing residual rainfall *(Figure 8.1)* gives a good indication of average percolation values where permeable rocks outcrop. This map indicates that average percolation into the Chalk in south-east England varies from more than 350 mm/year on Salisbury Plain to less than 100 mm/year in Essex.

Attempts have been made to relate percolation to rainfall in the chalk areas of south-east England[23,24]. Lapworth demonstrated the close relationship that percolation bears to average annual rainfall *(Figure 4.2)*. From an examination of data from percolation gauges, well fluctuations and spring flows, he derived the formula:

$$I = 0.9R - 343 \text{ mm} \tag{4.1}$$

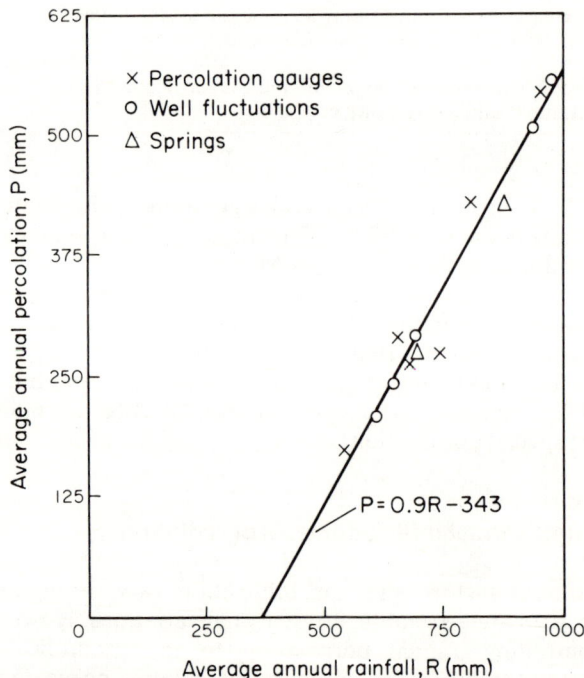

Figure 4.2   Relationship between average annual percolation and average annual rainfall (from Lapworth[24], courtesy Institution of Water Engineers)

where I = average annual percolation and R = average annual rainfall. As evaporation equals the difference between rainfall and percolation in permeable catchments, the formula can also be expressed in the form:

$$E = 0.1R + 343 \text{ mm} \qquad (4.2)$$

where E is average annual evaporation.

This formula gives realistic results where short grass covers very permeable ground but it should not be used for areas outside those for which it was derived without further confirmatory evidence.

More recently Wright[25] developed a formula that gave percolation to the Chalk in the Great Ouse Basin as:

$$I = 0.81R - 308 \text{ mm} \qquad (4.3)$$

The aquifer is partly overlain by boulder clay and the formula was derived by multiple regression procedures that related rainfall, baseflow and the geology.

Of course, in most catchments the surface geology consists of many soil and rock types with differing permeabilities. The total percolation to the catchment area may be determined from the baseflow component of total runoff from the area. From a knowledge of the catchment geology the total percolation derived in this manner can often be divided into the proportions entering the various rock types outcropping in the catchment. By this means the percolation through the boulder clay of Essex has been estimated to be between 25 and 50 mm/year, contrasting with about 100 mm/year for the more permeable boulder-clay cover in Norfolk[26].

Multiple regression procedures have also been used by Wright[25] to develop the following formulas for percolation through two different boulder clays in the Great Ouse Basin:

Chalky Boulder Clay $\qquad I = 0.202R - 70 \text{ mm} \qquad (4.4)$

Chalky Jurassic Boulder Clay $\qquad I = 0.202R - 77 \text{ mm} \qquad (4.5)$

The Chalky Jurassic Clay is the more tenacious of the two, and therefore allows somewhat less water to infiltrate. Percolation through the two boulder clays, with typical rainfall values for the area to which the formulas apply, is about 60 to 70 mm/year.

Lapworth[24] also estimated percolation in a drought year with a rainfall equal to 70 per cent of the average; such a year might occur once in 40 years over much of southern England. His figures are given in *Table 4.7* where they may be compared with later estimates made by alternative techniques for a once in 50 years occurrence.

Several experimental methods have been devised for investigating the percolation of water through the soil. Tests of fluorescent dyes[27] showed that Pyranine conc. was the most suitable of those examined: it suffered least degradation on contact with the soil and it could be detected several months after its application. Excavation of plots irrigated with this dye have shown a very non-uniform percolation pattern even over small areas. Tensiometers operated in groups of twenty or thirty buried at different depths can be employed to indicate moisture tension gradients in the soil and the direction of water movement. This is the approach being adopted in an experiment in

Table 4.7   ESTIMATES OF PERCOLATION DURING DROUGHT YEARS

Average annual rainfall (mm)	Drought year return period (*approximate*)	Infiltration in drought year (mm)	Percentage of average infiltration	Location
625	1 in 40*	55	25 ⎫	
750	1 in 40*	135	40 ⎪	General in
875	1 in 40*	215	47 ⎬	Southern England[24]
1000	1 in 40*	295	52 ⎭	
600	1 in 50	35	28	Ely Ouse[33]
740	1 in 50	65	22	Cirencester[92]
820	1 in 50	130	34	Beachy Head[66]
760	1 in 50	61	22	Lambourn*
760	1 in 50	75	30	Rockley Well*

*Unpublished data by courtesy of R. J. Mander (Thames Water Authority), from Lambourn soil water balance and Rockley well level response

Thetford Forest[28] where a system of recording tensiometers is being used to monitor the gradient in order to determine the quantity of drainage to the underlying Chalk and the evaporation from the surface.

Water movement in the unsaturated zone

The movement of water in the unsaturated zone is due to both gravity and capillary forces. The total potential at any point is the sum of the two, and water moves from areas of high to low potential. As gravity may not be the dominant force, water can move in any direction, although usually in a vertical direction, either upwards or downwards. The physics of water movement and consequent changes in moisture state have been the subject of considerable study. Notable contributions have been made by Childs[29] and Youngs[30] of the Unit of Soil Physics at Cambridge.

As the name implies, the unsaturated zone contains air or water vapour as well as water. The unsaturated hydraulic conductivity, often referred to as capillary conductivity, is a function of moisture content; the lower the moisture, the smaller the value, while maximum values are attained at near saturation. However, this property also depends upon such factors as pore sizes and their continuity, packing of the soil media and its texture. Thus the unsaturated hydraulic conductivity for a sand will eventually become less than that for a silt and a clay as moisture contents decrease, because the sand has less water-filled pores and the pores are too large to allow capillary movement. A fall in moisture content quickly reduces the unsaturated hydraulic conductivity; only a few centimetres of dry soil near roots may almost entirely restrict capillary conduction of water to them.

The movement of rainfall into a dry soil is controlled by both capillary and gravitational forces. Initially the capillary force is dominant, but the influence of gravity increases with time. The distribution of moisture in the soil profile as water moves downwards has been divided into four zones[31]:
1. Saturation zone
2. Transmission zone

3. Wetting zone
4. Wetting front

The saturation zone is a thin surface layer (about 15 mm thick) that, as the name implies, is saturated with water. It passes down, with a marked decline in moisture content, into the transmission zone. This zone has a fairly uniform moisture content. The wetting zone lies below the transmission zone, the lower boundary being referred to as the wetting front. The moisture gradient in the wetting zone is steep.

If infiltration continues the wetting front moves downwards and the transmission zone becomes longer. When infiltration stops, water is redistributed in the profile. The upper zones begin to drain and water continues to move down until all the zones reach field capacity. The process is illustrated in *Figure 4.3*, the two profiles indicating that the rate of redistribution depends upon the initial depth of infiltration.

Infiltration rates are influenced by two factors—storage capacity in the near surface layers and the rate that water can move downwards through the unsaturated zone. The limiting factor may be a thin relatively impermeable

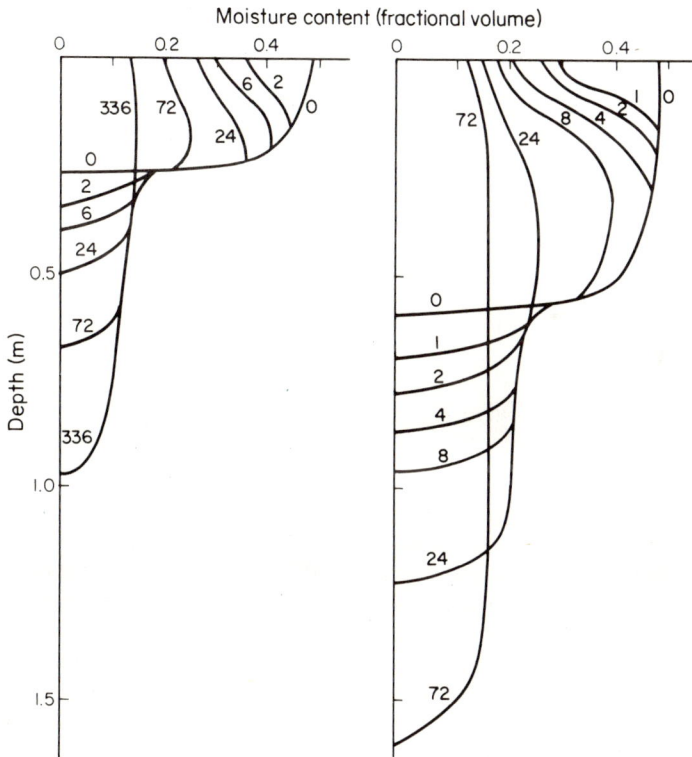

Figure 4.3   Moisture profiles during redistribution of moisture after infiltration; the numbers by each profile refer to the time, in hours, after the cessation of infiltration (from Youngs[94], courtesy The Williams Wilkins Co.)

layer such as a clay band or hardpan. If infiltration continues for long periods, 'perched' water tables can develop above such impeding layers.

When rainfall ceases, evaporation begins to dry out the surface layer, and water rises in the soil profile when capillary forces exceed gravitational forces.

If the water table is near the surface, water may be drawn up to the soil layer to replenish losses caused by evaporation and transpiration. The extent to which this is possible has always been an important factor when assessing the effect that drainage of groundwater may have on vegetation growth. Although early investigators claimed that water could rise considerable distances from deep water tables by capillarity, it is now accepted that such claims are not justified[32]. The rise of water due to capillary forces in a sand is about 0.3 m and in a clay about 1 m. In clays the upward movement is limited by the low hydraulic conductivity due to the fine pores.

The problem was given close attention during the course of the Great Ouse Groundwater Pilot Scheme[33]. The scheme involved extensive regional development of groundwater in the Chalk, and concern was expressed by the farming community that lowering the water table in the aquifer might reduce moisture contents in the soil zone to the detriment of crop yields. Preliminary assessments of the extent to which water would rise by capillary conduction were based on an equation by Gardner[34]:

$$d = \sqrt{(ka/E)} \qquad (4.6)$$

where d = the depth to the water table (from root zone or soil surface) that will permit an evaporation rate E by capillary conduction

       k = a constant (2.46)

       a = a factor relating unsaturated conductivity to moisture tension near the root zone; the value varies from about 400 for light sandy loam to 1500 for a clay loam

Applying the equation gave the following typical steady-state figures:

	Depth to water table (m)	
Evaporation rate	3 mm/day	5 mm/day
Light sandy soil	0.60	0.44
Heavy loam	1.20	0.86

As such high water tables are not normal during the summer growing season, drying of soil layers would soon limit capillary conductivity.

Only at two of the twenty sites in the Pilot Area at which detailed regular measurements of soil moisture were made were there any reductions of soil moisture in the upper two to three metres due to pumping groundwater. As would be expected from soil-water theory, these sites are within the riparian zone and adjacent to pumped wells. As the groundwater level declined in the Chalk, there followed, after some delay, drainage of a saturated sub-soil layer, sand in one case and peat in the other.

The general conclusion was that future groundwater operations on the same scale in similar areas would not affect crop yields, except possibly in the riparian zone where any reduction in soil moisture could even be beneficial.

The relationships between water-table depth, evaporation and crop growth have been studied in the East Anglian Fens[35] and Somerset moors, where sub-surface irrigation is practised. In these areas the water table is maintained close to the root zone by controlling water levels in field ditches from local rivers. Normally drought effects on growth can be minimised by keeping the water table within 0.9–1.2 m of the soil surface for arable crops and within 0.5–0.7 m for grass pasture.

The downward velocity of water molecules in the unsaturated zone is not the same as the movement of the wetting front. In a medium where intergranular flow predominates, movement is likely to be by displacement—that is, a water droplet added at the top displaces water throughout the column, and each year's input represents a layer in the profile; this mechanism is referred to as *piston flow*. However, as the velocity of water flow through pores of differing sizes is not uniform, some mixing must occur; this is due to *hydrodynamic dispersion*. It also seems unlikely that perfect piston flow operates in nature, as cracks and fissures in rocks allow water to move rapidly from the soil zone to deeper layers.

The flow mechanism through the unsaturated zone has been investigated in recent years using tritium pulses, produced in rainfall by thermonuclear explosions, as the tracer. The tritium concentration of water in the unsaturated zone of the Chalk has been examined at two sites, one in Berkshire and the other in Dorset[36,37]. In both cases the tritium profile reflected the tritium variations in rainfall. In *Figure 4.4* the peak due to the 1963–64 input is clearly visible, and a few metres lower down the 1958–59

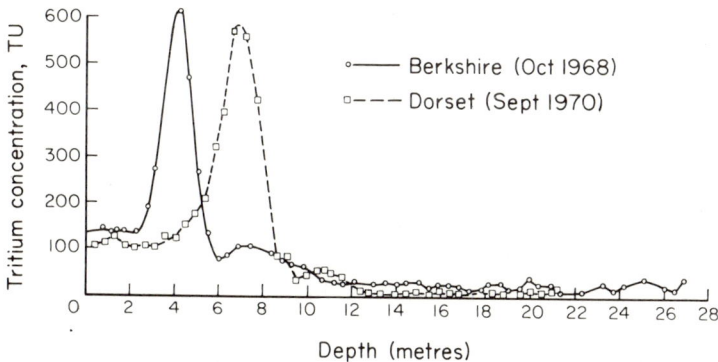

Figure 4.4 *Tritium concentrations in the unsaturated zone of the Chalk (after Smith and Richards[37])*

input is also discernible. The depth of the peaks indicates a mean downward velocity of water at both sites of about 0.88 m/year. Tritium concentrations at greater depths imply that some of the water input moved through fissures and cracks; at the Berkshire site this was about 15 per cent of the input, but in Dorset it was less than that.

These results imply that water movement in the Chalk is mainly by downward displacement through the pores of the rock, and the relatively low tritium concentration of groundwater in the Chalk as a whole suggests that this is probably the dominant process, although the proportion due to fissure flow on a regional scale is as yet unresolved.

Changes in groundwater quality, which can be related to rain periods even where the saturated zone is at some depth, affirm the occurrence of fissure flow[38]. There is some evidence that the greater the infiltration volume in any storm event the quicker the water table responds, and it may be that fissure flow is more dominant after extensive periods of rain when a saturated soil profile releases water to fissures. This process has been observed in the Chalk using fluorescein as a tracer, the water taking preferred fissure routes through the aquifer, ignoring other fissures. At other times water probably moves downwards more slowly and the proportion contributed to groundwater storage by rapid direct fissure flow is overall relatively small. Similar studies, taking advantage of tritium and nitrate marking of recharge water, have been conducted in the Chalk of east Yorkshire[39].

An interesting feature of the results obtained for the Chalk is that the classical percolation model based on rainfall less actual evaporation does not fit the observed total tritium input. The results suggest that infiltration is actually higher than would be expected from the classical model.

In contrast to the overall slow movement of water molecules in the Chalk, the actual effective rate of water transit, as measured by the response of the water table, is typically about 0.67 m/day. Thus the majority of infiltration in excess of soil moisture storage passes downwards to a water table at a depth of 6 m in 9 days, but takes 45 days if it is 30 m below the surface—although as already mentioned the first response may come earlier due to a direct component passing down fissures. In Hampshire the effective rate of water transit has been observed to be about 4 m/day[40].

Interflow or throughflow?

Horton's theory of the infiltration process[41] included the concept that, when infiltration capacity falls below rainfall intensity, overland flow occurs all over a hill-slope. But this only seems to be applicable to the infiltration process on bare hillsides with only a thin soil cover.

In Britain, where relatively thick soil and vegetation covers are more typical and where rain intensity is generally less than infiltration capacity, surface runoff (or overland flow) is not usual and a more important process is the slower lateral movement of water through the soil layer, referred to as *throughflow* by Kirkby and Chorley[42]. In this view the permeability of the soil zone tends to reduce with depth, which places a limit upon the amount of deep infiltration. Throughflow occurs above the level of reduced permeability. Soil moisture content increases down a hill-slope but only approaches saturation in a zone adjacent to the stream channels. During the course of a rainstorm the saturated area extends further up the hill-slope, but overland flow only occurs over the lower parts of the hill-slope where the soil is saturated.

Kirkby and Chorley[42] pointed out that Horton's view represents one extreme of the infiltration process while the concept of throughflow represents the other; at any particular time in a rain period variations of both extremes may be dominant on different parts of a hill-slope.

The term 'throughflow' appears to be a synonym of 'interflow', which has been defined as that part of infiltration that moves through the 'soil zone' without penetrating to the underlying zone of saturation. As such it embraces all water discharged from the unsaturated zone including that from perched water tables. Interflow may be 'thrown out' by impermeable soil layers as shallow springs or seepages; it may be augmented by tile drainage or controlled by the state of drainage ditches. Drainage works near the water table are an important factor determining the limits of moisture profiles and the pattern of soil-water flow to streams. Where tile-drains are laid close together, the water table is controlled and the ground no longer has such a large reservoir effect because infiltrating water is routed through the drains.

The term interflow tends to be associated with the conventional method of hydrograph analysis into the three components, storm runoff, interflow and baseflow (or groundwater discharge). From the above discussion it will be evident that processes giving rise to the storm runoff component depend upon the local physical conditions. Horton's view is that the entire basin contributes to storm runoff, but where throughflow is the dominant process this must produce storm runoff by providing water for overland flow in the saturated zone near the stream. Another view is that throughflow is not an important process but that the storm-flow component is produced from overland flow caused by direct precipitation on a limited area of saturated soil and also groundwater discharge where the water table reaches the surface[43]. The hill-slopes act as reservoirs which give rise to baseflow, the velocity of flow through the soil zone being too slow to produce storm-flow.

The importance of the various processes discussed will depend upon local factors such as geology, thickness and permeability of the soil zone and slope of the valley side. For example, in a study of a small impermeable catchment in Somerset underlain by Devonian rocks[44,45], two sub-basins were identified each with a distinctive soil and topographical character. Within the lower basin, hill-slope discharge was measured by a bank of lateral troughs, each one metre wide, set into the soil profile at the base of the hill-slope. These measured the flow from four distinct horizons, at depths of 0–100, 100–250, 250–450 and 450–750 mm. The soil was a free-draining brown earth up to 0.75 m deep and the ground slope was about 12°. Storm runoff was generated by direct channel precipitation over some 2 per cent of the basin. Although throughflow produced the main response of the basin to rainfall it did not produce storm-flow. On the other hand a true storm hydrograph was generated in the upper basin by a number of infiltration processes. 'Infiltration-excess' overland flow resulted from the surface peaty-gleyed podsol that covered extensive parts of the basin, and there was also a contribution from an extensive system of natural 'pipes' about 50 mm in diameter at the base of the peaty soil. Throughflow also contributed on the lower slopes.

Detailed examination of very localised areas (as, for example, in soil drainage problems) reveals the complicated relationship between the main processes. In one study carried out in Devon, a small area (0.85 hectare) near a hilltop was examined where the bed-rock was the Culm Measures. The soil was a surface water gley, waterlogging being caused by an impermeable layer in the profile, which impeded the movement of water.

The sub-soil was underlain by a highly permeable zone of weathered bed-rock. The minimum infiltration rate exceeded the majority of rainfall intensities, which discounted direct precipitation as the cause of the waterlogging. Water movement was restricted to the weathered rock zone, and the waterlogging was due to water being forced to the surface when flow was prevented by discontinuities in the soil profile[46].

Pipes such as those described by Weyman[44,45] in the Twin catchment appear to be quite widespread in Britain[47]. They play a significant role in the runoff processes in adjacent catchments on Plynlimon, which form the head-waters of the Severn and Wye and are covered with coniferous forest and grass respectively[48]. The soil strata are typically 100–150 mm of peat above a 50–100 mm clay/silt horizon overlying 500 mm of loose shaly mudstone flakes which grade into bed-rock. The hillside peat soils transmit water rapidly through a natural network of drainage paths or 'pipes'. Some, as much as 500 mm in diameter, are flowing permanently; they are in effect enclosed first order streams with slopes as low as half a degree. Their upstream sections carry only intermittent flow but may not dry up until several weeks have elapsed without significant rainfall.

Ephemeral pipes are the most common form. These occur at depths of less than 300 mm and have diameters of 20–400 mm. They usually run normal to the contours of the hill-slope and have been traced for several hundred metres. By definition they carry water only during and immediately after rain.

Large volumes of water appear to be lost to storage from pipes as flow proceeds down the hill-slope, and flow may not reach the bottom of the slope. The deep-seated permanent and intermittent pipe systems are the obvious source of base-flow in these Plynlimon catchments, and they also give rise to minor floods following moderate rainfall when the ephemeral system does not carry water. Similar pipe systems have been recorded in other moorland peat areas. Sometimes artificial drainage channels have cut and largely replaced the natural pipe network.

Measurements of stream flow in the chalk valleys of Dorset and Hampshire demonstrate the absence of surface runoff or rapid shallow drainage. However, examples of interflow superimposed on baseflow do occur. In the Sydling Valley, Dorset, infiltration rates in excess of about 2.75 mm/day result in an interflow component, as this rate is apparently the limit to deep vertical drainage in this catchment.

It seems clear that, whichever processes actually contribute to overland flow, the concept that only part of a catchment contributes—the so-called contributing area[49]—is valid under the conditions generally prevailing in Britain. The size of the contributing area is likely to vary during the course of a storm, particularly as the original saturated zone bordering the river expands[50,51].

Storage of water in the unsaturated zone

Water is retained in the soil largely by surface tension forces; a skin-like boundary exists at the air–water interface. On the hydrophilic surfaces that most soils possess an adhesive force is developed as a result of the very low angle of contact between the water and the surface. Salts in the soil water can cause osmotic forces to develop, and these can be important in some soils.

As the soil water content decreases it can be imagined that the curvature of the air–water meniscus between two soil particles increases. This increasing curvature means a decrease in free energy of the water molecules, fewer molecules escaping to the atmosphere and a lowering of the vapour pressure for a given temperature[52]. There are a number of ways of expressing this decrease in free energy, but most frequently it is considered as the capillary rise that would be required to reproduce the existing soil-moisture tension. For convenience this capillary rise is taken to be the logarithm of the suction force *(h)*, in centimetres of water, with which the soil moisture is in equilibrium. This force is termed the pF of the soil[53], where:

$$pF = \log_{10}h \qquad (4.7)$$

Table 4.8 shows the pF values for some of the important soil moisture states and the equivalent force expressed in other ways. For each soil series and each horizon within that series there are usually different sets of soil moisture relationships. A further complication is the difference that exists in the same sample of soil between its wetting and drying characteristics—the curve showing how water content varies with tension usually forms a hysteresis loop.

Table 4.8   pF VALUES FOR SOME IMPORTANT POINTS AND THEIR EQUIVALENTS

	pF	Tension (cm) Water	Tension (cm) Mercury	Atmospheres	Ergs
Oven-dry	7	10^7	76×10^4	10^4	98×10^8
Air-dry	6	10^6	76×10^3	10^3	98×10^7
Wilting point	4.2	1.58×10^4	1.14×10^3	15	147×10^5
Field capacity	2.7 to 1.8	517 to 63	38 to 4.6	0.5 to 0.06	49×10^4 to 6×10^4
Saturation	0	1	0.076	0.001	98

To determine these relationships, undisturbed core samples are subjected to test in the laboratory in apparatus that can recreate the conditions existing in the soil at different stages of wetting or drying. For example, simple sand-bath tension tables can be used over the range pF 1 to 3 and vacuum desiccators for the range pF 5 to 7. For the intervening part of the scale a pressure membrane apparatus[54] is usually employed, the relationship between applied air pressure (π) and pF being:

$$pF = 1.85 + \log_{10}\pi \qquad (4.8)$$

Soil samples are exposed at particular pF values in these devices until weighing shows that equilibrium has been reached. From the results of these tests pF curves can be determined *(Figure 4.5)* and can be seen to differ for the different soil types. One common feature is the constancy of the wilting point: pF 4.2 seems to be the maximum suction a plant root can exert to withdraw soil water, irrespective of the plant type and the soil type. Field

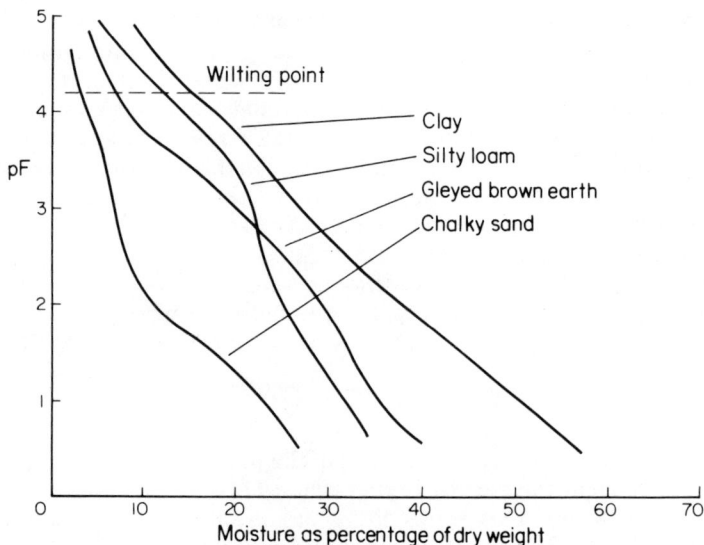

Figure 4.5   pF curves for some contrasting soils

capacity varies from soil to soil, as *Table 4.4* shows for the soils of the Dee. For plant growth and transpiration the amount of water held between wilting point and field capacity is important; this is often termed the *available water capacity* or the *available water*. For infiltration and runoff the capacity above field capacity is probably more important.

Both the storage and the movement of water in the soil are dependent upon the texture of the soil, its structure and several other factors[55]. The size of particles and the size of the pores they form determine the texture of a soil, while its structure is dependent on how the particles are arranged. Pore space increases from sands to clays, but the size and shape of pores and the presence of organic material influences both the storage and movement of water. Where the particle size distribution is known for a soil, Salter and Williams[56-58] have shown how texture can be related to moisture characteristics. They provide a graphical method for finding the available water capacity *(Figure 4.6)* based on laboratory tests covering a range from 60 mm/m in 1 m depth of sand to about 200 mm/m in 1 m depth of silt. Using values estimated from this diagram they obtained a mean error of ±10 per cent of the real values of available water capacity obtained from 39 soils by experiment. The method is known to produce an under-estimate where the soil has a high organic content.

A frequently used term related to the available water capacity is the *root constant*. This is defined[59] as 'the amount of easily accessible water held in the soil within the depth of rooting, and an extra inch added to allow for the water that can be extracted from the soil below the roots gives the total available water'. Chapter 3 discusses the estimation of the root constant and gives values *(Table 3.10)* for various crops. Of course root growth varies from year to year and during the growing season, and there are also the characteristics of the soil itself to take into account, yet despite these factors the concept is one that appears to work.

Several studies of roots have been undertaken that are significant in hydrological terms. For instance it has been found that some 90 per cent of the underground material of herbage plants is contained in the top 150 mm of the soil, and that approximately 99 per cent comes in the top 600 mm[60]. Reynolds has conducted studies of the spatial variations of root systems[61], and as part of the examination of evaporation in Thetford Forest the extension of new roots has been followed by soil coring and by measurements made through root observation windows[28].

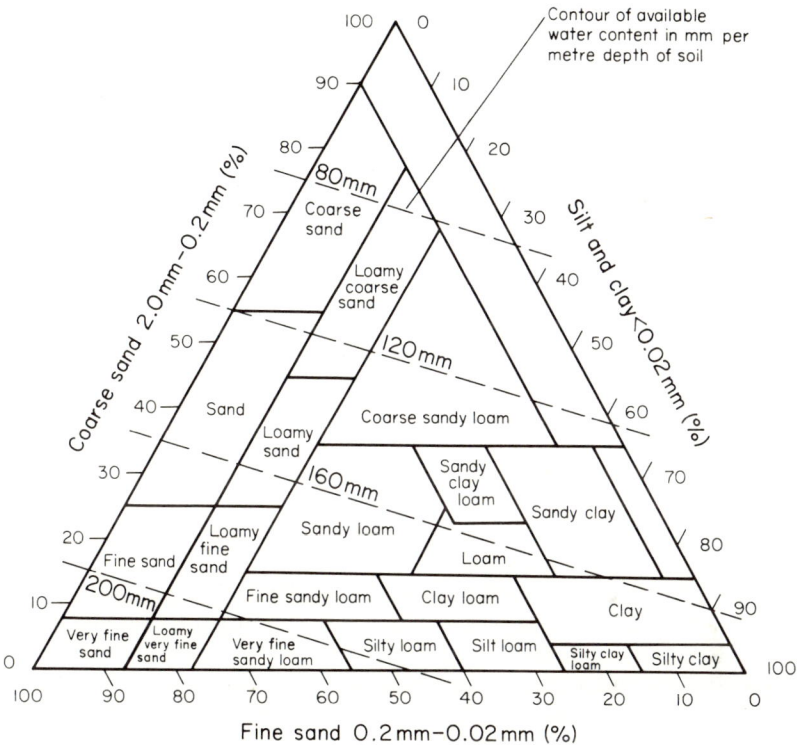

Figure 4.6 Graphical method for determining available water capacity (from Salter and Williams[58], courtesy Journal of Soil Science)

In most models simulating catchment behaviour, soil moisture storage is an important component. For example, in the model of the Dee[62], the measured physical characteristics of the soils were taken into account. Barton[63] classified soils in a similar way to Painter and to each group he allocated five interrelated soil moisture storage values. For the Cam, where the soils can be divided between those developed on boulder clay and those on the Chalk, it was found that a lumped model predicted the baseflow of the catchment adequately[64], when it might have been anticipated that some division of the basin would be necessary.

Measurement of soil moisture

It is probable that obtaining reliable measurements of soil moisture is one of the most intractable problems in hydrology[65]. Not only are there considerable difficulties in determining the water content of the soil at a particular point, but the spatial heterogeneity of soils is so enormous that areal assessments are considered by some to be virtually impossible to make. In fact there has been a tendency to avoid making soil moisture measurements and, wherever possible, to rely on estimates.

Some of the problems encountered have more to do with the nature of the soil itself than with the actual technique employed, such as how to deal with stones and organic matter. For the methods that rely on a buried sensor, there are the problems of ensuring proper contact between it and the soil, and how to emplace a sensor to avoid preferential movement of water into the hole in which it is buried. There are also the difficulties of producing a satisfactory calibration.

Gravimetric sampling

This method involves the extraction of a sample of soil from the required depth, enclosing it in an air-tight container and returning it to the laboratory, weighing, drying in an oven at 105°C for 24 hours and reweighing. The sample may be taken with a screw auger, but if an undisturbed core sample is obtained the moisture volume fraction (MVF) can be determined. Other methods are usually calibrated against this one, but although it has the advantage of being relatively cheap it is very time-consuming. It also has the disadvantage of destroying the site so that it is difficult to use on a routine basis.

By sampling on the third day after rain has saturated the profile it is possible to determine the field capacity of a soil[66].

Despite its drawbacks, gravimetric sampling has been employed effectively in a considerable number of other studies. For example, Edgell[67] used the method on Cader Idris to show that soil moisture content is related to land slope, while Goode[68] demonstrated that initially during the summer a larger soil moisture deficit is developed under long grass than short grass, until a mulch develops.

Tensiometers

The tension or suction developed at a particular point in the soil (also referred to as the negative pore-water pressure) may be measured by means of a tensiometer. This device consists of a porous ceramic pot, or alternatively a funnel with a sintered glass base, which is filled with water and connected by a water column to a manometer or vacuum gauge. The porous pot is buried in the soil at the desired depth, and water enters or leaves the pot until an equilibrium is set up with the soil water. This is registered by the vacuum gauge on the surface. The pore size of the pot can be selected to match the soil and the needs of the study—a small pore size giving a large range of tension, a large pore size a rapid response to changes in moisture content. While moist the pot is porous to water but not to air, but at tensions approaching pF 3 the water column breaks and air enters the system. Hence the use of tensiometers is confined to the wet end of the scale of soil moisture, and of course in converting from pF to moisture content, because of the hysteresis effect, it is necessary to know whether the soil is wetting or drying. Frost damage is another problem, but these cheap devices, which are relatively simple to make[69], have been widely used particularly for the control of irrigation systems[70] and in highway engineering research.

Electrical resistance units

Various types of electrical resistance unit have been constructed since they were first employed in Britain[71]. A pair of electrodes is embedded in a porous material, which takes up water from the soil until a state of equilibrium is reached. The resistance between the electrodes is governed by the moisture present in the porous material, which in turn is determined by the soil moisture content. Resistance is measured by a Wheatstone Bridge operating on a low voltage a.c. current of high frequency[72].

Some units have been made from cement, others from fibre-glass, but the most common types are manufactured from plaster of paris or nylon and stainless steel. Although units of the latter type are not buffered against the changes in conductivity of the soil water, they do not gradually decompose like the plaster units. Impregnating the plaster units with nylon resin prolongs their life, but it is difficult to build two units that react to soil water changes in an identical fashion. In addition to the moisture content–pF hysteresis, there is a second pF–gauge hysteresis between wetting and drying, so that calibration of these units is very difficult. However, plaster units are most reliable in drier soils and those of nylon and stainless steel in the wetter ones.

Rutter[73] employed plaster units in his study of water use of pine forest undertaken at Crowthorne in 1958. He installed these units at six depths between 150 mm and 2.4 m at four sites, first digging a pit and augering upward-sloping holes into one wall. The units were slightly smaller than the auger (120 mm diameter) and were chosen for their similar characteristics. They were calibrated against soil samples from the horizons where they were installed. Nylon and stainless steel units were installed in the Ray Catchment[74] at six depths from 150 mm to 1.7 m.

Because of calibration difficulties, the difficulty of maintaining contact with the soil, the influence of temperature, the effects of aging and several other drawbacks, most workers agree that electrical resistance units offer a cheap means of obtaining an index of soil moisture content but little else.

Neutron probes

This technique has been widely used and literature on it is considerable. A radioactive source emits fast neutrons into the soil where they are slowed down, principally by collision with hydrogen nuclei. Collision produces a cloud of slow 'thermal' neutrons; the density of this cloud is very largely a function of soil moisture content. The major components of the system needed to utilise this effect are:

1. a source—generally a capsule of between 30 and 100 millicuries of Americium-Beryllium (half life 450 years);
2. a detector—a boron trifluoride (BF_3) tube is the most common means of sampling the cloud of slow neutrons;
3. a counter—a device for counting the pulses emitted by the BF_3 tube in response to the slow neutrons.

The source and detector are housed close together in a probe (normally 38 mm in diameter), which is lowered down an access tube to the required depth *(Figure 4.7)*. The access tube, usually aluminium (44 mm in diameter,

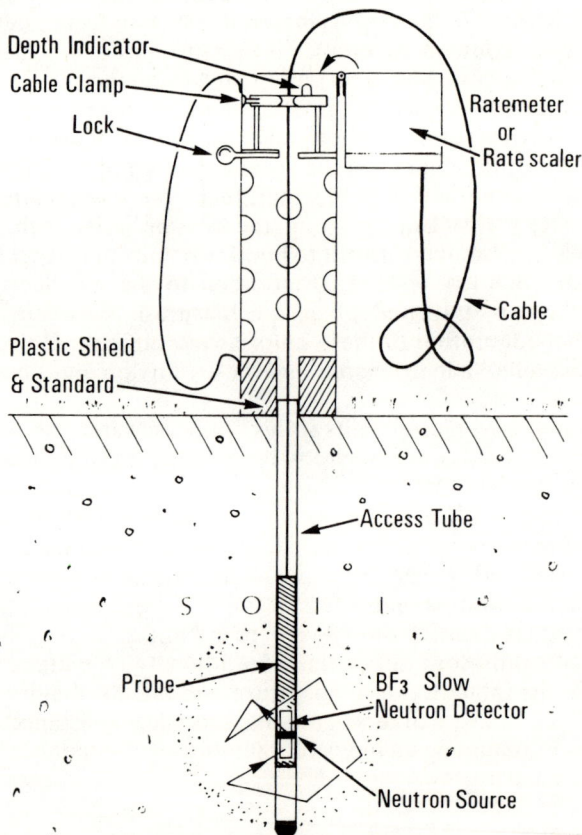

Figure 4.7 Diagrammatic representation of a neutron probe in use (not to scale) (courtesy Institute of Hydrology)

16 s.w.g. wall thickness), with a conical nose, is inserted into a tightly fitting hole augered to produce the minimum disturbance of the soil[75]. The cable carrying the probe is connected to the counter on the surface. This cable passes through the probe housing, so that when the probe is withdrawn from the soil for transport the operator is shielded from radiation emitted by the source. The counter incorporates the battery that powers the system, the electronic circuitry that controls it and a means of displaying the count.

The calibration of a probe can be based on theory, but usually it is determined by laboratory or field work. In the laboratory an access tube is placed centrally in a large drum of air-dry soil that has been packed to field density. The count rate is measured in this initially dry soil, then the soil is saturated and the amount of water this requires is measured so that a second point on the calibration curve can be determined. A field calibration is obtained by sampling with an auger close to an access tube for various soil moisture states and relating count rate to MVF. This method is lengthy and a wide scatter of calibration points frequently results.

The different devices available in the UK have been compared on several occasions by a group of users[76,77]. One outcome of these meetings was the development of a lightweight, rugged, one-piece instrument by the Institute of Hydrology. The 'Wallingford' probe[78] is now in widespread use and is available with three different types of counter. Statutory regulations cover its use, and it is designed so that an operator can be trained to use one in about half an hour. Use of the probe at fixed sites means that the same soil is sampled each time; as a consequence moisture differences can be determined very precisely.

In the field access tubes are closed at the top by rubber bungs and containers of silica gel inside prevent condensation. Because the probe samples a spherical body of soil, a false reading will be obtained where this sphere intersects the soil surface. In order to take measurements close to the surface a shallow tray (100 mm deep) of surface soil with perforations in the bottom is lifted temporarily over the top of the access tube to give a double thickness of soil.

Techniques have been developed[79] for processing the field data by computer, and several different methods of displaying the processed data are available. *Table 4.9* shows the daily variations in moisture content over part of the profile for one site in the Plynlimon catchments.

Neutron probes are in regular use in a number of catchment studies and at other sites across the country. The basins investigated include the River Ray[74], the Brenig[80], the Plynlimon catchments[81], the Cam[64], and the Great Ouse Pilot Scheme area[82]. Few studies of changes in areal averages have been made and published. The problem is to decide on the network and the sampling procedure to provide sufficiently representative observations for a whole basin. A choice has to be made, as Bell says[2], between measuring the maximum number of sites, each with a relatively high random error, and measuring a small number of sites, each with a relatively low random error. The first means that a short time (say 5 minutes) is spent on taking readings at each site; the second that a longer time is spent (15 minutes or more). Ideally observations should be taken simultaneously across a catchment, but in practice the network is confined to the number of tubes that can be visited in one day, depending upon which of the above strategies is chosen.

The Unsaturated Zone

Table 4.9 DAY-TO-TABLE VARIATIONS IN SOIL MOISTURE VOLUME FRACTIONS AND WATER CONTENT FOR A SITE IN THE PLYNLIMON CATCHMENTS IN 1971 (FROM ROBERTS[79], COURTESY INSTITUTE OF HYDROLOGY)

Day No. in year	Soil moisture volume fractions at various depths						Water content of total profile (mm)
	200 mm	300 mm	400 mm	500 mm	600 mm	700 mm	
195	0.804	0.826	0.745	0.592	0.412	0.345	49.3
196	0.809	0.837	0.745	0.584	0.408	0.345	49.4
197	0.797	0.826	0.730	0.588	0.398	0.341	48.8
198	0.797	0.821	0.743	0.589	0.401	0.338	48.8
199	0.806	0.816	0.724	0.587	0.398	0.342	48.8
200	0.798	0.818	0.745	0.575	0.393	0.331	48.6
202	0.777	0.806	0.739	0.600	0.412	0.338	48.4
204	0.781	0.814	0.715	0.582	0.392	0.331	47.9
205	0.783	0.805	0.746	0.599	0.412	0.324	48.4
206	0.785	0.809	0.758	0.610	0.429	0.326	49.0
207	0.785	0.814	0.729	0.573	0.387	0.329	47.9
208	0.788	0.811	0.741	0.582	0.412	0.322	48.4
209	0.787	0.806	0.729	0.593	0.395	0.326	48.2
210	0.778	0.796	0.740	0.591	0.401	0.323	48.0
213	0.819	0.868	0.772	0.672	0.536	0.429	53.2
214	0.814	0.854	0.769	0.627	0.503	0.442	52.3
215	0.840	0.849	0.773	0.615	0.461	0.433	52.3
216	0.859	0.888	0.804	0.696	0.557	0.438	55.3
217	0.925	0.920	0.813	0.700	0.559	0.435	57.4
218	0.878	0.897	0.816	0.697	0.568	0.443	56.2
220	0.836	0.879	0.814	0.690	0.550	0.441	54.6
221	0.918	0.935	0.836	0.700	0.569	0.460	57.9
222	0.880	0.854	0.769	0.664	0.513	0.424	54.2
223	0.930	0.933	0.776	0.681	0.543	0.442	57.0
224	0.947	0.923	0.805	0.688	0.552	0.436	57.7
225	0.942	0.930	0.819	0.700	0.559	0.447	58.1
226	0.955	0.933	0.843	0.703	0.555	0.444	58.6
Mean	0.838	0.854	0.768	0.633	0.469	0.384	52.0
Std. dev.	0.060	0.048	0.037	0.050	0.072	0.054	3.9

Other instrumental methods

Changes in moisture content alter several of a soil's characteristics and these alterations have been employed to obtain measurements of moisture content; for example, changes in resistance to twisting and shearing forces, changes in electrical capacitance, variations in thermal conductivity, and differences in the attenuation of gamma rays and radio signals. Some of these methods are employed for specialised purposes but none is in widespread use.

Estimation of soil moisture

Countrywide maps of the estimated state of soil moisture are issued about once a fortnight by the Meteorological Office *(Figure 4.8)*, together with

Figure 4.8   Estimated soil moisture deficit, 29 October 1969 (from Grindley[84], courtesy Hydrological Sciences Bulletin)

data in tabular form for certain river basins. These maps indicate areas where there is a flood risk during the winter; while the build-up and extent of the soil moisture deficit are indicated by the summer maps. The method used in constructing these maps has been described[83,84], but in brief it depends on estimates of potential transpiration and assumptions about root constants for a number of representative sites. It also relies upon the Penman[85] procedure for allowing for the divergence of actual and potential transpira-

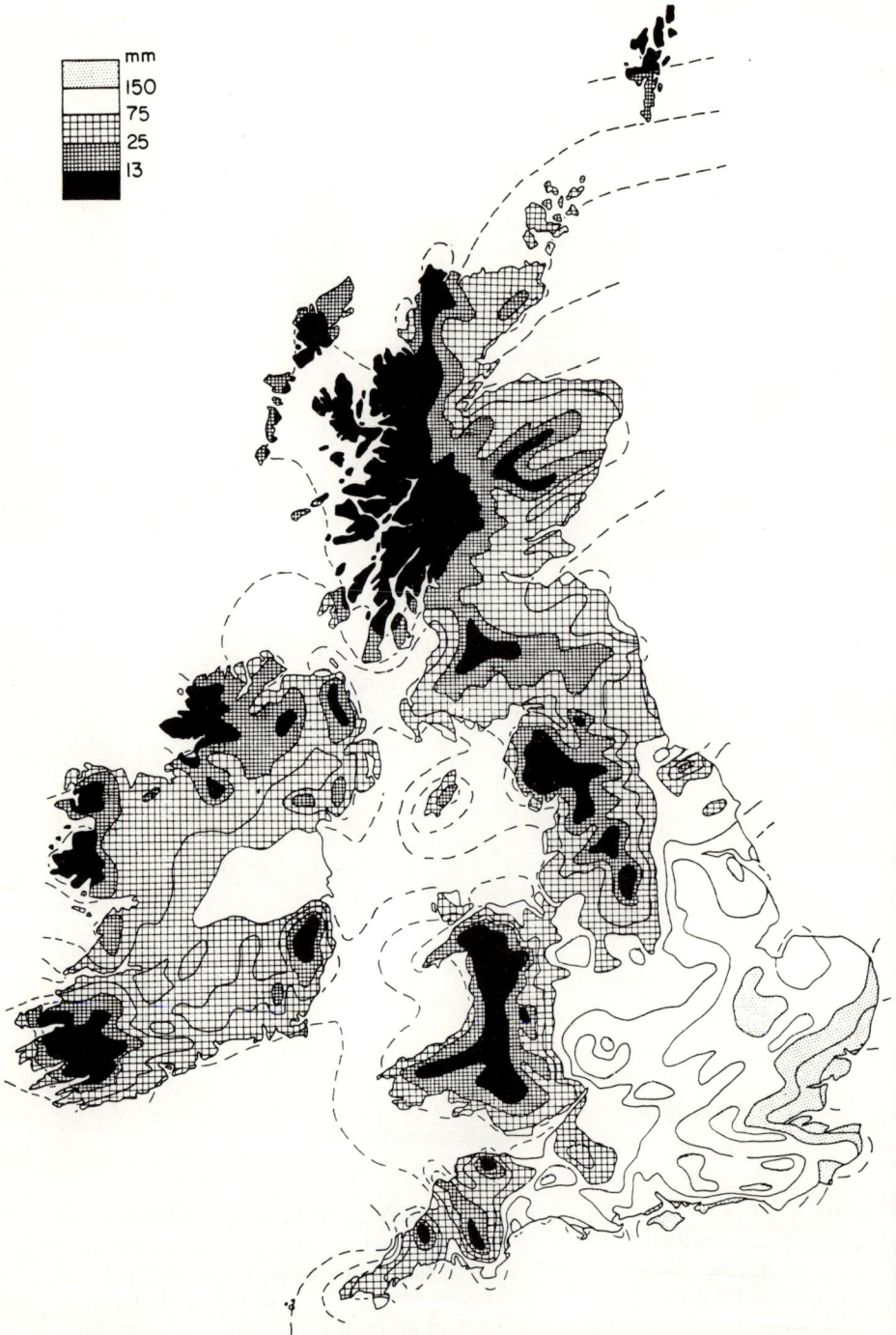

Figure 4.9   Annual average soil moisture deficit (from Green[86], courtesy Blackwell Scientific Publications Ltd)

tion and on the assumption that, when rainfall exceeds potential transpiration, the soil moisture deficit is reduced by the difference between the two amounts. The estimates are made on a daily basis and the soil moisture state is continually updated. The average annual potential soil moisture deficit *(Figure 4.9)* has been estimated by Green[86] by summing the difference between average potential transpiration and average rainfall for the six summer months for a large number of sites.

Applications

Some of the ways in which knowledge of soil moisture, infiltration and percolation have been employed for particular purposes have already been described. Further examples of the use of measured or estimated values of these factors are discussed in the remainder of this chapter.

Soil state and flood runoff

For many years the 'rational' method has been used for obtaining the peak flood flow from:

$$Q = kAi \qquad (4.9)$$

where A = the catchment area
         k = the impermeable fraction of area A
         i = the design rainfall intensity on the catchment averaged over the 'time of concentration'

In this context it has become usual to talk about 'percentage runoff' or 'impermeability factors' for catchments. Hempsall[87] suggests the following values for k:

Catchment	Large area	Small, steep area
Rocky and impermeable	0.8	1.0
Slightly permeable, bare country	0.6	0.8
Slightly permeable but with vegetation or cultivated areas	0.4	0.6
Cultivated absorbent soil	0.3	0.4
Deep sandy absorbent soil	0.2	0.3
Heavy mature forest on permeable rock	0.1	0.2

The RRL hydrograph method[88], which was developed by the then Road Research Laboratory for the design of urban sewer systems, and which employs an advanced form of this type of approach, uses a time–area method to form a hydrograph. In this method the grassed areas in a catchment are assumed to be totally pervious but the hard surfaces may be grouped according to their relative impermeabilities.

As a first step in the development of methods of predicting flow, a number of studies have employed the antecedent precipitation index (API) as an indication of the wetness of a catchment. This followed Linsley's[89] work in developing an API for the Thames. The method has been employed subsequently for estimating both flood discharges[90] and discharges for longer periods[91] for a number of catchments. The basis of the approach is to allot weights *(a)* to the rainfall amounts *(R)* recorded before the time in question, in the following manner:

$$\text{API} = aR_t + a_1R_{t-1} + a_2R_{t-2} \quad \ldots \quad a_nR_{t-n} \tag{4.10}$$

When an API is devised for flood prediction purposes, the values of the weights usually decrease in magnitude with increasing time prior to the event to be estimated, e.g. $a = 0.9, a_1 = (0.9)^2 \ldots a_n = (0.9)^{n+1}$. When mean monthly flows are being estimated, it is frequently found that the weights do not decrease uniformly with time but reflect the monthly distribution of rainfall amounts.

Once the API has been calculated it can be updated daily with ease when the gauge rainfalls that are used to determine the catchment value are known. As the API is an index of the water stored in the catchment that has not yet drained away, it might be expected to be related to the catchment's recession constant k.

The estimated soil moisture maps issued by the Meteorological Office have already been described. The success of such procedures for predicting the onset of the winter lowland flood season can be seen from the comparison *(Table 4.10)* with soil conditions observed by the Middle Level Commissioners in their East Anglian Fens[93].

A special case exists when the ground freezes. The normal ability of the ground to absorb water is halted, with consequent dangers of enhanced runoff in a subsequent wet spell. Some of the very worst floods in lowland England have been caused by spring rain melting a snow cover at a rate much in excess of the capacity of the soil to accept it. March 1947 is a well-known example. A great deal of low-lying agricultural land was inundated, especially in the Fens where the embanked main rivers were breached. The case of rain falling directly on to frozen ground clear of snow is much rarer; to be a serious threat the ground must be frozen to a depth that will be unaffected by the relatively warm air mass associated with the rain. Such conditions appeared to be likely early in 1963 when record depths of frozen ground (about 1 metre) were experienced. Fortunately the spring was quite dry and the snow cover melted slowly.

Another unusual cause of flooding can be the 'panning' of the soil crust during a major drought. Not only can clay become brick-hard but also peat may dry to the point where it cannot be rapidly resaturated. Thus rain, falling typically at high intensity in a thunderstorm that climaxes a long sultry dry spell, may cause higher flows than the ordinary river flood channel can cope with. A good example is the Foston, Derbyshire, flood of 1959[87].

The first comprehensive study of the importance of soil wetness on flood magnitudes was carried through by the Flood Studies Team[10]. They showed that it was possible to predict (with a multiple regression coefficient of 0.657 from 1447 observations) the proportion of storm runoff *(Q)* to storm rainfall *(P)*:

$$Q/P \times 100\% = 0.22 \, (\text{CWI} - 125) + 0.10(P - 10) + 95.5 \, \text{SI} + 0.12 \, \text{URB} \tag{4.11}$$

Table 4.10 WINTER FLOOD SEASONS IN THE FENS
(MIDDLE LEVEL, CAMBRIDGESHIRE)

Winter	Meteorological Office's date of saturation by calculation	Date of saturation by observation	Calculation error Days early	Calculation error Days late
1939–40	18 Nov	16 Oct	—	33
1940–41	17 Nov	7 Nov	—	10
1941–42	13 Jan	Not attained	50+	—
1942–43	16 Dec	Not attained	80+	—
1943–44	Not attained	Not attained	Correct	—
1944–45	17 Nov	16 Nov	—	1
1945–46	27 Dec	28 Dec	1	—
1946–47	14 Nov	13 Nov	—	1
1947–48	17 Jan	Not attained	50+	—
1948–49	13 Dec	Not attained	80+	—
1949–50	24 Nov	19 Dec	25	—
1950–51	19 Nov	11 Nov	—	8
1951–52	24 Dec	8 Nov	—	46
1952–53	19 Nov	19 Nov	Correct	—
1953–54	9 Feb	1 Jan	—	39
1954–55	5 Nov	5 Nov	Correct	—
1955–56	22 Jan	11 Jan	—	11
1956–57	28 Dec	18 Dec	—	10
1957–58	6 Dec	21 Oct	—	46
1958–59	4 Oct	1 Oct	—	3
1959–60	11 Jan	4 Jan	—	7
1960–61	1 Nov	19 Oct	—	13
1961–62	10 Dec	15 Dec	5	—
1962–63	3 Feb	28 Jan	—	4
1963–64	30 Dec	28 Nov	—	32
1964–65	25 March	18 Jan	—	56

where CWI (catchment wetness index) $= 125 + API5 - SMD$

API5 (antecedent precipitation index) is the first five terms of the series defined on page 154, adopting $a = (0.5)^{1/2}, a_1 = (0.5)^{3/2} \ldots a_n = (0.5)^{n+1/2}$; API5 can be found approximately from $P0.5^{D/24}$, where D = design storm duration (hours) for which the commencing API value is required, and P = precipitation over $2D$ hours prior to start of design storm.

SMD = prevailing soil moisture deficit (if any) at start of API5 calculations

SI (soil index) $= (0.15S_1 + 0.3S_2 + 0.4S_3 + 0.45S_4 + 0.5S_5) \div 1 - S_u$

$S_1, S_2 \ldots S_5$ = percentage of area covered by soil types 1 to 5 (see *Table 4.3*)

S_u = unclassified areas of water or urban development (as percentage of catchment)

URB = percentage of catchment covered by urban development

The standard error of estimate is 15 per cent. Likely values of Q/P range from 30 per cent in deep soils in the south-east during typical storms to 75 per cent in the mountainous uplands of the west in extreme floods. However, it must be stressed that this equation is not designed to explain every event. Instead it is to assist in estimating a flood of desired probability from the appropriate rainfall via the unit hydrograph technique.

REFERENCES

1. Jacks, G. V., *Soil,* Nelson, London, 215 (1958)
2. Bell, J. P., 'Neutron probe practice', *Institute of Hydrology Report No. 19* (1973)
3. Soil Survey of England and Wales, *Annual Report,* HMSO (1973)
4. Agricultural Advisory Council, *Modern farming and the soil,* HMSO (1971)
5. Russell, Sir E. J., *The world of the soil,* Collins (1957)
6. Taylor, J. A., 'Soils and vegetation', in *Wales: a physical, historical and regional geography* (ed. E. G. Bowen), Methuen (1957)
7. Edmonds, D. T., Painter, R. B. and Ashley, G. D., 'A semi-quantitative hydrological classification of soils in north-east England', *J. Soil Science,* **21,** 256–264 (1970)
8. Painter, R. B., 'A hydrological classification of the soils of England and Wales', *Proc. Instn Civil Engrs,* **48,** 93–95 (1971)
9. United States Bureau of Reclamation, *Design of small dams,* US Dept. of Interior, 413–424 (1965)
10. *Flood Studies Report,* Volume 1, Chapter 4, Natural Environment Research Council (1975)
11. Rudeforth, C. C. and Thomasson, A. J., 'Hydrological properties of soils in the River Dee catchment', *Soil Survey Special Survey No. 4,* 12 (1970)
12. Taylor, J. A. and Tucker, R. B., 'The peat deposits of Wales: an inventory and interpretation', *Proc. Third International Peat Congress,* 163–173 (1968)
13. Bower, M. M., 'The distribution of erosion in Blanket Peat in the Pennines', *Trans. Inst. Brit. Geographers,* **29,** 17–30 (1961)
14. Hill Farming Research Organisation, *Third Report 1961–1964,* 75–78 (1964)
15. Conway, V. M. and Millar, A., 'The hydrology of some small peat-covered catchments in the northern Pennines', *J. Instn Water Engrs,* **14,** 415–424 (1960)
16. Hills, R. C., 'The influence of land management and soil characteristics on infiltration and the occurrence of overland flow', *J. Hydrology,* **13,** 163–181 (1971)
17. Hills, R. C., 'Lateral flow under cylinder infiltrometers: a graphical correction procedure', *J. Hydrology,* **13,** 153–162 (1971)
18. Field Drainage Experimental Unit, Minist. Agric. Fish. and Food, *Annual Report 1970*
19. Fenwick, I. M., 'Some problems in soil permeability measurement', *Essays in Geography for Austin Miller,* Reading University Press (1965)
20. Shockley, D. C., 'The influence of soil moisture holding characteristics upon sprinkler irrigation design, in *Planned irrigation,* Wright Rain Ltd (1956)
21. Hoare, E. R., 'Soil, moisture and plant growth', in *Planned irrigation,* Wright Rain Ltd (1956)
22. Satchell, R. L. H. and Edworthy, K. J., 'Recharge of the Nottinghamshire Bunter Sandstone', *Proc. Symp. on advanced techniques in river basin management: Trent Model Research Programme,* Instn Water Engrs, 75–93 (1972)
23. Lewis, W. V., 'Some aspects of percolation in south-east England', *Proc. Geol. Assoc.,* **54,** 171–184 (1943)
24. Lapworth, C. F., 'Percolation in the Chalk', *J. Instn Water Engrs,* **2,** 97–120 (1948)
25. Wright, C. E., *Combined use of surface and groundwater in the Ely Ouse and Nar catchments,* Water Resources Board, Reading, 43 (1974)
26. Ineson, J. and Downing, R. A., 'Some hydrogeological factors in permeable catchment studies', *J. Instn Water Engrs,* **19,** 59–80 (1965)
27. Reynolds, E. R. C., 'The percolation of rainwater through the soil demonstrated by fluorescent dyes', *J. Soil Science,* **17,** 127–132 (1966)
28. Institute of Hydrology, *Research 1972–73,* Natural Environment Research Council, 66 (1973)
29. Childs, E. C., *An introduction to the physical basis of soil water phenomena,* John Wiley & Sons, London, 495 (1969)
30. Youngs, E. G., 'Basic laws of soil water', *Peat Hydrology Report No. 16,* Institute of Hydrology, Wallingford, 4–6, (1972)
31. Bodman, O. B. and Colman, E. A., 'Moisture and energy conditions during downward entry of water into soils', *Proc. Soil Survey Soc. Amer.,* **8,** 116–122 (1943)
32. Baver, L. D., *Soil Physics,* John Wiley & Sons, London, 489 (1956)
33. Drennan, D. S. H., 'Agricultural investigations in the pilot scheme area 1969–71', Appendix A of *Great Ouse Groundwater Pilot Scheme Final Report,* Great Ouse River Authority, Cambridge (1972)

34. Gardner, W. R., 'Some steady-state solutions of the unsaturated moisture flow equation with application to evaporation from a water table', *Soil Science,* **85,** 228 (1958)
35. Nicholson, H. H., Firth, D. H., Eden, A., Alderman, G., Baker, C. J. L. and Heimberg, M., 'The effect of groundwater level upon productivity and composition of Fenland Grass (II)', *J. Agric. Science,* **43,** 265–274 (1953)
36. Smith, D. B., Wearn, P. L., Richards, H. J. and Rowe, P. C., 'Water movement in the unsaturated zone of high and low permeability strata by measuring tritium', *Proc. Symp. isotope hydrology,* Int. Atomic Energy Agency, Vienna, 73–87 (1970)
37. Smith, D. B. and Richards, H. J., 'Selected environmental studies using radioactive isotopes', *Peaceful uses of atomic energy, Vol. 14,* Int. Atomic Energy Agency, Vienna, 469–480 (1972)
38. Warren, S. C., 'Chemical aspects of controlled pumping and automation', *Proc. Soc. Water Treat. Exam.,* **13,** 7–11 (1964)
39. Foster, S. S. D. and Crease, R. I., 'Nitrate pollution of Chalk groundwater in east Yorkshire—a hydrogeological appraisal', *J. Instn Water Engrs,* **28,** 178–194 (1974)
40. Headworth, H. G., 'The analysis of natural groundwater level fluctuations in the Chalk of Hampshire', *J. Instn Water Engrs.* **26,** 107–124 (1972)
41. Horton, R. E., 'Erosional development of streams and their drainage basins: hydrophysical approach to quantitative morphology, *Geol. Soc. Amer. Bull.,* **56,** 275–370 (1945)
42. Kirkby, M. J. and Chorley, R. J., 'Throughflow, overland flow and erosion', *Bull. IASH,* **12,** 5–21 (1967)
43. Dunne, T. and Black, R. D., 'An experimental investigation of runoff production in permeable soils', *Water Resources Research,* **6,** 478–490 (1970)
44. Weyman, D. R., 'Throughflow on hill-slopes and its relation to the stream hydrograph', *Bull. IASH,* **15,** 25–33 (1970)
45. Weyman, D. R., 'Runoff process, contributing area and streamflow in a small upland catchment', in *Fluvial processes in instrumented watersheds,* Inst. Brit. Geographers, 33–43 (1974)
46. Rycroft, D., 'Drainage investigations in the south-west', in *Annual Report 1971,* Field Drainage Experimental Unit, Minist. Agric. Fish. and Food, 7–15 (1972)
47. Jones, A., 'Soil piping and stream channel initiation', *Water Resources Research,* **7,** 602–610 (1971)
48. Institute of Hydrology, *Research 1971–72,* Natural Environment Research Council, 67 (1972)
49. Betson, R. P., 'What is watershed runoff?', *J. Geophys. Research,* **69,** 1541–1551 (1964)
50. Tennessee Valley Authority, 'Area-stream factor correlation, a pilot study in the Elk River basin', *Bull. IASH,* **10,** 22–37 (1965)
51. Hewlett, J. D. and Hibbert, R. A., 'Factors affecting the response of small watersheds to precipitation in humid areas', *Proc. Int. Symp. on forest hydrology* (ed. W. Sopper and H. Lull), Pergamon Press, 275–290 (1965)
52. Ingram, J., 'Soil moisture', *Research,* **14,** 63–70 (1961)
53. Schofield, R. M., 'The pF of the water in soil', *Trans. Third Int. Cong. Soil Science,* **2,** 37–48 (1935)
54. Coleman, J. D. and Marsh, A. D., 'An investigation of the pressure-membrane method for measuring the suction properties of soil', *J. Soil Science,* **12,** 343–362 (1961)
55. Stewart, V. I., 'Water in the soil', *Welsh Soils Discussion Group Report No. 4,* 1–10 (1963)
56. Salter, P. J. and Williams, J. B., 'The influence of texture on the moisture characteristics of soils: II Available water capacity and moisture release characteristics', *J. Soil Science,* **16,** 310–317 (1965)
57. Salter, P. J., Berry, G. and Williams, J. B., 'III Quantitative relationships between particle size composition and available water capacity', *J. Soil Science,* **17,** 93–98 (1966)
58. Salter, P. J. and Williams, J. B., 'IV A method of estimating the available water capacities of profiles in the field', *J. Soil Science,* **18,** 173–181 (1967)
59. Penman, H. L., 'Vegetation and hydrology', *Technical Communication No. 53,* Commonwealth Bureau of Soils, Harpenden (1963)
60. Troughton, A., 'Studies on the roots and storage organs of herbage plants', *J. British Grassland Soc.,* **6,** 197–206 (1951)
61. Reynolds, E. R. C., 'Root distribution and the cause of its spatial variability in Pseudotsuga taxifolia (poir)', *Brit. Plant and Soil,* **32,** 501–517 (1970)

62. Jamieson, D. G. and Wilkinson, J. C., 'River Dee research programme 3: a short-term control strategy for multi-purpose reservoir systems', *Water Resources Research*, **8**, 911–920 (1972)
63. Barton, B. M. J., 'Synthetic monthly runoff records for ungauged British catchments', *J. Instn Water Engrs*, **27**, 149–162 (1973)
64. Dickinson, W. T. and Douglas, J. R., 'A conceptual runoff model for the Cam catchment', *Institute of Hydrology Report No. 17*, 44 (1972)
65. McCulloch, J. S. G., 'Hydrological networks for measurement of evaporation and soil moisture', *Proc. Symp. on design of hydrological networks*, Quebec, IASH Pub. No. 68, 579–588 (1965)
66. Binnie and Partners, *Report of the effects of Friston Forest on the yield of Friston Pumping Station*, Eastbourne Waterworks Company (1966)
67. Edgell, M. C. R., 'A preliminary study of some environmental variables in an upland ecosystem: Cader Idris, Merionethshire', *J. Ecology*, **59**, 193 (1971)
68. Goode, J. E., 'Soil moisture deficits developed under long and short grass', *Annual Report, East Malling Research Station* (1955)
69. Jacobs, J. C. and West, G., 'A field tensiometer for soil moisture studies', *Road Research Laboratory Note No. RN/4085/JCJ.GW* (Nov. 1967)
70. Wells, D. A. and Soffe, R., 'A tensiometer for the control of glasshouse irrigation', *J. Agric. Engng Research*, **6**, 16–26 (1961)
71. Croney, D., Coleman, J. D. and Currer, E. W. H., 'The determination of the pore water pressure and moisture content of soil using electrical resistance gauges', *Road Research Laboratory Note No. RN/1455/DC.JDC.EWHC* (Nov 1950)
72. Croney, D., Coleman, J. D. and Currer, E. W. H., 'The electrical resistance method of measuring soil moisture', *British J. Applied Physics*, **2**, 85–91 (1951)
73. Rutter, A. J., 'Studies in the water relations of pinus sylvestris in plantation conditions: II The annual cycle of soil moisture change and derived estimates of evaporation', *J. Applied Ecology*, **1**, 29–44 (1964)
74. Edwards, K. A. and Rodda, J. C., 'A preliminary study of the water balance of a small clay catchment', *J. Hydrology (NZ)*, **9**, 202–218 (1970)
75. Eeles, C. W. O., 'Installation of access tubes and the calibration of neutron moisture probes', *Institute of Hydrology Report No. 7* (1969)
76. Bell, J. P. and McCulloch, J. S. G., 'Soil moisture estimation by the neutron method in Britain', *J. Hydrology*, **4**, 254–263 (1966)
77. Bell, J. P. and McCulloch, J. S. G., 'Soil moisture estimation by the neutron method in Britain—a further report', *J. Hydrology*, **7**, 415–433 (1969)
78. Bell, J. P., 'A new design principle for neutron soil moisture gauges: the "Wallingford" neutron probe', *Soil Science*, **108**, 160–164 (1969)
79. Roberts, G., 'The processing of soil moisture data', *Institute of Hydrology Report No. 18* (1972)
80. Cole, J. A. and Green, M. J., 'Measuring soil moisture in the Brenig catchment: problems of using neutron scatter equipment in soil with peaty layers', *Proc. Symp. on water in the unsaturated zone*, Wageningen, IASH Pub. No. 82 (1966)
81. Rodda, J. C., 'Progress at Plynlimon—problems of investigating the effect of land use on the hydrological cycle', *British Association Meeting*, Swansea (1971)
82. Binnie and Partners, *Water balance report for 1972*, Ouse Groundwater Pilot Scheme, Great Ouse River Authority, Cambridge (1973)
83. Grindley, J., 'The estimation of soil moisture deficits', *Met. Mag.*, **96**, 97–108 (1967)
84. Grindley, J., 'Estimating and mapping of evaporation', *Proc. Symp. on world water balance*, IASH Pub. No. 92, **1**, 200–213 (1970)
85. Penman, H. L., 'The dependence of transpiration on weather and soil conditions', *J. Soil Science*, **1**, 74–89 (1949)
86. Green, F. H. W., 'A map of annual average potential water deficit in the British Isles', *J. Applied Ecology*, **1**, 151–158 (1964)
87. Hempsall, M. S., 'The assessment of flood risk in the United Kingdom', *J. Chartered Insurance Institute*, **59**, 115–137 (1962)
88. Watkins, L. H., 'The design of urban sewer systems', *Technical Paper No. 55*, Road Research Laboratory, HMSO (1962)
89. Linsley, R. K., 'River forecasting in the United States', *Conference on forecasting river floods*, Dept. of Civil Engineering, Imperial College, London (1958)
90. Lambert, A. O., 'An investigation into infiltration and interception rates during storm

rainfalls and their application to flood prediction', *J. Instn Water Engrs,* **21,** 525–535 (1967)

91. Andrews, F. M., 'Some aspects of the hydrology of the Thames Basin', *Proc. Instn Civil Engrs,* **21,** 55–90 (1962)
92. Burton, A. R., 'Hydrological study of the Latton groundwater source', *J. Instn Water Engrs,* **22,** 287–293 (1968)
93. Fillenham, L. F. and Jack, W. L., 'Evaluation of flood prevention schemes for the Middle Level, Cambridgeshire', *J. Instn Water Engrs,* **29,** 297–304 (1975)
94. Youngs, E. G., 'Redistribution of moisture in porous materials after infiltration, 2', *Soil Science,* **86,** 202–207 (1958)

Chapter 5

Groundwater

Introduction

Groundwater is the part of sub-surface water that occurs in the voids of rocks in the zone of saturation. It can be developed for water supply providing the rocks are sufficiently permeable to allow flow to collecting points, such as wells or springs, in quantities considered to be economic.

Although groundwater is widely distributed in the UK and, therefore, available for small domestic supplies, its availability on a scale to provide for public water supply or the larger industrial demands is mainly restricted to those areas in England where the more permeable sedimentary rock formations occur. Wherever groundwater is available in sufficient quantities it is generally preferred as a source of water because of its low cost, good quality, relatively constant temperature and chemical composition, and the fact that it is not so susceptible to pollution as a surface source.

In England during the early stage of industrialisation and the associated urbanisation, considerable advantage was taken of the presence of permeable rocks to develop groundwater at sites or in areas where it was required. Examples are the London Basin, Merseyside, Birmingham, the coastal towns of southern and eastern England, among many others. Early developments were primarily taking advantage of the permeability of aquifers, which allowed water to flow to the well or wells at the desired rates; in other words the aquifer was being used as a distribution system. As the extent of development increased the natural storage of water in the rocks began to be depleted to a significant extent in many areas. Undesirable consequences became apparent in the form of declining groundwater levels, reduction of river flows, particularly during the summer, and saline intrusion near coastlines. Consideration of the amount of storage available was gradually becoming a more critical factor. This has been accentuated further in the last decade following the greater emphasis placed on the more effective use of the large volumes of storage available in aquifers. Over the years there has been a trend from local development of individual sites to the regional development of aquifers. This requires the management of aquifers to meet all the demands made upon them, for it is now accepted that aquifers provide water not only for water supply but also to maintain river flows for the preservation of amenity and public health, fisheries and navigation rights. Therefore, management of a groundwater reservoir must

160

be considered in the context of the integrated use of surface water and groundwater, and the optimum use of both surface and groundwater storage to conserve water.

The efficient management of a groundwater reservoir is more complex than that of a surface reservoir. Information is required about variations in the permeability and the storage capacity of the rocks, the sources of inflow to the aquifer and the positions of the various outlets, as well as the effect of groundwater abstraction upon both the inflow and the outflow. It is important to appreciate that water levels have to be lowered before an aquifer can be developed efficiently and that a wide range of water levels is necessary to derive the greatest benefits from the storage available.

Use of groundwater in the UK

The total groundwater abstraction in England and Wales in 1972 was 2434 million m³. This represented about 16 per cent of the total water use in that year but the relatively low percentage was due to the large volume of surface water used for cooling. A better indication of the role of groundwater is given by the fact that it provides 35 per cent of the water used for public supply.

The abstraction from individual aquifers has been analysed for the period 1948–1963[1]. This information, based on returns made under Section 6 of the Water Act, 1945, indicated that for this period, which is probably representative of the current situation, 60 per cent of the total abstraction was from the Chalk and Triassic sandstones and 20 per cent, but mainly for mine-drainage, from the Coal Measures. (When groundwater pumped for mine-drainage is excluded, the Chalk and Triassic sandstones contribute about 75 per cent of the total amount of groundwater used for water supply.) Of the remaining aquifers the Lower Greensand, Permian, Inferior Oolite and superficial deposits were of approximately equal importance, the Lower Greensand accounting for about 3.5 per cent of the total use and the remainder about 2 per cent each. In marked contrast to the rest of the world, in the UK river-deposited sands and gravels are not important aquifers.

In Scotland and Northern Ireland groundwater resources are not of regional significance but they are important in some areas for local development[2-4]. In both countries groundwater contributes about 5 per cent to public supply. The available yield from underground sources for public supply in Scotland is only about 100 Ml/d, while private abstractors pump about 20 Ml/d. Groundwater sources provide some 25 Ml/d towards the public supply demand in Northern Ireland. The fact that a relatively high proportion of the total groundwater use in both countries is derived from springs reflects the speculative nature of drilling in the less permeable rocks that predominate in these countries.

In the past groundwater has usually been developed by pumping directly into pipe-distribution systems to meet the demand, but, as already mentioned, in the future there will be a greater emphasis on the use of aquifers as storage reservoirs. Attention will be directed to the optimum use of this storage in combination with surface water resources, as discussed in Chapter 8.

The main sources of groundwater in the UK will obviously continue to be the Chalk and Triassic sandstones. The Chalk crops out, or is overlain by relatively small thicknesses of superficial deposits, within some 17 000 km² in England. This is about twice the area within which the Triassic sandstones either outcrop or are covered by superficial deposits. In general terms these figures indicate the relative importance of the resources of these aquifers that are derived from infiltration of rainfall. However, they do not indicate the full potential of these rocks as aquifers or their relative value as groundwater reservoirs; for this it is necessary to know their storage coefficients and their effective water-bearing thicknesses. The storage coefficient of the Chalk is typically 2 per cent or less, whereas that of the Triassic sandstones is commonly 20 per cent. Furthermore, the effective thickness of the Triassic sandstones is greater than that of the Chalk, which may have an upper limit of 50–100 m. These figures give some indication of the relative importance of the two aquifers as reservoirs and emphasise the potential of the Triassic sandstones—and, in fact, sandstones in general. This should not be interpreted as saying that the Chalk is not a large reservoir itself. To give but one illustration of the storage capacity of aquifers, the upper 50 m of the saturated Bunter Sandstone in Nottingham-shire, outcropping over some 1000 km², contains about 10 000 million m³ of available water, a figure that may be compared with 1800 million m³, the total storage in 1970 in surface reservoirs in the United Kingdom used for water supply.

Thus the total water storage in aquifers is very large, and considerable effort is now being applied towards developing this resource to a greater extent without incurring undesirable side effects. It must be realised that, unless the water is to be 'mined' (which is not usually acceptable in the long term), only part of the storage need be used to balance uneven variations in the long-term and short-term distribution of rainfall and to store surplus surface runoff underground by artificial recharge. Some of the problems and constraints are discussed later in this chapter and in Chapter 8.

Techniques and instrumentation

Assessment of groundwater resources

The groundwater resources of an aquifer can be divided conveniently into 'replenishment resources' and 'storage resources'. Replenishment resources are those dependent upon annual infiltration to the aquifer and are equated with long-term mean infiltration. Storage resources on the other hand represent water lying in an aquifer below the principal natural outlets. These resources are not depleted in a drought. They are estimated from a knowledge of the thickness of the saturated aquifer and the appropriate coefficient of storage. As already remarked, the volume is often consider-able[5,6]. However, all the water permanently stored in aquifers cannot be developed, for constraints invariably exist such as deterioration of quality or reduction of hydraulic conductivity with increasing depth.

Infiltration to an aquifer is commonly estimated from the difference between mean annual rainfall and actual evaporation over the aquifer outcrop. Where the outcrop does not support direct runoff the difference between the two parameters is taken as the natural recharge. If the aquifer is overlain by superficial deposits the infiltration is reduced by a factor, often arbitrarily selected but of course related to the nature of the superficial deposits. A further allowance may be necessary if direct runoff from the superficial deposits can recharge the aquifer at the periphery of their outcrop.

Sometimes the infiltration through superficial deposits can be estimated from more reliable data in adjacent similar catchments, as in East Anglia where the infiltration derived from base flow analyses has been assumed to apply to larger areas of similar basic geology.

In many studies in the UK[7-9], groundwater resources have been estimated from measurements of rainfall and actual evaporation. The difference between the two measurements is often small, for example less than 100 mm/year in parts of south-east England, and in these circumstances an appreciable error can arise in the value derived for groundwater resources.

With the increasing number of river flow gauging stations and the length of flow records available, the average groundwater component in river flow can be assessed readily by hydrograph analysis. The discharge can be equated with infiltration into the groundwater catchment above the gauge. Allowance has to be made for any underflow beneath the gauge[10].

Not uncommonly the surface and groundwater divides do not coincide; this has to be taken into account when relating groundwater discharge to infiltration expressed as depth of water over the appropriate infiltration area. As an example the groundwater catchment of the River Itchen to Allbrook is some 35 per cent larger than the surface catchment[11]. Similarly in the London Basin rain infiltrating into the surface catchment of the River Lee actually discharges into the Great Ouse basin[7].

The assessment of groundwater resources is often associated with completing a water balance for the catchment area, of the form:

$$P = E + S_R + G_R \pm U \pm S \qquad (5.1)$$

where
P = precipitation
E = actual evaporation
S_R = direct runoff (including interflow)
G_R = groundwater discharge
U = net groundwater flow through the aquifer (i.e. inflow less outflow)
S = change in storage, groundwater and soil moisture

The equation is usually solved for a period of years so that changes in storage (which are difficult to measure) are of less significance. In some circumstances the net groundwater flow through the aquifer (U) can be appreciable and may represent the main part of the groundwater resources. This can occur, for example, if the water balance is made for an aquifer unit at outcrop, but the flow represented by U is from the outcrop area into a confined area and represents the main replenishment resources of the confined area. A case in point is the Lincolnshire Limestone in south Lincolnshire where most of the infiltration into the limestone flows in an easterly direction through the aquifer below confining Middle and Upper Jurassic beds. This flow has been estimated[6] by solving the equation for U.

A similar application occurs when assessing the amount of direct groundwater flow to the sea from a coastal aquifer. A water balance is made for a number of years and the equation again solved for U, which represents in this case discharge to the sea.

Other methods are available for estimating infiltration and hence groundwater resources[12]. Infiltration can be measured with lysimeters or infiltration gauges but the results are unlikely to be representative on a regional scale. Equations have also been derived relating infiltration to rainfall[13,14] (see Chapter 4), and increasingly mathematical and analogue models of aquifer systems are being used to assess groundwater resources.

The groundwater flow through an aquifer is given by:

$$Q = TiW = KiA = KiWm \qquad (5.2)$$

where
- Q = quantity of flow through the aquifer in a given time
- T = transmissivity
- i = hydraulic gradient
- W = width through which flow takes place
- K = hydraulic conductivity
- A = cross sectional area through which flow takes place
- m = saturated thickness of the aquifer.

If water is abstracted from an aquifer at different rates over a number of years and the aquifer is confined and therefore remains fully saturated (or if it is unconfined and the fall of water level due to abstraction is small in relation to the total saturated thickness), the hydraulic gradient will be related to the quantity abstracted. A linear relationship can be obtained between groundwater level in the aquifer and abstraction, from which an estimate can be made of the 'safe yield' of the aquifer[15] *(Figure 5.1)*. This figure illustrates the point that water levels have to be lowered to develop groundwater resources; the problem is to decide when further lowering will cause deleterious consequences.

This section has considered various approaches to the assessment of groundwater resources but it is important to realise that groundwater storage and river flow are intimately related, groundwater development usually being at the expense of river flow. For this reason the extent to which the total available groundwater resources can be developed requires careful study of all the implications on the remainder of the hydrological cycle.

Pumping tests

A pumping test involves abstracting water from a well at a controlled rate and observing the rate of change of water level in the well and in nearby observation wells. The purpose of the test should be to determine one or more of the following:
1. Hydraulic properties of the aquifer and associated deposits.
2. Yield–drawdown curve for the well.
3. Efficiency of the well.
4. Change of quality of the water with time.
5. Regional implications of developing the aquifer, for example the aquifer–river relationship.

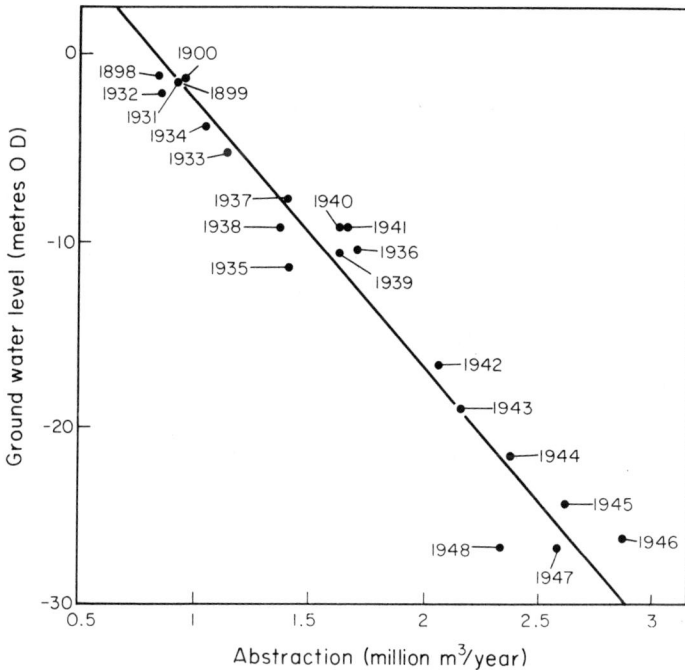

Figure 5.1 Relationship between groundwater levels in the Chalk and abstraction in the Colne Valley, Essex (from Slater[15], courtesy Institution of Water Engineers)

Traditionally major wells in the UK are tested after completion for 14 to 21 days and the tests are generally timed for the autumn, when water levels are low. This arose because it was considered that such a period gave the water engineer some confidence in the long-term reliability of the well and also allowed sufficient time for any significant interference effects to become obvious in nearby wells. However, the length of a test depends very much on its purpose, and in some cases a few hours or a few days will suffice.

Hydraulic properties of an aquifer

The properties of an aquifer that can be calculated from a pumping test are hydraulic conductivity, transmissivity and storage coefficient. In addition the relationship of the aquifer to overlying and underlying semi-permeable deposits can also be assessed and the position calculated of natural boundaries (such as rivers, faults or lithological changes) that influence the flow of water to the well.

To meet all these objectives water level changes due to pumping must be measured in observation boreholes. Generally the local physical features and geological relationships indicate possible boundaries, and observation boreholes should be sited to confirm their existence. For example if the pumped well is near a river, observation boreholes should be sited on both sides of the river, in lines at right angles to and parallel to the river front, with the object of defining the form of the cone of depression and determining whether or not the river is acting as a recharge source.

A first stage in the interpretation of the data from a pumping test is to make a log–log plot of water level drawdown against time since pumping commenced for each observation borehole, and compare the form of the curve with the 'Theis type curve'[16]. The form of the data curve will often indicate whether the aquifer is behaving as an infinite homogeneous aquifer, whether there is leakage from semi-permeable layers (thereby reducing the drawdown), or whether the curve is departing from the type curve because of an impermeable boundary that is increasing the drawdown.

Analysis of water level changes in an observation borehole near a pumped well indicates the average hydraulic characteristics of the aquifer, and the boundaries that exist, in the area within which drawdown of water levels occurs. This has the advantage that the properties are representative of a relatively large volume of rock.

Observation boreholes should not be sited too close to the pumped well otherwise local variations in the aquifer, which may be unrepresentative, could influence the results. On the other hand boreholes should be close enough to ensure that significant water level changes are induced by pumping and that the changes caused by pumping overshadow any small errors that may result from corrections for natural changes such as seasonal trends, and barometric or tidal effects.

Ideally observation boreholes should penetrate the full thickness of the aquifer, but corrections for partial penetration can be made. Where leakage through overlying semi-permeable beds is suspected boreholes should be sited to establish whether such leakage occurs. As a general rule, under unconfined conditions observation boreholes should be within 150 m of the pumped well if the abstraction rate is 50 l/s or more, and within 100 m if the pumping rate is less than this. In a confined aquifer the effects of pumping are propagated rapidly over a wide area and observation boreholes can be sited up to several hundred metres from the pumped well.

If the pumping test is for the calculation of aquifer properties, it may only be necessary to pump for some 12 to 72 hours. However, tests should be continued until the local hydrogeological factors affecting the flow to the well have been determined; this means that a preliminary analysis of the data must be made as the test proceeds.

Methods devised for analysing pumping tests under a variety of different conditions have been adequately summarised in the literature[17] and are not discussed here.

Before being tested a well should be developed to its maximum capacity. This involves increasing the permeability and porosity in the vicinity of the well by removing fine-grained material from the aquifer or fine-grained material, produced during drilling, from the walls of the well where it may be blocking fissures leading into the well. This is done by surging (for example alternately starting and stopping the pump, which allows water to fall back into the well and thereby reverses the flow in the immediate vicinity of the well), by using a jetting tool to clean the well face or screen or, in limestones or calcareous aquifers, by treating with acid.

Natural water level trends must be established before the test by taking measurements in observation boreholes daily for at least seven days prior to the test. More frequent measurements, or the use of automatic recorders, may be necessary if corrections to water levels are required because of variations caused by barometric pressure changes or induced by tidal movements.

At least one observation borehole should be outside the radius of influence of the pumped well, for use as a control. Care should be taken to ensure that the site of the control borehole(s) is representative of conditions in the vicinity of the pumped well.

Water abstracted from the well should be disposed of outside the radius of influence of the pumped well, and if possible into a watercourse, to avoid recirculation of water during the test. In the absence of a watercourse, temporary pipes are necessary to transport the water 500–1000 m before discharging on to the ground. The distance depends very much on the length of the test.

Yield–drawdown curve

The yield–drawdown curve of a well is defined by pumping the well at three or more rates, each rate being continued until an 'equilibrium' pumping water level is attained or until such a level can be estimated by extrapolation of the data. From such a curve the permanent pumping plant can be designed and an estimate made of pumping costs.

The form of the yield–drawdown curve will vary at different times if the seasonal fluctuation of the natural groundwater level is large in relation to the thickness of the aquifer. This is because the saturated thickness of the aquifer (and hence the yield) will be less when the water level is low. In such situations the pumping test should be carried out at a time of minimum water levels. Changes in the shape of the yield–drawdown curve with time may indicate deterioration in the capacity of a well, or perhaps improvement in its capacity following surging or other well development techniques.

Yield–drawdown curves for Chalk wells conform to a family of type curves[18] *(Figure 5.2)*. At low inflow velocities the relationship between yield and drawdown may be linear (as would be expected from Darcy's Law), but the curves become non-linear at high rates of abstraction when the flow in the vicinity of the well becomes turbulent. Departure of the yield–drawdown curve from the type curve indicates the maximum capacity of the well. Above this rate there is little or no increase of yield for a large increase of drawdown. This stage is reached because of excessive head losses at the higher rates of abstraction.

A family of type curves also exists for sandstone aquifers including the Triassic sandstones, Lower Greensand and Spilsby Sandstone[18]. In these aquifers departure from the type curve occurs more gradually than is often the case for wells in the Chalk, probably because intergranular flow is more important.

Values for transmissivity around a producing well in the Chalk can be estimated from the yield–drawdown curve of the well by using a relationship derived between the transmissivity and the yield of the well at an equilibrium drawdown of 3 m[18].

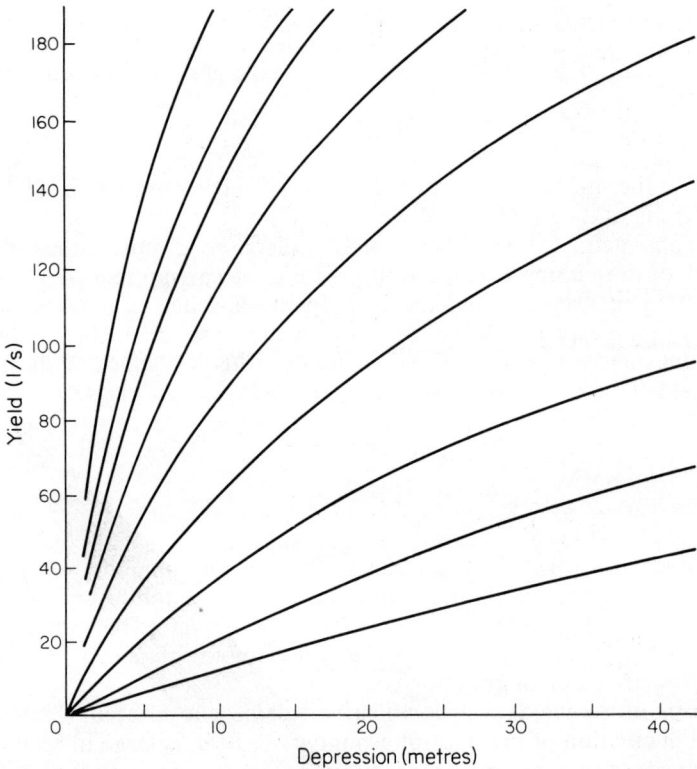

Figure 5.2   Type yield–depression curves for Chalk wells (from Ineson[18], courtesy Institution of Water Engineers)

Well efficiency

The water level drawdown in a pumped well that fully penetrates an infinite confined aquifer has two components—formation loss and well loss. Formation loss is a function of both pumping rate and length of pumping. It is determined from standard formulae for the calculation of drawdown by inserting hydraulic properties for the aquifer derived from a constant rate pumping test.

Well loss is a function of pumping rate only and does not vary with time. It is caused by resistance to flow in the well and through the well-face and well-screen (if present), and by the occurrence of turbulent flow in the vicinity of the well. Under turbulent flow conditions the well loss increases rapidly with the pumping rate[19]. Well efficiency is defined as the ratio of the theoretical formation loss to the total drawdown in the well. Under less than ideal conditions drawdown may be increased by:

1. partial penetration of the pumped well;
2. dewatering of the upper part of the aquifer;
3. impermeable boundaries to the aquifer.

These factors reduce the well efficiency ratio. Drawdown may be reduced by recharge boundaries to the aquifer.

Jacob[20] showed how the formation and well-loss components could be calculated from a step-drawdown pumping test. In his interpretation he assumed well loss was proportional to the square of the pumping rate. However, at low rates of abstraction the flow in the vicinity of the well can remain laminar, and in these circumstances the well loss is directly proportional to the pumping rate[21]. Rorabaugh[21] suggested that when turbulent flow occurs the well loss is more accurately given if the pumping rate is raised to a power greater than 2 and suggested 2.5 as a representative value, but a value as high as 3.5 has been quoted[19].

According to Rorabaugh[21] the equations representing the drawdown in a well are:

$$S_w = B_t Q + C'Q \text{ for } Q < Q_c$$
$$S_w = B_t Q + CQ^p \text{ for } Q > Q_c$$

where

S_w = drawdown
Q = pumping rate
B_t = formation loss factor
$C', C \text{ and } p$ = constants relating to the well loss term (Jacob assumed $p = 2$)
Q_c = critical rate at which the flow near the well becomes turbulent

A method for differentiating the well loss caused by laminar and turbulent flow and for calculating the constants has been given by Rorabaugh[21]; its application has been reviewed by Lennox[19], who has also reviewed the stages involved in an analysis of step-drawdown data.

Maintaining well yields involves frequent assessments of well efficiency. Yield-drawdown tests should be made regularly to detect any deterioration in well performance.

Change in water quality

During a pumping test water samples should be collected for chemical and bacteriological analysis. The chemical examination of a sequence of samples indicates whether quality changes with time during pumping. The chemical nature of the water indicates whether incrustation and corrosion are likely to be problems.

Regional analysis of aquifer systems

Proposals made in the UK for the intensive development of groundwater resources on a regional scale, in conjunction with surface runoff, have led to the need for pumping tests to assess the consequences of such developments on the hydrological regime of a catchment. These tests are usually made to determine the effects of pumping from a number of wells on groundwater levels, baseflow to rivers, and induced leakage from rivers in response to head changes caused by pumping. The emphasis is on the regional analysis of an aquifer/river system, and river gauging stations and many observation boreholes are necessary for adequate interpretation. Such tests often continue for 100–150 days or even longer. As an example the Great Ouse Groundwater Scheme in the Thet catchment of Norfolk involved pumping 70 Ml/d from 18 wells for up to 250 days. The consequences were measured using over eighty deep observation boreholes, some fifty shallow 75 mm diameter tube wells, four river flow gauging stations and some twenty sites for evaluating changes of soil moisture[22,23].

Design of sand screens and gravel packs

Well screens and properly designed gravel packs are necessary in wells drawing water from unconsolidated sands, if optimum yields are to be obtained over long periods of time. The aquifers in which such precautions are usually necessary include the Pleistocene Crag, Tertiary sands, Upper and Lower Greensand and the Ashdown Beds. The design of screens and packs has been adequately covered in the literature, and reference should be made to standard texts for details[17,24]. For convenience, the various design stages are briefly summarised below:

1. Collect adequate representative samples of the aquifer.
2. Determine the grain-size distribution, throughout the sequence if the aquifer is not homogeneous, and obtain the effective grain-size (the sieve size that retains 90 per cent of the aquifer particles), the mean grain-size and the uniformity coefficient (the sieve size that retains 40 per cent of the aquifer divided by the effective grain-size).
3. Decide upon the type of pack necessary—that is, either natural or artificial (either uniform or graded)[25]. Very often an artificial pack is used; if the uniformity coefficient of the aquifer is less than 3 a uniform pack is appropriate, but if it is greater than 3 a graded pack should be inserted. A natural pack can be developed from the aquifer itself by surging and pumping the well, providing the effective grain-size of the aquifer is greater than 0.25 mm, the 90 per cent size is less than 10.0 mm and the uniformity coefficient is between 3 and 10. It involves selecting a slot size for the screen that allows about 60 per cent of the finer constituents in the aquifer in the vicinity of the screen to pass through, while retaining the coarser particles. An artificial pack is a material, coarser than the aquifer, that is placed around the screen to retain the aquifer. An artificial pack should be at least 70–80 mm thick but less than 200 mm. It is usually about 150 mm thick. Preformed artificial packs, of appropriate size, glued to a slotted screen and ready for insertion in a well, are available.
4. Design the pack (if one is required) from grain-size distributions of the aquifer.
5. Decide upon the slot size of the screen from grain-size distributions of the aquifer or pack.
6. Decide upon the length of the screen from the thickness of the aquifer, position of permeable zones and probable pumping water level. Generally in a confined aquifer only 70–80 per cent of the aquifer thickness is screened, while under water-table conditions only the lower third or half may be screened because of the drawdown.

7. Decide upon the well diameter and the diameter of the screen to ensure adequate yield at a suitable entrance velocity through the screen.

Wells in unconsolidated sands should be carefully developed by surging up to a maximum yield. During service they should be pumped at less than the maximum developed yield and sudden large changes in rate should be avoided. Lack of attention to well design and development is responsible for the short life of many wells in arenaceous aquifers. On the other hand, careful design of packs and attention to well development and maintenance should give well efficiencies greater than 80 per cent.

Application of geophysical methods

Geophysical methods are being applied increasingly in exploration for groundwater and in its subsequent development. Exploration aspects are concerned with the location of groundwater, the determination of geological conditions favourable to its occurrence, and the differentiation between fresh and brackish or saline water within aquifers. Surface methods are applied to these objectives.

Once groundwater has been developed by means of wells or boreholes, sub-surface methods are used to determine the nature and properties of the strata exposed in the borehole walls and of the fluids that the strata contain. These methods can also be used to determine the physical and chemical properties of the water column, and to record the constructional details of the borehole[12,26,27].

Surface methods

The principal ones are electrical resistivity, seismic, gravity and magnetic methods. Surface methods have not been used extensively for groundwater exploration in the UK, since the geology of the country is known in some detail, but electrical resistivity and seismic methods have been applied to give more detailed insight into geological conditions in local areas.

In the electrical resistivity method the average resistivity of the ground is measured with a series of electrodes arranged in line. Normally only four electrodes are used, and lateral changes in resistivity are detected by traversing with an electrode array in which the spacing between the various electrodes is kept constant. If vertical changes in lithology are to be investigated, the effective depth of penetration of the measurements is progressively increased by expanding the distance between the electrodes without altering their pattern. If good resistivity contrasts exist between successive formations, the depths at which the lithological changes occur may be estimated by plotting the resistivity values obtained at different electrode spacings against the corresponding spacings. However in many cases, and especially when more than two layers are involved, the interpretation of resistivity data requires detailed analysis and the use of theoretical curves.

Surface resistivity measurements can sometimes be used to determine the depth of a saline groundwater interface.

In the seismic method, shock waves are generated in the ground by means of small explosive charges or mechanical vibration, and the velocities in the formation of selected waves are measured by timing their arrival at sensitive detectors placed at carefully-chosen distances from the signal source. As in the resistivity method, the effective depth of penetration of the measurements can be varied, in this case by changing the spacing between signal source and detector.

Resistivity and shallow-penetration seismic methods are often employed in groundwater investigations to estimate the thickness and lateral extent of water-bearing gravels, and to locate buried channels in bedrock. On the other hand, magnetic and gravity methods, which involve the measurement of small local variations in the earth's magnetic and gravitational fields, have not been employed to any extent in groundwater investigations in the UK.

Sub-surface methods

These involve lowering probes mounted on electric cable into a borehole to measure a number of different properties of the rock and its contained fluids. Some of these methods are designed to measure rock or rock-fluid properties exclusively, disregarding as far as possible the influence of the borehole, while other methods are restricted to the borehole fluid. In each case the variations in the measured parameter are recorded against depth, and the resulting borehole log is compared with similar logs recording other parameters in order to arrive at a balanced interpretation.

The logs most commonly employed in groundwater studies at the present time measure:

    electrical resistivity (of the formation)
    electrical resistivity (of the borehole fluid)
    spontaneous potential
    natural gamma radiation
    temperature gradient (in the borehole fluid)
    vertical flowrate (of the borehole fluid)

Other logs that are useful but perhaps less commonly used include:

    caliper and casing collar locator
    neutron
    density (of the formation)
    acoustic velocity (in the formation)

Electrical resistivity logs measure the resistivity of the formation when this is saturated with an electrically conductive fluid. The measurements are made between an earthed electrode, which may be situated at the surface, and an array of electrodes on the borehole logging cable. As in surface resistivity surveying, the effective depth of penetration of the measurements can be varied by changing the electrode geometry.

Borehole resistivity devices are of two main types, namely those designed to measure the resistivity of the formation close to the borehole (shallow-penetration tools) and those intended to measure formation resistivity at points beyond the influence of the borehole and any fluid escaping from it (deep-penetration tools). Shallow-penetration tools average their measurements over a small depth interval and therefore have good vertical definition, but their response is affected by the presence of the borehole and by any borehole fluid that may have invaded the formation. Deep-penetration tools average their measurements over a greater depth interval, but they are less affected by the borehole and they record values that are probably closer to the true resistivity of the formation. So-called 'focused-current' devices combine good vertical definition with deep lateral penetration, but they are more expensive than the simple resistivity probes and are not widely used in groundwater work at present.

When resistivity logs are run under the right borehole conditions with both shallow-penetration and focused-current tools, information about the relative permeability of the formation at different levels can sometimes be derived by careful interpretation. In most groundwater investigations only the simple types of resistivity log are likely to be available. However, these logs can give useful information about relative porosities in the formation, and resistivity profiles that are characteristic of particular lithological zones may be used for correlation between different boreholes.

Changes in resistivity profiles have been related to known horizons in the Chalk[28] *(Figure 5.3)*, and this has facilitated correlation between boreholes throughout southern England and made possible the more effective control of drilling programmes.

Standard curves have been produced for some of the most widely used electrode arrangements, and for the proprietary tools run by geophysical service companies, to permit quantitative interpretation of resistivity logs in

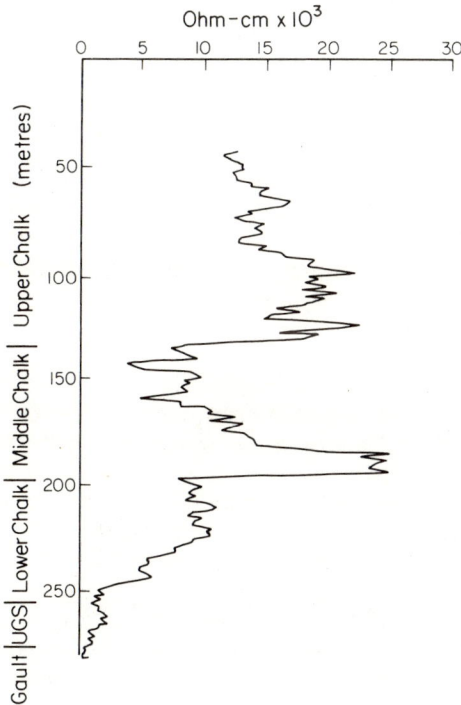

Figure 5.3   Electrical resistivity log of the Chalk in a borehole at Leatherhead (from Gray[28], reproduced by permission of the Controller, Her Majesty's Stationery Office)

terms of formation porosity. These curves are useful when applied with the proper corrections, but when accurate porosity data are required then more modern methods using nuclear interactions[29] and acoustic velocities may be preferred.

Special resistivity tools are available for logging dry boreholes or those that contain electrically non-conductive muds, but these devices do not have much direct application in groundwater studies.

While formation resistivity logs are intended to ignore as far as possible the nature of the fluid in the borehole, fluid resistivity logs are confined to the borehole fluid itself. The method of measurement is essentially the same as that used for formation resistivity, but the downhole electrodes are screened from the formation. Changes in fluid resistivity with depth can be related to changes in groundwater quality, and since the ionic content of waters from different aquifers may vary, fluid resistivity logging can sometimes be used to distinguish between waters derived from different aquifers in multi-aquifer wells. Changes in resistivity may also indicate levels of water inflow to a well.

If sufficient background data are available, empirical relationships may be established in a particular groundwater province between fluid resistivity and the concentration of the predominant ions in that particular ground-water mixture. However, the main value of fluid resistivity logging lies in the location of saline water in aquifers that are subject to seawater intrusion, or to contamination by connate water at depth[30].

The spontaneous potential (SP) log records the sum of small electrical potentials that may occur in a borehole when it is filled with an electrically conductive mud. These potentials are set up mainly by interaction between the drilling mud, the formation water and shale in the formation, but measurable potentials can also be generated by movement of fluid through the borehole walls.

Changes in SP are measured between a moving electrode in the borehole and an earthed electrode at the surface. Because the SP log responds mainly to the presence of shales it can be used to distinguish between these and the relatively more permeable sands. SP values taken from the log can also be used to calculate the electrical resistivity of the formation water, but this technique may be unreliable when applied to freshwater sands, and in most aquifers the resistivity of the groundwater can be found by more direct means.

The SP log is nearly always run in conjunction with resistivity logs as part of the standard electrical survey, but in water wells the SP log is often of very poor quality since the borehole conditions required for good SP development may not be present. When a good SP is seen in these conditions it is usually due to fluid movement rather than to the presence of shale. The presence of shale-type formations is more reliably demonstrated by the gamma-ray log.

The gamma-ray log measures the natural gamma radioactivity of the formation. Most of this activity comes from radioactive elements in the uranium-radium and thorium series and from radioactive potassium. The highest levels of radioactivity in sedimentary rocks are found in clays and shales, and consequently these rocks display relatively high gamma activity while limestones and most sands show much lower activity. Because of this effect the gamma-ray log can be used to distinguish between beds that are likely to be permeable and water-productive, and those that are more likely to form barriers to groundwater movement. For example it has been used effectively to correlate thick sandstone/mudstone sequences in the Coal Measures and Millstone Grit in the East Midlands[31]. A gradual rise has been recorded in the level of gamma activity with increasing depth in the Lower Chalk; this is probably due to an increasing clay content[32].

The gamma-ray log is made by lowering a geiger or scintillation-type detector into the borehole, then raising it at a constant speed while measurements of radiation intensity are made and automatically averaged over a preset time interval or 'time constant'. The best quality logs are obtained by using a slow hoisting speed combined with a long time-constant, since this procedure minimises statistical variations in measured radio-activity that might otherwise be mistaken for significant features on the log. Most gamma-ray logs carry a separate record of this statistical variation, made by holding the detector stationary in the borehole for several minutes, and this check is an important aid in interpretation.

The temperature of the ground to a depth of 20–30 metres is similar to the mean air temperature in the locality. Below this depth ground temperatures increase at an average rate, in the UK, of 1°C for 30 m, the actual rate of increase being related to the thermal conductivity of the rocks. This basic thermal gradient may be disturbed by the movement of groundwater, and a temperature log may identify points of groundwater entry to or loss from a well. Temperature measurements in the borehole are also required for the quantitative interpretation of formation and fluid resistivity logs.

Simple temperature logs are made by lowering a thermistor probe into the borehole and recording the temperature gradient. Differential temperature logs are made by recording the difference in temperature between two thermistors spaced a short distance apart on the logging cable[33]. When the slope of the temperature log is constant the differential log will be a straight line, but when the temperature gradient changes the differential log will show a deflection. The differential temperature log can be made highly sensitive and is capable of detecting very small temperature changes, but care must be exercised in its interpretation since the thermal regime in a water-filled borehole may be affected by factors that are not necessarily connected with significant groundwater movement.

Vertical groundwater movement in a borehole may be measured with a sensitive flowmeter, or by injecting a tracer at a known depth and time and following its movement with a suitable detector. Mechanical flowmeters are adequate for measuring high rates of flow in pumped wells, but tracers are usually needed to measure the low velocities (often less than 0.1 m/s) which are more typical of natural groundwater flows[33]. Brine is often employed as a tracer in water wells because the small amounts used do not affect the quality of the water, and the brine itself is easily detected with a fluid resistivity probe[30,34,35].

When a profile of calibrated flow measurements is available for a well it is possible to assess the contributions made by different aquifers or producing zones to the total yield of that well. For accurate interpretation, however, a caliper log is needed.

The caliper logging tool is equipped with spring-loaded arms that press against the borehole walls as the tool is drawn slowly up the hole. The movement of the arms is recorded against depth and the resulting log provides a continuous measurement of the average diameter of the borehole. Caliper logs are essential for the proper interpretation of most of the other geophysical logs when these are being used quantitatively. They also give information about the nature of the formation when washouts and other cavities, and sometimes fractures, are shown on the borehole diameter profile. Caliper logs also show constructional features of the borehole such as changes in bit and casing size, while the casing collar locator, a simple magnetic device often combined with other logging tools, shows the depth of joints in the casing string. These features may be of particular interest in old wells, especially if the original records have been lost.

Although relative porosities can be estimated from conventional resistivity logs, and measured quantitatively when good logs and correction data are available, porosity measurements are more reliably made by using a combination of nuclear and acoustic techniques. The parameters studied here are fast neutron moderation, gamma ray scattering, and acoustic wave velocity in the formation.

Neutron logs measure the moderating effect of light atoms, chiefly hydrogen, on a flux of fast neutrons emitted by an isotopic source or neutron generator in the logging tool. In water wells most of the hydrogen within the sphere of investigation of the tool is assumed to be in the form of water, thus the neutron response can be related to the water-filled porosity of the formation.

The scattered gamma-ray or formation density log measures the effect of the formation matrix on gamma rays emitted from a collimated source in the logging tool. Gamma rays that have been scattered by the formation matrix are counted by one or more detectors, which are placed in the same tool but shielded from the radioactive source, so that only gamma rays that have been scattered by the formation can reach the detectors. The degree of scatter achieved is proportional to the electron density of the matrix, and from measurement of this effect the bulk density and porosity of the formation can be derived.

The acoustic log[36] records the travel time of a sonic signal that is generated in the logging tool, passes through the formation parallel to the borehole, and is picked up by receivers placed further along the tool. The travel time of the sonic signal is dependent on lithology and porosity, and if the former is known the porosity can be determined.

All three porosity logs are affected by changes in borehole diameter. However, these undesirable effects can be kept to a minimum by the use of sidewall tools or multiple detector arrays[36].

Maximum benefit can only be obtained from geophysical borehole logging when the various logs are interrelated and studied with the rest of the available well data. When unexplained features are noted on individual logs, or when composite interpretation reveals inconsistencies among the measurements, the direct inspection of the borehole walls with an underwater television camera may provide the additional information that enables a final interpretation to be made.

Instruments for measuring groundwater levels

The depth to water in boreholes is usually measured with a simple electrical instrument consisting of a dry battery, an indicator such as a milliameter, a length of twin-core cable, and a twin-electrode probe that is lowered into the borehole. When the probe reaches the water surface the circuit is completed and this is shown by the indicator at the surface. Non-electrical instruments that operate on the principle of the float and inertial brake are also available.

If measurements are required when the well is overflowing a pressure gauge can be used, or a piezometer can be made from transparent plastic tubing. The well must of course be capped, with suitable provision for attaching the gauge or tubing[37].

When a continuous record is required of water level fluctuations, a float-operated recorder is used. On this instrument time is recorded along the length of the horizontal chart drum; the drum is rotated by the movement of a float, acting through a gear train that determines the vertical recording scale. Other types of recorder use a vertical drum, or a horizontal drum with the water level recorded along the horizontal axis. The motive power for the time drive may be supplied by a spring clock, a weight-driven clock or an electric clock powered by battery or mains supply. A range of different time scales can be provided, and special strip-chart recorders, which will run for several months unattended, are available.

Chart recorders are preferred for use during pumping tests when immediate visual interpretation may be needed, but punched-tape recorders are often used for hydrometric networks since this method of recording is more suitable for automatic data processing. Recorders of this type make readings at pre-set time intervals only, but they will record on paper tape in a variety of codes, some of which are suitable for direct input to a computer.

A float-operated instrument has been developed[38] that can be attached to the top of the casing of a small diameter borehole (50–100 mm diameter). The water level is recorded by a four-figure counter calibrated in metres and centimetres.

Float-operated water level recorders cannot easily be installed in wells containing pumps and other pipework, and in the future instruments using pressure transducers may be used for this type of installation. Measurement of water level by sonic reflection is also a possibility[33].

Groundwater observation-well networks were established by the river authorities, in the main aquifers in England and Wales, to monitor changes in groundwater storage and quality. A suitable borehole distribution for such networks is one borehole per 25 km². Measurements are generally made manually, usually at monthly intervals, but one borehole every 100–250 km² should be fitted with an automatic water level recorder. However, the ideal density depends very much on the shape of the aquifer outcrop, the extent to which it is dissected by river systems and the importance of the aquifer for water supply; fewer boreholes are required in a confined aquifer, unless it is being actively developed. In 1950 the Norfolk County Council developed a network in the Chalk of Norfolk comprising some 72 boreholes in an area of 5180 km² (equivalent to one borehole per 73 km²). This was one of the earliest networks and measurements at monthly intervals have been maintained since 1950.

More detailed studies, relating to specific groundwater problems, require much denser networks. The preparation and updating of groundwater level maps must also be based on measurements made in many more wells and boreholes, although measurements may be made at less frequent intervals, possibly twice a year or even annually. Reference has been made to the fact that during the Great Ouse Groundwater Pilot Scheme changes in groundwater storage were measured using over eighty deep observation boreholes and some fifty shallow tube wells within an area of 71 km^2.

In designing future groundwater networks consideration has to be given to the fact that telemetry systems will be incorporated to a greater extent.

The occurrence of groundwater in the UK

The fundamental factor controlling the occurrence of groundwater in the UK is the distribution of the relatively impermeable Pre-Cambrian and Palaeozoic rocks in the north and west and the restriction of the main outcrops of the more permeable Mesozoic and Tertiary sequences to the south and east of England. It is convenient to consider the occurrence of groundwater in relation to the outcrops of pre-Carboniferous, Carboniferous and post-Carboniferous rocks *(Table 5.1* and *Figure 5.4)*.

Pre-Carboniferous rocks occur in six regions:
1. Northern Scotland
2. Southern Uplands of Scotland

Table 5.1 THE GEOLOGICAL SEQUENCE IN BRITAIN WITH PARTICULAR REFERENCE TO HYDROGEOLOGY

Era	System	Sub-system	Typical sequence
Cainozoic	Quaternary	Holocene[4]	
		Pleistocene	*Upper and Middle Pleistocene*[5] Lower { *Weybourne Crag** *Norwich Crag** *Red Crag**
	Tertiary	Pliocene	Coralline Crag*
		Oligocene	
		Eocene	

3. Lake District
4. Wales
5. South-west peninsula of England
6. Northern Ireland

A wide variety of sedimentary, igneous and metamorphic rocks are found in these areas but they are essentially hard and compact and have been extensively folded and faulted. The available evidence is that they do not yield water in any quantity except in favourable situations where fracture or fissure zones are well developed. Such zones are often indicated by topographical features. The average yield from boreholes in igneous rocks, metamorphic rocks (other than virtually impermeable cleaved mudstones) and Palaeozoic sandstones and limestones varies from about 0.25 to 5 l/s, but individual wells can be failures. It is more usual to obtain small amounts of groundwater from these rocks by developing springs. Potential sources of water supply for local development in areas of such compact rocks are alluvial deposits and, except in the south-west peninsula of England, fluvio-glacial sands and gravels in the main valleys.

Carboniferous rocks also occur mainly in six regions:

1. Midland Valley of Scotland
2. Northern England
3. Welsh Borderland including the Dee Valley
4. Midland coalfields of England
5. South Wales
6. Devon and Cornwall

Details of local sequences of hydrogeological importance

HAMPSHIRE BASIN	LONDON BASIN	
Hampstead Beds§		
Bembridge Marls		
Bembridge Limestone		
Osborne Beds‡		
Upper & Middle Headon Beds‡		
Lower Headon Beds‡		
Barton Beds§		
Bracklesham Beds§	Bagshot Beds*	
Lower Bagshot Beds*		
London Clay	London Clay	
	*Blackheath and Oldhaven Beds**	In the London Basin these deposits are referred to as the Lower London Tertiaries
Reading Beds‡	*Woolwich* and Reading Beds§	
	*Thanet Beds**	

Table 5.1 THE GEOLOGICAL SEQUENCE IN BRITAIN WITH PARTICULAR
REFERENCE TO HYDROGEOLOGY—*continued*

Era	System	Sub-system	Typical sequence
Mesozoic	Cretaceous	Upper Cretaceous	CHALK
		Lower Cretaceous	*Upper Greensand* Gault‡
			LOWER GREENSAND
			Weald Clay
			Hastings Beds
	Jurassic		Purbeck Beds†‡[2]
		Upper Jurassic	Portland Beds[6] Kimmeridge Clay CORALLIAN BEDS†[7] Oxford Clay Kellaways Beds*‡ Upper Cornbrash†
		Middle Jurassic	Lower Cornbrash†
			Great Oolite Series
			Inferior Oolite Series
		Lower Jurassic	Lias‡[9]
	Triassic	Upper Triassic	Rhaetic‡
			Keuper
		Lower Triassic	BUNTER

Details of local sequences of hydrogeological importance

HAMPSHIRE BASIN	LONDON BASIN	NORFOLK	LINCOLNSHIRE
CHALK	CHALK	CHALK	CHALK
Upper Greensand *Gault‡*	*Upper Greensand* *Gault‡*	Red Chalk† Carstone*	Red Chalk† Carstone*
SURREY & KENT FOLKESTONE BEDS Sandgate Beds§ HYTHE BEDS†* Atherfield Clay	**S. ENGLAND** Carstone* Sandrock \ *Ferruginous* | *Sands* Atherfield Clay		Sutterby Marl Fulletby Beds§
Weald Clay	Weald Clay		Tealby Clay and Limestone
Tunbridge Wells Sand Wadhurst Clay *Ashdown Beds** **SW. ENGLAND** { Portland Stone† { Portland Sand§		*Sandringham* *Sands* including Snettisham Clay	Claxby Ironstone Hundleby Clay *Upper Spilsby* *Sandstone* *Lower* *Spilsby* *Sandstone*

BATH
Forest Marble[8]
Bradford Clay
GREAT OOLITE
    LIMESTONE
Upper Fuller's Earth‡
Fuller's Earth Rock†
Lower Fuller's Earth‡
	N. COTSWOLDS
	Forest Marble[8]
	
	GREAT OOLITE
	    LIMESTONE
	
	Lower Fuller's Earth‡
		LINCOLNSHIRE
		Great Oolite Clay
		Great Oolite Lst
		
		Upper Estuarine
		    Series §
			YORKSHIRE
			
			Upper
			    Estuarine
			    Series §

INFERIOR
OOLITE†
	INFERIOR
	OOLITE†
		LINCOLNSHIRE
		    LST
		Lower Estuarine Series§
		Northampton Sand
			Scarborough Lst
			Middle Estuarine
			    Series§
			Millepore Bed†
			Lower Estuarine
			    Series§
			Dogger*†

MIDLANDS & **NOTTS.** Keuper Marl *Keuper Sandstone*	**S.W. ENGLAND** Keuper Marl	**S. LANCASHIRE** Keuper Marl *Keuper Sandstone*	**N.W. ENGLAND** Keuper Marl *Kirklinton Sandstone*

UPPER MOTTLED
    SANDSTONE
PEBBLE BEDS
LOWER MOTTLED
    SANDSTONE
	UPPER
	    SANDSTONE[10]
	PEBBLE BEDS
		BUNTER
		    SANDSTONE
			ST. BEES
			    SANDSTONE

Table 5.1   THE GEOLOGICAL SEQUENCE IN BRITAIN WITH PARTICULAR
REFERENCE TO HYDROGEOLOGY—*continued*

Era	System	Sub-system	Typical sequence
Upper Palaeozoic	Permian[1]		Upper Permian Marl[3] UPPER MAGNESIAN LST MIDDLE MAGNESIAN LST LOWER MAGNESIAN LST Marl Slate YELLOW SANDS
	Carboniferous	Upper Carboniferous	Upper Coal Measures[11]
			Middle Coal Measures[11] Lower Coal Measures[11]
			Millstone Grit[12]
		Lower Carboniferous	CARBONIFEROUS LST[13]
Lower Palae-ozoic	Devonian including Old Red Sst.		
	Silurian		
	Ordovician		
	Cambrian		
	Pre-Cambrian		

Notes:
1. The Permian includes thick evaporite sequences in Durham and north-east Yorkshire. Because of their nature and depth they are of no value for water supply and are not included in the table.
2. The boundary between the Cretaceous and the Jurassic lies within the Purbeck beds
3. The Permian rocks are very variable. The sequence taken as a 'typical' is that found in Durham
4. Alluvial clays and silts, blown sand, peat
5. Sands, boulder clays and river gravels
6. Portland Beds are waterbearing but their distribution and ready drainage make them unimportant aquifers
7. Corallian Beds are mainly lsts in northern and southern England but clays (Ampthill Clay) in central England
8. Forest Marble comprises limestones and clays
9. In SW. England the U. Lias includes Cotswold, Midford, Yeovil and Bridport Sands
10. In SW. England the U. Sandstone may be partly of Keuper age
11. Coal Measures are mainly alternations of sandstone and mudstone. Sandstones are important in the U. Coal Measures

In many of these areas the Carboniferous sequence consists predominantly of a series of either limestones and mudstones or sandstones and mudstones, but from the hydrogeological point of view the massive limestones of the Carboniferous Limestones series, well developed in the Peak District of Derbyshire, the Mendip Hills, South Wales and parts of northern England, are in a distinct category. In general terms the Carboniferous rocks yield moderate supplies of water derived essentially from fissures. As a result of karstic features, groundwater flow is concentrated in major fissure systems in the massive limestones beds.

The outcrops of the post-Carboniferous rocks are mainly restricted to eastern and southern England and the English Midlands, with an extension into north-west England. They contain the principal aquifers, namely the Magnesian Limestone, the Permo-Triassic sandstones, Jurassic limestones, Lower Greensand and the Chalk. The post-Triassic rocks and the

Details of local sequences of hydrogeological importance

YORKS & NOTTS	S.W. ENGLAND	S. LANCASHIRE	N.W. ENGLAND
Upper Permian Marl	Lower Marls		Eden Shales
Upper Magnesian Lst	*Lower Sandstone*	Manchester Marls	Marls
Middle Permian Marl	*Breccias and*		Magnesian Lst
Lower Magnesian Lst	*Conglomerates*	COLLYHURST SST	Hilton Plant Beds‡
Lower Permian Marl	Watcombe Clay		PENRITH SST (incl.
Basal Sands and			Upper Brockram*)
Breccias			Lower Brockram*

MIDLANDS
Clent Group*‡
Enville or Con-
glomerate Group*‡
Keele Group‡
Halesowen Group*
Etruria Group‡

12. Millstone Grit is alternation of sandstone and mudstone
13. In many areas massive lst but otherwise variable including lsts, ssts and mudstones

The following symbols indicate the general nature of deposits when this is not apparent from the name:
      * Sand or sandstone
      † Limestone
      ‡ Clay or mudstone
      § Interbedded sand, silt and clay
The use of two symbols indicates both lithologies are important

BLOCK capitals indicate major aquifers; *italics* secondary aquifers
Other sandstones and limestones, and also thin sandstones and limestones interbedded in argillaceous rocks, may yield water but they are of only local significance
Beds with the same name or shown as equivalents are not necessarily of exactly the same age

Permo-Trias of north-east England have a general regional dip to the east or south, but the remaining Triassic rocks lie in a series of structural basins.

A range of different types of calcareous aquifers is well represented in England and Wales. The Chalk is a soft fine-grained porous limestone characterised by diffuse groundwater flow. At the other extreme, the massive Carboniferous Limestone is a hard, compact limestone with well developed karstic features. The Jurassic limestones are mid-way between the two, being more compact and less porous than the Chalk but not to the same extent as the Carboniferous rocks. Major fissure systems are much more extensively developed than in the Chalk.

The main sandstone aquifers are the Permian sandstones, the Bunter and Keuper sandstones (referred to as the Triassic sandstones) and the Cretaceous Lower Greensand.

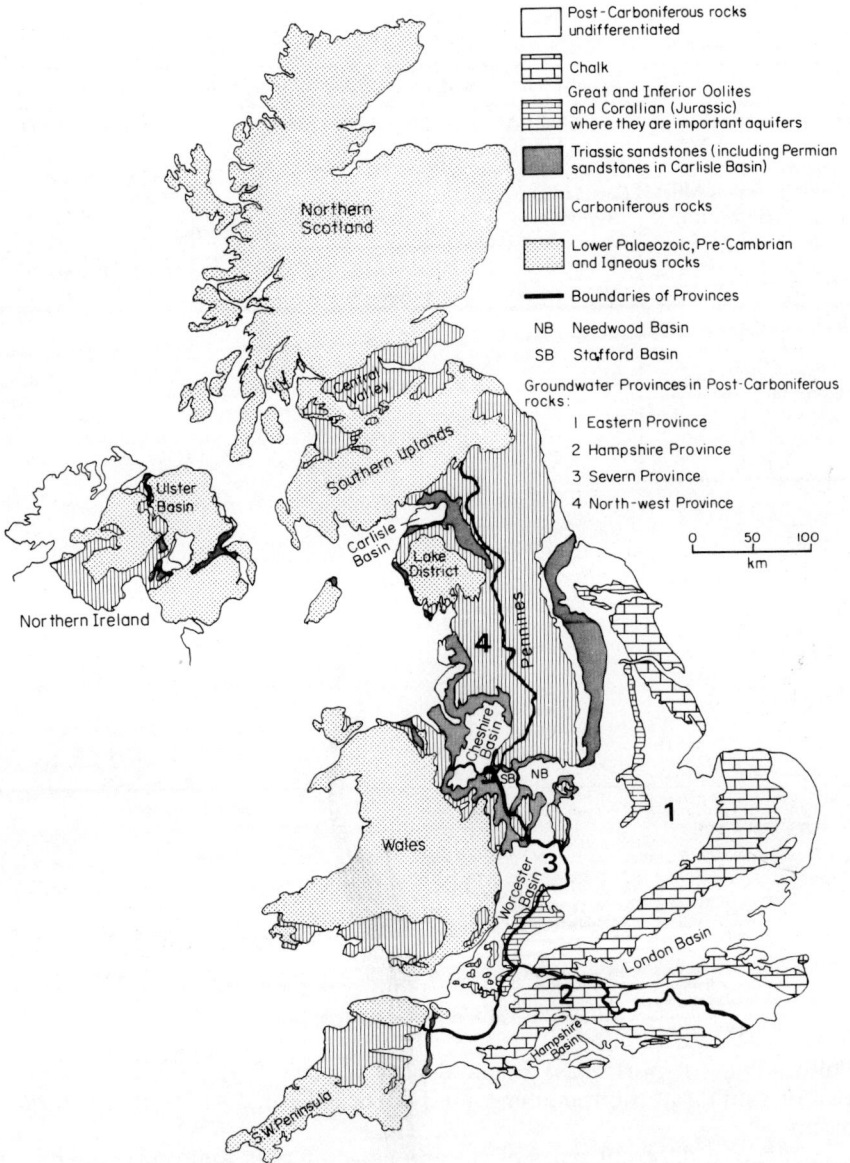

Figure 5.4 *Groundwater provinces and outcrops of major aquifers*

Groundwater provinces in post-Carboniferous rocks

The post-Carboniferous rocks can be divided conveniently into a number of groundwater basins or 'provinces' based upon the probable form of regional flow systems, each leading to a common drainage outlet. (*Note.* 'Basin' is the more suitable term but 'province' is preferred here to avoid confusion with river basin and structural basin, as used in the geological sense.) The existence of regional hydraulic continuity, which recognises that water moves through deposits of widely differing permeabilities, including those normally considered to be impermeable, has to be taken into account in defining such natural provinces. Recent studies[39-41] of theoretical flow models and their application to real situations have indicated that flow lines are strongly influenced by topography and geology. Dominant recharge areas correspond to the principal uplands while major lowland areas are discharge zones. The upper limit of a flow system is the water table and the lower limit an impermeable or virtually impermeable boundary at depth—for example Pre-Cambrian and some Palaeozoic rocks. Local, intermediate and regional flow systems may occur[39]. These views can be applied in general terms to the post-Carboniferous rocks in England, and in many areas Carboniferous outcrops can also be readily incorporated. The form of such regional flow patterns may be important in assessing the consequences of deep sub-surface waste disposal.

The post-Carboniferous rocks of eastern and southern England consist typically of an alternating sequence of aquifers and aquicludes, the aquifers forming a series of escarpments. Most of the water infiltrating into the outcrops of the aquifers is discharged to the rivers draining the outcrop areas and represents local flow systems. The remainder moves through the aquifers below overlying impermeable beds, or leaks from the aquifers to deeper zones, ultimately finding its way to outlets by moving, if necessary, through relatively impermeable beds, the direction of flow being related to fluid potential gradients. This water is moving in a regional flow system. The principles are illustrated in *Figure 5.5,* a cross-section from the Severn valley to the lower Thames valley showing possible flow-lines in post-Permian rocks. The true picture will be considerably more complex, if only because interpretation of flow in pre-Triassic rocks has been excluded.

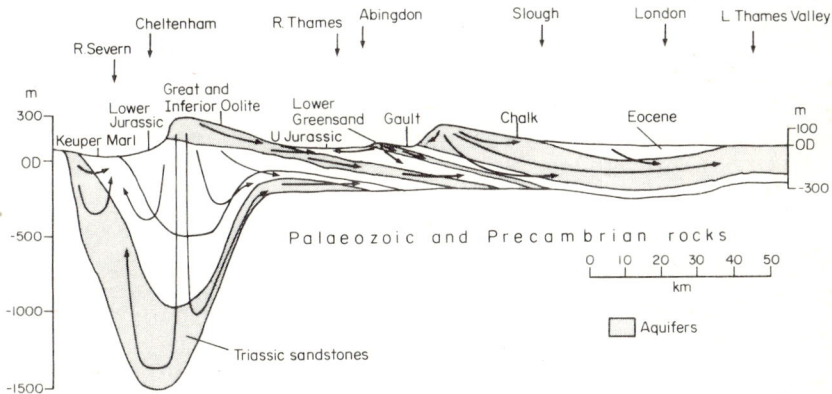

Figure 5.5   West–east section across England showing possible flow lines in post-Permian rocks

It is necessary to keep in perspective the rates of flow in the different flow systems. At depths of some 1000 m or more the permeability of aquifers may be only of the order of 10^{-3} m/d, while values for confining argillaceous deposits may be 10^{-6} m/d. With such low permeabilities cross-formational flow will be by preferred routes of greater permeability, such as faults or fracture zones. In a thick isotropic, homogeneous aquifer the flow direction in the saturated zone below recharge areas is vertically downwards, but in many aquifers, because of the much higher horizontal hydraulic conductivity, flow has a strong horizontal component. This is particularly so in carbonate rocks and in sandstones with a soluble cement. For this reason water tends to flow whenever possible in a horizontal direction towards natural outlets, and vertical flow in the less permeable parts of an aquifer is limited and inter-aquifer transfer relatively small. But the important point is that it does occur and must be taken into account when defining the limits of a natural groundwater province. In the London Basin large volumes of water flow in local flow systems through the Chalk to the rivers draining the outcrop. Less water is moving more slowly through the aquifer below the Eocene in a regional system, and most of this is moving in the upper 50 m of the aquifer. At still greater depths, in the Lower Greensand below the Chalk, the water is probably moving even more slowly, ultimately to outlets that necessitate upward leakage through the Gault and the Chalk[7].

By bearing in mind the probable distribution of regional fluid potentials and the resultant regional directions of flow, the post-Carboniferous rocks may be placed in the following four natural groundwater basins or provinces *(Figure 5.4)*:

1. *Eastern Province* draining to the North Sea
2. *Hampshire Province* draining to the English Channel
3. *Severn Province* draining to the Bristol Channel
4. *North-west Province* draining to the Irish Sea

The Eastern Province includes a Permian to Pleistocene sequence with a general regional flow to the east to outlets in the North Sea. The regional water level divide actually coincides with the Pennine watershed, and the Carboniferous rocks on the eastern side of the Pennines form part of this province; some of the Carboniferous outcrops in the Midlands are also included. Chemical variations of groundwater in the Carboniferous Limestone, Millstone Grit and Coal Measures at depth below the Permo-Trias imply movement in an easterly direction in these deposits, probably under natural conditions to outlets via the Triassic sandstones[31]. Part of this flow must now be to mine-drainage centres in the Coal Measures.

Infiltration into the Triassic sandstones forming the western margin of the Stafford structural basin flows beneath that basin and the Needwood Basin, probably partly to outlets in the Lower Trent valley but also to outlets further east.

Variations in the geological sequences are responsible for local modifications of flow within the regional system. For example, under natural conditions deep percolation through the Lincolnshire Limestone in south Lincolnshire may have flowed to the North Sea through buried glacial channels centred on the Wash and part probably discharged through overlying Jurassic clays to the Fens[6].

In the London Basin, the Lower Thames is the natural outlet not only for flow in the Chalk but also for the Lower Greensand and for Jurassic rocks deriving their water from infiltration in the Cotswold Hills *(Figure 5.5)*. Groundwater flowing at depth from the Triassic sandstones outcrop east of the Warwickshire Coalfield, which actually lies in the Severn river basin, may flow to an eastern outlet rather than to the Bristol Channel.

The Hampshire Province includes Jurassic to Oligocene rocks draining to the English Channel. The Permo-Triassic sequence of Devon should be included in this province.

Regional groundwater flow in post-Carboniferous rocks in the Severn Province is through Triassic sandstones below the Keuper Marl of the Worcester structural basin to outlets in the Bristol Channel by upward leakage through less permeable deposits. Local flow systems in Carboniferous rocks are centred on the Mendip Hills and also in post-Triassic rocks around the margin of the province.

In the North-west Province the regional flow is in Triassic rocks draining to the Irish Sea from the Vale of Clwyd Trough, the Cheshire Basin and the outcrops to the north and south of Morecambe Bay. Further north the Carlisle Basin also drains to the Irish Sea. The Carboniferous rocks west of the Pennines form part of the North-west Province. Water flows through the Carboniferous sequence to outlets via the Triassic sandstones of the Cheshire and Carlisle Basins just as it does east of the Pennine watershed. A similar situation must exist in the Clwyd and lower Dee valleys. The groundwater in the Triassic sandstones of the Ulster Basin drains to the North Channel between Ireland and Scotland.

The natural regional flow patterns outlined above have been disturbed by the development of groundwater resources and by pumping for mine-drainage purposes. The main outlet for groundwater flow in the London Basin is now wells that penetrate the London Clay to the Chalk. Drainage of the Kent coalfield has produced a downward potential in the Lower Greensand and presumably induced leakage from the Chalk into the Lower Greensand. The ultimate outlet for this water is mine-drainage from the collieries in the coalfield. Similar situations exist in other coalfields where downward potentials have developed in the Permian and Triassic rocks.

Chalk

The Chalk is the principal aquifer in the UK, supplying about 50 per cent of the total use of groundwater. It is a soft white limestone, which underlies eastern and southern England from Yorkshire to the south Devon coast, except in the Weald where it has been removed by erosion. The rock consists of two fine powders that differ both in nature and texture[42]. The coarser consists of shell debris and foraminifera, while the finer is composed of minute calcareous shells (coccoliths) of plankton and their disintegration products. The Chalk is therefore predominantly of organic origin and, although the shells of the small organisms may be infilled with fine material, many are hollow and the pores of the rock do not contain cement, except as mentioned later.

A high proportion of shell debris gives a gritty, friable chalk, abundant foraminifera tend to give a hard, nodular chalk, while soft chalk consists almost entirely of coccolith material. The particle size of the coarser fraction varies between 10 and 100 μm in diameter and the finer fraction between 0.5 and 4 μm. These particle sizes are equivalent to coarse silt and clay, respectively.

The Middle and Upper Chalks consist almost entirely of calcium carbonate, 96–99 per cent being usual although the figure is less in marl bands. The Lower Chalk is more marly and may contain as little as 40 per cent calcium carbonate, but usually between 65 and 80 per cent. The insoluble fraction is mainly clay. It was examined in some detail in Chalk cores from a borehole at Leatherhead and found to consist primarily of smectite and illite in the Upper and Middle Chalk, but in the upper part of the Lower Chalk smectite was of less importance and an appreciable fraction was kaolinite[43]. These clay minerals influence the chemical composition of groundwater through ion exchange. Smectite has a particularly high cation exchange capacity, but illite and kaolinite are involved in such reactions to a lesser extent.

Porosity and development of permeability

The porosity of the Chalk generally exceeds 20 per cent. Values based on 250 samples from southern England are summarised below[44]:

	Upper and Middle Chalk	*Lower Chalk*
Normal range	41–50 per cent	21–30 per cent
Extreme range	34–53 per cent	17–41 per cent

The porosity estimated from geophysical logging in two boreholes in south-east England varied between 30 and 40 per cent. Such high values are due to the presence of hollow micro-organisms, and to the complex shape and the loose packing of the particles. The harder chalks have a lower porosity, generally less than 20 per cent. On a regional scale the Chalk of Yorkshire has a lower porosity than the Chalk of southern England, while the White Limestone of Northern Ireland is a more extreme example with a porosity of about 5 per cent[45]. The reduction in porosity is due to secondary calcite cementation.

The porosity varies throughout the Chalk sequence. This is indicated by electrical resistivity measurements, which depend partly upon the amount of water in the interstices and, therefore, upon the percentage volume of the voids; the more porous the aquifer the lower the resistivity. The harder rock bands in the Chalk, such as the Totternhoe Stone and the Melbourn and Chalk rocks, which have lower porosities than the softer chalks, are the better water-yielding horizons. This is because, although they are more cemented, they are also more fissured.

An aquifer has to be porous to provide storage space for water but the pores must be interconnected to allow the passage of the water, thereby imparting permeability to the rock. Although the soft chalks have a high primary porosity the rock is extremely fine grained. The typical pore diameter is less than one micrometre and as a result water tends to be held by capillary forces.

The primary permeability of the Chalk is small and the value of the rock as an aquifer, as with most limestones, depends upon secondary porosity and permeability due to fissures and joints. These are formed by tectonic movements, differential changes in volume on consolidation together with associated mineralogical changes, and the release of pressure as erosion gradually removes overburden. Once fissures and fractures have been formed they are enlarged by percolating water. Fissures in the Chalk are of three types: 'pipe' or cylindrical fissures, vertical or high angle cracks or joints, and bedding plane fissures (which are often associated with tabular flints layers). The formation of the horizontal fissures may be facilitated by the gradual reduction of overburden pressure by erosion, which produces planes of incipient weakness at right angles to the relief of pressure. As erosion has been more extensive along the lines of valleys the development of horizontal fissures would be expected to be greater in these areas. Frost action during the glacial period probably increased the extent of fissuring in the upper part of the Chalk. The frequency of fissures tends to decrease with increasing depth.

Groundwater flow through the Chalk is diffuse; that is, it is through numerous interconnected small cracks and fissures. The flow is laminar and Darcy's Law is generally obeyed. Weathering and climatic factors and the development of regional topographical features have been instrumental in determining the permeability pattern. Fissures have developed, as in most limestones, by solution of calcium carbonate near and within the fluctuation range of the regional water table. In the lower parts of the main valleys and in coastal areas they were enlarged below the present minimum water level during the glacial period when groundwater circulation was controlled by a sea level up to 100 m below the present level, although at the maximum extents of the various ice advances groundwater circulation must have been restricted by permafrost. Because of the form of groundwater flow lines, solution fissures do occur below the minimum water level *(Figure 5.10)*, but at depth fissures of tectonic origin or those due to relief of vertical pressure are probably more important.

The valleys are discharge areas for groundwater, and the flow per unit cross-sectional area is concentrated along these lines. Generally discharge is through small springs and seepages, although individual large springs are known; for example, the Chadwell Spring in the Lee valley in Hertfordshire had an average flow of 9–13 Ml/d under natural conditions before groundwater was developed in the London Basin.

Specific yield

Because the Chalk is an extremely fine-grained rock, the aquifer retains water by capillary attraction and the specific yield is relatively small. Specific yield may be determined by analysing movements of groundwater levels, and values for Hampshire and Sussex derived by this method range from about 1 to 4 per cent[46,47].

Specific yield may also be calculated by analysing water level changes in observation boreholes caused by nearby pumping wells. This method gave average values of about 1 and 3 per cent from extensive pumping tests in the Lambourn valley of Berkshire[48] and Thet valley of south-west Norfolk[22,23] respectively. Aquifer tests in south-east England generally give values ranging from 1 to 5 per cent.

The average specific yield over a catchment can be assessed by analysing river hydrographs to give the groundwater component and relating this to changes in groundwater level for a given period. Sufficient wells must be available to give an adequate measure of the change in volume of saturated rock. If the method is applied during a period of water level recession, the difference between groundwater levels after a given time represents the volume of aquifer drained during that time, which when related to groundwater discharge gives the specific yield:

$$\text{Specific yield} = \frac{\text{Volume of groundwater discharge}}{\text{Change in aquifer storage}}$$

The method has been applied to the Test and Itchen catchments in Hampshire, giving values of 3.3 and 3.4 per cent respectively[47]. When applied to the Wissey catchment in Norfolk an average value of 1.8 per cent[49] was probably influenced by an extensive cover of boulder clay; the true value for the Chalk is likely to be higher and probably of the order of 3 per cent. A similar result was obtained by Wright[50] for the Wissey and three other catchments in East Anglia. Areal values of the coefficient of storage were derived from estimates of the dry weather recession constant and the relationship between mean groundwater levels in the catchments and baseflow. The mean value for the four catchments was 1.5 per cent, again reflecting the presence of boulder clay as well as Chalk.

Permeability

The permeability of the Chalk is related to the extent of fissuring, because the mass of the aquifer has a very low hydraulic conductivity. For example, the mean values for intergranular horizontal and vertical hydraulic conductivity $(K_H$ and $K_V)$ of samples from the Upper and Middle Chalk from a borehole in Berkshire are [51]:

	Upper Chalk (m/d)	*Middle Chalk* (m/d)
K_H	1.48×10^{-3}	6.47×10^{-4}
K_V	1.58×10^{-3}	7.22×10^{-4}

A conspicuous feature is the essentially isotropic nature of the material. The overall permeability of the aquifer in the vicinity of the borehole was estimated to be about 30 m/d from a pumping test, and this difference between intergranular and total permeability must be typical of the aquifer.

Ineson[18] obtained a relationship between the transmissivity of the Chalk in the vicinity of a well and the yield for a drawdown of 3.08 m (10 ft). Applying this on an areal basis he was able to illustrate variations in the permeability of the aquifer[52]. Reference has been made to the close relationship between permeability and topography, the more permeable Chalk occurring in the valleys with a rapid reduction in values away from the valley lines *(Figure 5.6)*. The reduction is so marked that in many areas the Chalk could be regarded as two aquifers—linear zones of highly permeable rock flanked by rock of much lower permeability. Transmissivities in the main valleys, including dry valleys, are typically 1500–3000 m²/d declining to 10–20 m²/d under the higher ground. This relationship between permeability and topography is clearly seen in East Anglia, the Chiltern Hills and Salisbury Plain. It may persist where the valleys extend on to the overlying

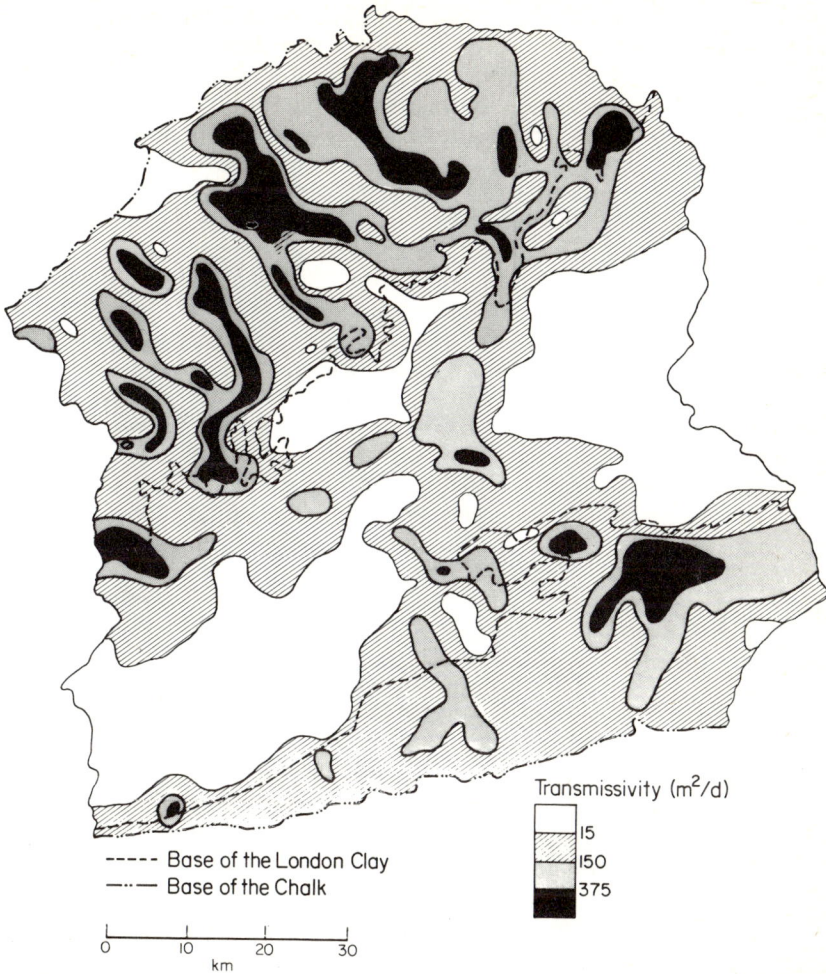

Figure 5.6 Variations in the transmissivity of the Chalk in the London Basin (from WRB publication[7])

Eocene, possibly because the alignment of the valley is controlled by structural features or because greater erosion along the valley causes differential loading of the Chalk and a tendency for the rock to fracture along the valley line. Whatever the cause, the flow lines in the aquifer beneath the Eocene will tend to converge below the valleys and accentuate by solution any incipient weakness in the aquifer responsible for the permeability pattern.

The permeability in Yorkshire and Lincolnshire is not closely related to topography, and structure plays a more dominant role. The rock is harder in these areas than in south-east England but generally less permeable.

Permeability can be related to the structure of the aquifer. This is clearly illustrated in the London Basin[7], where synclines are associated with lower transmissivity, the reverse being the case for anticlines. The Chalk in the shallow syncline extending north-east from Stowmarket has a low permeability and contains saline connate water[53]. In Hampshire the regional structural pattern of east-west folds is reflected in variations in transmissivity, but in this area high values occur along synclines[54].

The dominant factor influencing the permeability of the Upper and Middle Chalk is undoubtedly weathering: the base level of erosion controls the position of groundwater outlets and hence the form of the flow lines. Certain types of Chalk are more susceptible to fissuring but incipient fractures require moving water to develop them. Recent data suggest that the Chalk is generally fissured to depths of about 50–100 m below the water table. As already mentioned, the fissured depth is related to water-level fluctuations and, in some areas, to low sea levels during the Pleistocene period. Certain horizons in the Chalk such as the Totternhoe Stone in Cambridgeshire and the Melbourn Rock in Kent are particularly permeable. The Melbourn Rock has yielded appreciable supplies from depths of 130 m in Kent. Generally it is worth extending a borehole to intersect these rock bands if they are reasonably close to the surface, say within 80 m (or possibly more if local experience suggests they are productive), but neither the Melbourn Rock nor the Chalk Rock in the Lambourn[48] and Thet valleys[55] respectively, proved to be productive during the recent extensive groundwater investigations.

In the course of sinking mine shafts to the Coal Measures in east Kent, the Chalk was found to be dry at depths of about 100–130 m[56], although small yields have been reported from 180 m at Tilmanstone Colliery and 150 m at Betteshanger Colliery[57]. Most of the water derived from the Chalk below the Eocene in the London Basin is obtained from the upper 60 m[7].

As a general rule it can be accepted that the upper 50–60 m of the Chalk is the principal aquifer.

The Lower Chalk contains more marl than the Middle and Upper divisions but the proportion is less in north Norfolk, Lincolnshire and Yorkshire where this division can yield significant amounts of water.

The actual rate of groundwater flow through the Chalk is extremely variable. Although diffuse flow predominates, relatively rapid flow occurs through well-defined fissure systems. Experiments with tracers showed that water entering swallow holes in the Chalk in the Colne Valley near Potters Bar emerges three days later 16 km to the north-east near Ware in the Lee Valley[58]. This and other examples indicate that flow velocities in such well-defined fissures can attain 2000–5000 m/d.

Development of groundwater in the Chalk

Nowadays groundwater is developed in the Chalk by wells up to 1 m in diameter and as much as 120–150 m deep. Such wells can yield in excess of 13 Ml/d, while 4 Ml/d is quite common from sites in valleys. Away from the valleys yields are lower and at present these areas are usually only exploited for domestic or agricultural purposes, or where small supplies for industry are required and choice of site is limited. Only exceptionally are yields of more than 5–10 l/s obtained in such situations from 250–300 mm diameter boreholes. Boreholes of 100 mm diameter are generally large enough to provide about 1 l/s required for agricultural purposes other than spray irrigation.

In many parts of East Anglia the upper 10–15 m of the Chalk, where overlain by Pleistocene deposits, consists of small fragments of chalk in a matrix of fine chalk. This layer, which was formed during the glacial period by frost action, is soft and plastic when saturated with water and has a low permeability. It limits well yields and its plastic nature causes problems during drilling. The layer should be lined out and drilling continued to firmer chalk at greater depths.

Where the Chalk is overlain by Eocene clays, supplies are less than at outcrop. Near the margin of the Eocene some 10 l/s may be obtained, but as the thickness of cover increases yields decline and the average is about 5 l/s, although the drawdown in such situations may exceed 30 m. Because of the heterogeneous nature of the Chalk, yields of individual wells can vary widely even in limited areas.

In the past the yields of large diameter wells were supplemented by driving adits, often for considerable distances in different directions, from different levels in the well. Sometimes individual wells were connected by adits. Generally adits were driven to intersect the principal fissure or joint directions.

Adit systems, which are typically 1.2 m wide and 1.8 m high, are common in the North and South Downs where large supplies of water are required to supply centres such as Brighton, Eastbourne, the coastal towns of Kent and parts of London. Brighton Water Department (now part of the Southern Water Authority) derives 60 per cent of its supply from 13 wells and 13.6 km of adits (the remaining 40 per cent comes from 15 wells without adits), the Southern Water Authority has eight pumping stations and 14 km of adits on the Isle of Thanet, while the Thames Water Authority's underground sources in the Lee and Darent Valleys comprise 48 pumping stations and 18 km of adits.

Triassic sandstones

The Triassic sandstones comprise the Keuper Sandstone and the Bunter Series, the latter being divided into the Upper Mottled Sandstone, Pebble Beds and Lower Mottled Sandstone. These rocks are predominantly fine- to medium-grained sandstones but include pebble beds (in places cemented into conglomerates), coarse sandstones, siltstones and mudstones. Mudstones form an appreciable proportion of the upper part of the Keuper Sandstone in some areas; the name Waterstones is given to these deposits, which are more appropriately grouped, from the hydrogeological point of view, with the overlying Keuper Marl. The Keuper sandstones tend to be finer grained than those of the Bunter Series. In parts of north-west England the Permian rocks below the Trias are also sandstones and, as it is difficult to differentiate the two formations and as they form a single aquifer, they are generally referred to as the Permo-Triassic sandstones.

The Triassic sandstones, and the Permian sandstones of north-west England referred to above, provide about 25 per cent of the total groundwater used in England and Wales.

Aquifer properties

The porosity of the Triassic sandstones, like that of other arenaceous rocks, depends upon the packing, sorting and orientation of the grains and the degree of cementation. The Bunter Sandstone tends to be a soft compact rock and is not usually well-cemented, although bands of hard cemented sandstone do occur. The Keuper Sandstone is generally more cemented.

The typical Bunter Sandstone of the Midlands has a porosity of about 30 per cent and a specific yield between 20 and 25 per cent[59] *(Figure 5.7)*.

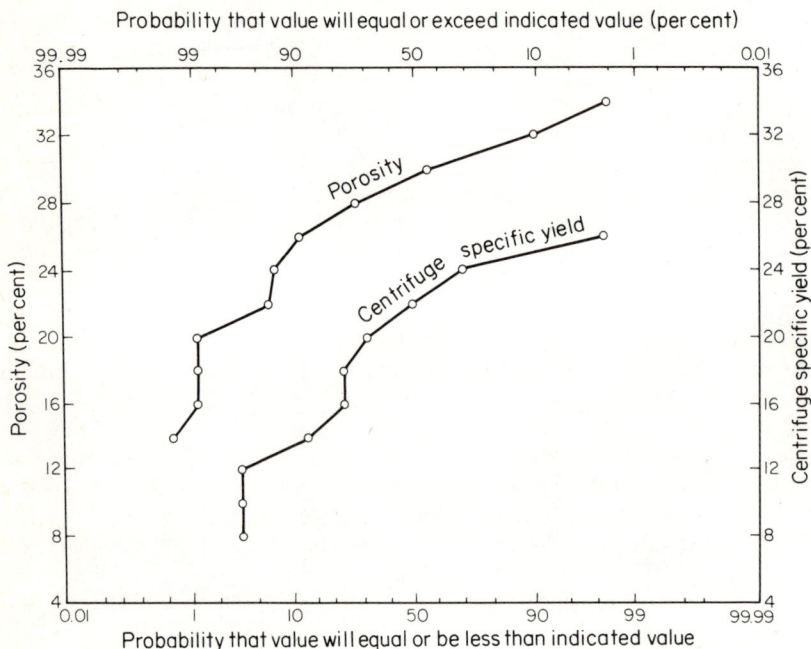

Figure 5.7   Probability distribution of porosity and centrifuge specific yield in the Bunter Sandstone of Nottinghamshire (after Lovelock, reproduced from Williams et al.[59])

Values are somewhat lower in the Clwyd valley, where porosity is between 15 and 30 per cent and the specific yield between 4 and 17 per cent[60]. Porosity values are similar in south Lancashire and Cheshire, varying from 12 to 30 per cent[61], but specific yields are typically about 20 per cent[62]. As the degree of cementation increases the porosity and specific yield of unfissured rock are reduced, but in practice in such circumstances secondary effective porosity due to fissures may significantly increase field values. However, some cemented sandstones in Somerset have porosities as low as 5 per cent and specific yields of less than 1 per cent[63].

Lovelock[64] examined the nature of groundwater flow in the Triassic sandstones by comparing transmissivity values based on laboratory permeability studies with values obtained from field pumping tests. He introduced the term 'intergranular transmissivity', T_i, to describe the intergranular component of total transmissivity, T_t. According to Lovelock, T_i is the product of the geometric mean intergranular permeability of the aquifer and its saturated thickness, and is based upon analyses of core samples taken at closely spaced intervals from an observation borehole drilled in the vicinity of a production well. T_i is determined by standard analysis of the drawdown curve in the observation borehole due to pumping from the production well. Departure of T_i from T_t is interpreted as a quantitative indication of the degree of groundwater movement in the aquifer through fissures. Significant differences were found *(Table 5.2)*, the data indicating that fissure flow is a widespread phenomenon in these aquifers.

Table 5.2 COMPARISON OF INTERGRANULAR AND TOTAL TRANSMISSIVITY IN PERMO-TRIASSIC SANDSTONES (AFTER LOVELOCK[64])

Locality	Formation	T_t/T_i ratio*
Kennel Bridge, Comber	Permian Sandstone	1.0
Upper Dunsforth, Yorks.	Bunter Sandstone	3.1
Edwinstowe, Notts.	Bunter Sandstone	4.4
Studforth, Yorks.	Bunter Sandstone	6.3
Ainderby Steeple, Yorks.	Bunter Sandstone	6.8
Claro House, Yorks.	Bunter Sandstone	7.1
Littleton Colliery, Staffs.	Bunter Pebble Beds	10.3
Vale of Clwyd	Bunter Sandstone	12.4
Marton, Yorks.	Bunter Sandstone	13.3
Dumfries	Permian Sandstone	20.4
Vale of Clwyd	Bunter Sandstone	21.3
Rainton, Yorks.	Bunter Sandstone	50.0
West Cumberland	St. Bees Sandstone	55.6
Haw Hill, Comber	Bunter Sandstone	153.5
West Cumberland	St. Bees Sandstone	8500

*T_i is the product of the geometric mean intergranular permeability of the aquifer and its saturated thickness

The aquifer properties of the Bunter Sandstone were examined in some detail at a site at Edwinstowe near Mansfield, Nottinghamshire[59]. The geometric mean horizontal *(K_H)* and vertical permeability *(K_V)* of core samples from two boreholes were 3.93 and 2.72 m/d. For the majority of the samples K_H exceeded K_V, the ratio varying between 1.33 and 2; the higher values are typical of finer grained sandstones due to the increasing importance of clay particles oriented parallel to the bedding in such rocks. The transmissivity of the aquifer at this locality, based on laboratory determinations, was estimated to be about 300 m²/d, but field pumping tests indicated it was actually about 1500 m²/d. In this particular case four times as much water moves through fissures as by intergranular flow. A similar situation was revealed from an investigation of the sandstones in the Vale of Clwyd[60] and the Fylde of Lancashire[65]; transmissivity values as high as 1000 m²/d are primarily due to fissure flow, but much lower values of 100–300 m²/d occur where intergranular flow predominates. The intergranular K_H to K_V ratio of the sandstones in the Clwyd valley varies between 4:1 and 8:1.

In contrast to the above examples, flow is considered to be predominantly intergranular through the Permo-Triassic sandstones of south Lancashire and north Cheshire[66]. Field pumping tests give hydraulic conductivities of between 0.1 and 10 m/d, a range very similar to that for laboratory determinations on unfissured core samples. Samples in the lower permeability range have K_H values 100 times greater than K_V but, as in Nottinghamshire, the sandstone becomes more isotropic when hydraulic conductivities are in the range 1–10 m/d, presumably because the clay content decreases as the grain size increases.

The transmissivity of the Triassic sandstones in south Lancashire and Cheshire ranges from 20 to 2000 m²/d, assuming a saturated thickness of 200 m and intergranular flow. Generally in the Midlands values[67] based on pumping tests vary from 350 to 750 m²/d, but where the sandstone tends to be cemented and the rock is not well fissured more typical values are 150 m²/d.

The Upper Sandstone of Devon has a hydraulic conductivity of about 5 m/d, which is estimated to be one-tenth that of the underlying Bunter Sandstone. In Somerset, where the Triassic sandstones are cemented, transmissivity values of less than 15–20 m²/d have been recorded but these are related to the more limited thicknesses of the aquifer in south-west England[63].

In the country as a whole the Keuper Sandstone is more cemented and tends to have a lower permeability than the Bunter Sandstone. This is indicated by lower specific capacities of wells, which are drilled mainly in the Keuper Sandstone[67].

There does not appear to be the same relationship between topography and permeability in the Triassic sandstones that is so typical of the Chalk. The factors controlling the regional variations in the permeability of the sandstones are more likely to be the environment under which they were originally deposited and the subsequent geological history, including cementation and the development of fissures and fractures. In some areas recent subsidence, following the mining of coal in the underlying Coal Measures, has increased the fissure permeability.

Occurrence of water

The Triassic sandstones crop out around the peripheries of structural basins, and in the centre of these basins they are overlain by the Keuper Marl which acts as a confining bed. Under natural conditions water in the sandstones overflows from discharge points at the margin of the marl outcrop in low-lying areas. Part flows below the marl and may re-emerge in rivers after moving upwards through the marl where the potential gradient is favourable. That this is so is indicated, for example, by the fact that water overflows from wells drilled to the Triassic sandstones through the Keuper Marl in the Trent valley and its tributary valleys[67].

The annual range of groundwater levels in the Triassic sandstones is small, and is usually of the order of 1–2 m. This is a reflection of the high storage coefficient and relatively high permeability.

The sandstones have been extensively developed for water supply, especially where the aquifer occurs in the vicinity of large centres of population. In some districts abstraction exceeds the available resources and water levels have fallen significantly, sometimes with undesirable changes in quality. Several case histories are briefly discussed below.

In *south Lancashire* and *north Cheshire* Permo-Triassic sandstones, up to 1000 m thick, form the northern margin of the Cheshire Basin. The sandstones have been extensively developed around the mouth of the Mersey and in Manchester, where water levels are now below mean sea level over appreciable areas and in places more than 20 m below sea level. Levels have fallen by more than 30 m since 1910 and as a result saline intrusion has occurred from the Mersey. Wells sited away from the estuary have been contaminated by upconing of connate saline water present in the sandstones at depth[62,66].

In *Birmingham*[8] groundwater storage was, until recently, being reduced by about 6 Ml/d and water levels had been falling for over 50 years. The rate of fall was greatest (exceeding 1 m/year in east Birmingham) to the east of the Birmingham-Hints Fault, where the sandstones are overlain by the Keuper Marl. The total fall east of the fault was over 75 m but west of the fault, where the sandstone lies beneath glacial deposits, it is only 40 m.

At *Burton-upon-Trent*[67] wells drilled before 1920 through the Keuper Marl to the Triassic sandstones encountered flowing artesian conditions but as a result of pumping the water level is now 15–30 m below mean sea level. Some wells in the sandstones at Burton have chloride concentrations in excess of 1000 or even 2000 mg/l. The source is probably connate water either from the sandstones themselves or derived from pre-Triassic rocks.

Yields from the Triassic sandstones

Wells of 600–900 mm diameter in the Midlands generally yield 5–10 Ml/d. The aquifer is often developed by pumping stations consisting of several wells yielding in total 15–20 Ml/d. Yields tend to be low in some areas, for example west of Derby, north of Cannock Chase and at Burton-upon-Trent[67]. These variations must be related to the physical properties of the aquifer, but unfortunately the regional variations of these are not well known and are usually inferred from well yields.

Typical yields on the northern limb of the Cheshire Basin are 2–4 Ml/d, while in the Fylde extensive development of groundwater by the Fylde Water Board has proved that individual wells yield 3–7 Ml/d, the actual amount depending upon the extent of fissuring in the aquifer[65]. Further north in the Vale of Eden and the Carlisle Basin, where the Triassic sandstones are referred to as the St Bees and Kirklinton sandstones, the average maximum yield is 0.5–1 Ml/d. About 2 Ml/d may be anticipated in the Vale of Clwyd, with maximum values of the order of 5–7 Ml/d[60]. Major wells in east Devon provide about 2 Ml/d but in west Somerset, where the sandstones are more cemented and less porous, 0.5 Ml/d is more typical[63].

Generally yields from the Keuper Sandstone are lower than from the Bunter Sandstone. In the east Midlands, where the Bunter Sandstone is absent and only the Keuper Sandstone is present, average yields from 300 mm wells amount to between 0.5–1 Ml/d. Colliery shafts passing through the Keuper Sandstone in the Leicestershire coalfield derive 1–1.5 Ml/d from this formation[67].

Lower Greensand

The Lower Greensand outcrops around the margin of the Weald, in the Isle of Wight and north-west of the London Basin in Bedfordshire and Cambridgeshire. In the Weald the formation includes two aquifers, the Folkestone Beds and the Hythe Beds. The Folkestone Beds are typically fine- to medium-grained quartzose sands. They are separated from the Hythe Beds by the Sandgate Beds, which are generally semi-permeable argillaceous sands or silty clays, although in some areas they are mainly sandstones and sandy limestones. The Hythe Beds consist of alternating layers of sandy limestone and calcareous sandstone, but in the western part of the Weald they are mainly sandstones. From the outcrop in Kent and Surrey the Lower Greensand thins northwards below the North Downs and dies out at a depth of 200–400 m below OD along an east-west line extending through south London to Richmond.

In the Isle of Wight and Dorset the sequence is represented by a thick series of sands. North-west of London, in Bedfordshire and Cambridgeshire, some 70 m of sands form a narrow outcrop, from which the formation extends south-east below the Chalk but dies out along a south-west to north-east line extending from St. Albans into East Anglia.

Typical values for the hydraulic conductivity of sands in the Lower Greensand are between 5 and 20 m/d and for specific yield between 10 and 20 per cent.

The sand facies south of London yields 10–25 l/s to large diameter wells and in favourable situations 35 l/s. Larger quantities tend to be obtained from the Hythe Beds, 600 mm wells giving 35 l/s and exceptionally over 100 l/s. In Bedfordshire and Cambridgeshire 25 l/s is a representative yield[68].

West of London, near Slough, the Lower Greensand occurs at depths of 300 m below the Gault, the Chalk and Tertiary deposits. Flowing artesian conditions were encountered when the first wells were drilled to the formation about the beginning of the century, but extensive development of the resources, particularly since the 1930s, has lowered water levels by as much as 30 m near Slough and by 5 m at distances up to 25 km from the town, although confined conditions still exist[69]. Wells in Slough yield up to about 25 l/s.

The Sandringham Sands in Norfolk, previously referred to the Lower Greensand, are now known to be equivalent to the Upper Jurassic and Hastings Beds and are considered under the latter heading (see page 207).

Aquifers in the Jurassic

The Jurassic includes a wide range of sedimentary rock types with individual lithologies showing considerable variation in thickness and lateral extent. A number of aquifers occur in the sequence, of which the limestones of the Inferior Oolite and Great Oolite series in the Middle Jurassic are the most important *(Table 5.1, Figures 5.4 and 5.8)*.

The Inferior Oolite is well developed in the Cotswold Hills, attaining a maximum thickness of over 100 m east of Cheltenham. It also occurs south of the Mendip Hills and in Lincolnshire, where it is known as the Lincolnshire Limestone *(Figure 5.8)*.

In Dorset and the Cotswold Hills, a series of sands up to 75 m thick, referred to in different areas as the Bridport, Yeovil, Midford or Cotswold

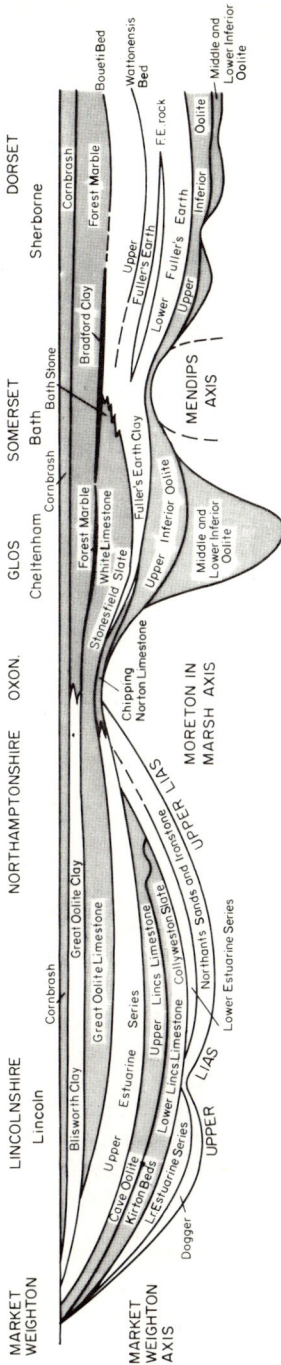

Figure 5.8 'Lateral variations in the Middle Jurassic between south Yorkshire and Dorset; the chief developments of Limestone are shaded, thicknesses are not accurately to scale (from Bennison and Wright[70], courtesy Edward Arnold Ltd)

Sands, forms part of the Upper Lias. These sands are in hydraulic continuity with the overlying Inferior Oolite. Well yields from the sands are generally small but many springs drain from them, deriving some of their supply from the Inferior Oolite. Springs issuing from the Midford Sands provide part of the water supply for Bath.

The Great Oolite Limestone is an important aquifer in the Cotswold Hills. In the south Cotswolds it is separated from the Inferior Oolite by the Fuller's Earth Clay, and the two limestones form distinct aquifers. Further north the clay passes laterally into limestone and the Inferior and Great oolites form a single aquifer, although the Inferior Oolite is much thinner than to the south and the sequence includes a number of clay layers *(Figure 5.8)*.

The Forest Marble, which includes limestones, and the Cornbrash, which is mainly limestone, are not important aquifers in their own right, but in the Cotswold Hills they supplement the yields of wells drilled to the Great and Inferior oolites and in some areas may be considered to form a single aquifer with the underlying limestones.

The Corallian consists of limestones and calcareous sandstones in Yorkshire and in southern England, south of Oxford. It is an aquifer of some significance in Yorkshire, where it is over 100 m thick giving maximum yields from individual wells in the range 25–50 l/s and occasionally in excess of this. In north Wiltshire and Berkshire it consists of 20 m of calcareous sands with interbedded hard grit bands. Typical well yields are up to 10 l/s.

Other limestones and sandstones in the Jurassic are of importance locally, for example the Marlstone in Northamptonshire, but most groundwater development from Jurassic rocks is from the limestones of the Great and Inferior oolite series and these are discussed further below.

Aquifer properties of the Middle Jurassic limestones

The transmissivity of these limestones is very dependent upon the extent of fissure systems, which are commonly related to a rectangular joint pattern. Massive permeable limestones contain fissures some 50–75 mm wide and spaced about 2 m apart. However, the limestones are not homogeneous and in consequence water level changes induced by pumped wells are different in different directions. The transmissivity of the Inferior and Great oolite limestones often exceeds 1500 m²/d and may be as high as 4000–4500 m²/d. In outcrop areas values change seasonally as the thickness of saturated aquifer varies, and extremely fissured limestones at outcrop may have a low transmissivity because of the relatively small thickness of saturated rock.

The Lincolnshire Limestone is more permeable in south Lincolnshire than in north Lincolnshire; this is probably due to some extent to the more argillaceous nature of the limestone in the north which has restricted the circulation of water through the rock[6].

Areal variations in the permeability of the Jurassic limestones are not well known but relative permeability may be indicated by the quality of the water. For example the saline connate waters have been flushed out of the more permeable Lincolnshire Limestone in south Lincolnshire but are still retained in the less permeable aquifer to the north[6] (Chapter 7).

The porosity of the Jurassic limestones is generally between 15 and 20 per cent while the specific yield is not likely to exceed 5 per cent; a value of 2 per cent was obtained during a study of the potential yield of the Great Oolite, Forest Marble and Cornbrash near Cirencester[71].

Because much of the storage in the limestones is in the form of extensive interconnected fissure systems, there is a considerable variation in storage between the winter and summer seasons and during and following individual periods of rain. The outcrops have been deeply dissected by valleys extending into the underlying impermeable beds, and this allows rapid drainage through the fissure systems.

Yields from the Middle Jurassic limestones

As the storage potential in the outcrop area is limited yields tend to be small. Reliable perennial sources are usually found where the limestones remain fully saturated under confined conditions. In such situations yields are prolific if extensive fissure systems have been naturally formed. Artesian flows in excess of 22 Ml/d have been recorded in south Lincolnshire, where some of the largest yields in Britain are obtained from individual wells[6]. In north Lincolnshire the majority of wells give less than 25 l/s although exceptions do occur; for example at one site four 300 mm wells gave a total yield of 200 l/s. Large capacity wells in the Inferior and Great oolites produce between 50 and 100 l/s and in some cases as much as 200 l/s.

Permian

Permian rocks occur in the north of England both east and west of the Pennines. East of the Pennines the lower part of the sequence consists of either sands or breccias and the upper part is formed by the Magnesian Limestone, which is predominantly a dolomite. The limestone is divided in Yorkshire into upper and lower divisions by marls, while further south in Nottinghamshire the upper part of the sequence passes laterally into Triassic sandstones.

The Yellow Sands occur at the base of the Permian in Durham. They are unconsolidated, permeable sands varying in thickness up to more than 50 m. Because of their nature cementation or freezing processes were necessary to control water inflows during the sinking of mine shafts through the Permian to the underlying Coal Measures. The inflow from 10 m of sand penetrated by a shaft at South Hetton Colliery was about 100 l/s. Because of the water-bearing nature of the sands colliery workings do not approach nearer the base of the Permian than 30–50 m. An indication of the permeability of the deposit was given by a 75 mm diameter borehole, drilled from a colliery heading to prove the position of the Permian, which yielded 125 l/s; the flow was ultimately developed for public supply[72]. Yields from medium- to large-diameter boreholes are generally at least 5–10 l/s.

The Magnesian Limestone is as much as 200 m thick where it crops out in Durham. Although it is a compact rock, extensive fissuring and (particularly in east Durham) its cavernous, cellular and brecciated nature have made it extremely permeable. Yields from individual boreholes exceed 50 l/s and the aquifer has been extensively developed near Sunderland and West Hartlepool. Individual fissures can give considerable supplies. For instance, in a shaft at Easington Colliery three fissures within a vertical distance of 5 m each yielded more than 150 l/s, and a total of 580 l/s, while the total inflow from 140 m of limestone was nearly 760 l/s[72].

Recent investigations[73] have increased knowledge of the hydrogeology of the Magnesian Limestone south of the West Hartlepool Fault (which trends east–west through Hartlepool in south Durham). In this area the Upper Magnesian Limestone is overlain and underlain by the Upper and Middle Permian Marls respectively. The Upper and Lower Magnesian Limestones are massive, dolomitic limestones of limited porosity that tend to yield water only where fissured. The Middle Magnesian Limestone, on the other hand, is very porous and generally provides about 3 Ml/d. However, less than 30 m of the upper part of the entire Magnesian Limestone sequence of 200 m acts as the main aquifer. This is probably because groundwater flow patterns are controlled by a base level of erosion. Because of the dip of the deposits, the permeable zone crosses the various formation units. The average transmissivity of the Magnesian Limestone is about 300 m²/d but can exceed 1500 m²/d.

In Yorkshire and Nottinghamshire the Magnesian Limestone is generally more compact than in Durham, but in Yorkshire it is in part cellular and contains cavernous hollows. Minutely cellular limestone occurs in the Lower Magnesian Limestone in Nottinghamshire but the porosity is due to the solution of calcareous ooliths and effective porosity is low[67]. The yields from the limestone in Yorkshire and Nottinghamshire are commonly less than 10 l/s and reflect the lower permeability of the formation.

The high hardness of the water from the Permian is a factor that tends to discourage development for water supply east of the Pennines.

West of the Pennines, in the Vale of Eden, the Permian is represented by the Penrith Sandstone, which is underlain in the south of the vale by breccias (the Brockram). The sequence has a maximum thickness of about 450 m with 200 m representing the Penrith Sandstone. The aquifer is virtually undeveloped but the coarse-grained nature of the sandstone indicates that the permeability is relatively high. The mean horizontal and vertical intergranular hydraulic conductivities are about 1.5 and 0.9 m/d[59,64].

The Collyhurst Sandstone forms the lower part of the Permian in Lancashire. It is separated from the Triassic sandstones by the Manchester Marl but the two aquifers are often in hydraulic continuity as a result of faults. The Collyhurst Sandstone is a good aquifer in its own right and yields of the order of 50 l/s can be anticipated from large-diameter wells.

In Somerset and Devon, the Permian consists of fine-grained sandstones and breccias overlain by marls, the arenaceous facies being as much as 130 m thick. Yields from wells are generally only about 5 l/s, although exceptionally 10 l/s or more[63].

Carboniferous Limestone

The Carboniferous Limestone Series includes a variety of sedimentary rocks but the rock type generally associated with the series is the massive limestone well developed in the Peak District of Derbyshire (with a thickness of more than 450 m), the Mendip Hills (up to 1000 m), south Wales (up to 800 m), north Wales (up to 1000 m), and north-west Yorkshire (over 200 m). Sandstones form significant parts of the series in some areas, as in the Forest of Dean and Anglesey, and predominantly shale sequences are found in other areas, for example south-east of the Peak District and in the Craven district, near Settle.

North of the Craven district the series comprises alternating limestone, sandstone and shale sequences (referred to as the Yoredale Series) in which individual limestones can attain considerable thicknesses, in some cases exceeding 200 m. Still further north, in Northumberland and the Midland Valley of Scotland, the sequence is more arenaceous but includes shales and thin limestones. In Northumberland the series includes the Fell Sandstone which is 200–350 m thick.

In general the Carboniferous limestones have a low porosity compared with Mesozoic limestones. They are compact, well-jointed rocks. The specific yield from the body of the rock is negligible and groundwater flow is entirely along joints and fissures that have been extensively enlarged by solution. The landform that develops on such limestones, where the drainage is entirely underground, is referred to as *karst,* although in the UK these landforms are actually classified as *fluvio-karst* because both fluvial and karst processes have been involved[74].

The principal areas of karst landforms in the UK are on the Carboniferous Limestone in north-west Yorkshire, the Peak District and the Mendip Hills[75], although the longest and deepest cave in the UK is found in the Carboniferous Limestone in south Wales. It is Ogof Ffynnon Ddu, which has 40 km of underground passages. Such karst landforms developed mainly in the Quaternary and to a lesser extent in the Tertiary. Karstification was particularly active during phases of ice-sheet recession. The development of swallow-holes in north-west Yorkshire can often be related to the extent of the boulder clay or the position of ice-fronts. In the Mendips and north-west Yorkshire, cave systems have formed at different levels that have been related to changes in local base-water levels during the Pleistocene[76,77].

A water table, as generally understood, does not occur in the Carboniferous Limestone. The fissure systems tend to be discrete underground basins of varying sizes. Studies of flow paths from swallow-holes to resurgences in the Mendips and north-west Yorkshire, using tracers, have demonstrated their complex nature[78-80]. Individual swallow-holes may feed more than one outlet, and flow lines cross but do not mix. At St Dunstan's Well in the eastern part of the Mendips two springs issue 2 m apart, each fed by separate systems of swallow-holes and each with a distinctive water chemistry[81]. The water in the limestone flows in the main conduits or channels at velocities of some 6000–7000 m/d[78].

In many areas groundwater in the limestone is derived from both surface runoff entering swallow-holes and direct infiltration of rain into the limestone or overlying soil (referred to as percolation input)[78]. In the Mendips direct infiltration accounts for 90–100 per cent of the spring discharges at times of low flow, over 65 per cent at high flows and 50 per cent during exceptional floods[82-84]. The distribution of rainfall between the

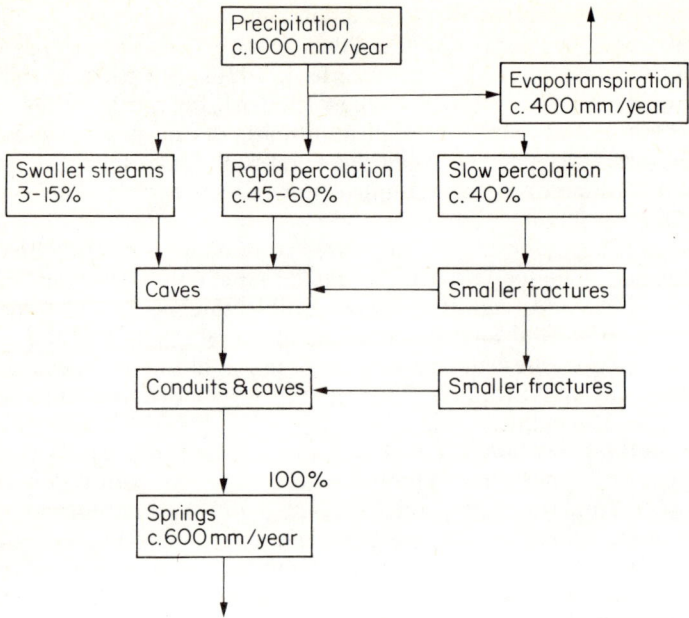

*Figure 5.9   Paths taken by water flowing through the Carboniferous Limestone in the Mendip
Hills (from Atkinson et al.[86], courtesy Somerset County Council)*

various channels and the flow paths through the limestone are illustrated in
Figure 5.9.

The proportion of direct infiltration depends to some extent on the size of
the limestone outcrop. For example, in south Wales, where this is limited
and there is considerable recharge of surface flow from impermeable
deposits overlying the limestone, the proportion of direct infiltration varies
between 20 and 40 per cent of the spring discharges[85].

The direct infiltration tends to flow in smaller channels than runoff
entering swallow-holes, which lead to the larger subterranean channels or
conduits because of the larger volumes of water involved. The percolation
water flows at velocities of 10–200 m/d—significantly lower than rates in the
conduits, where turbulent flow conditions obtain[78].

The storage in the limestone has been divided into three components: (a)
in conduits, (b) in fissures in the vadose or unsaturated zone and (c) in
flooded fissures, i.e. fissures in the saturated zone[78]. The volume represented
by the conduits is small, generally less than 1 per cent of the total. Although
the storage in the unsaturated zone is several times greater than that in the
conduits, both together represent only about 1 per cent of the storage in the
saturated zone.

Tentative groundwater contours have been drawn for the Carboniferous
Limestone in the Peak District[67,87], but they must be regarded as only a
general guide to the top of the saturated zone. They indicate that surface and
groundwater divides do not always coincide. Igneous rocks intruded into the
limestone form perched water tables and these give rise to surface discharges
in some areas, though this water recharges the limestone when the stream

passes off the igneous rock. Mining for metalliferous ores has occurred particularly in the eastern part of the outcrop. Mine-drainage by tunnels (referred to as soughs) has permanently lowered groundwater levels and these tunnels now represent the principal discharge points from the limestone. In places the natural groundwater divide has been displaced appreciably by artificial drainage[67,87].

Drilling for water is speculative in the limestones because of the negligible specific yield of the mass of the rock and the fact that water flows in discrete channels not closely spaced or extensively interconnected. Yields from wells vary considerably and there is a high probability of failure to obtain a supply.

In karst areas springs are often used for water supply, as for example in the Mendip Hills, Wales and the Peak District. The Bristol Waterworks Company makes extensive use of springs issuing from the limestone in the Mendip Hills. The flow rates respond rapidly to rainfall and the annual range of flow is considerable. For example, the various springs giving rise to the headwaters of the River Chew range in flow from as little as 0.9 to 90 Ml/d and the springs at Cheddar vary from 9 to 455 Ml/d. Because of this extreme range due to the limited storage capacity of the limestone, springs are developed by the Company in conjunction with several reservoirs at sites around the Mendip Hills, the water from the springs flowing into the reservoirs.

The Carboniferous Limestone is most effectively developed for water supply if fissures are intersected by tunnelling. This is generally uneconomic but where it has been necessary for mining, large volumes of water have been intercepted. For example pumping for mine drainage at a haematite mine near Barrow-in-Furness amounted to over 30 Ml/d; a further example is Meerbrook Sough in Derbyshire which discharges about 58 Ml/d to the River Derwent, the water being collected from some 8 km of tunnels, which are partly in the Millstone Grit. Part of the discharge is used for public water supply[67]. In 1879, during the construction of the Severn Tunnel, water under artesian pressure derived from the Carboniferous Limestone was intersected in fissures. The flow was estimated to be 90 Ml/d and pumping is necessary to this day to drain the tunnel, the water being used for local industry[88].

The manner in which fissure and cave systems develop in karstic limestones has been extensively debated[74], for it is fundamental to the development of groundwater resources in such aquifers. According to Rhoades and Sinacori[89] the initial flow paths in a limestone with a poorly developed but regular joint system would attempt to approach the idealised curve pattern *(Figure 5.10)*, described by Hubbert[90] and amplified subsequently by Toth[39]. However, solution would be more extensive along the shorter flow lines in the upper part of the saturated zone. This would eventually lead to a concentration of flow and solution at the higher level with consequently a reduction of flow at greater depths, a tendency concentrating groundwater discharge from a few main springs to a greater extent than occurs in aquifers where diffuse groundwater flow predominates[89]. During the course of such a sequence high level springs are ultimately captured by fissures leading to outlets at lower levels *(Figure 5.10)*. The idealised sequence probably explains the development of flow systems in relatively low-dipping limestones such as generally occur in the

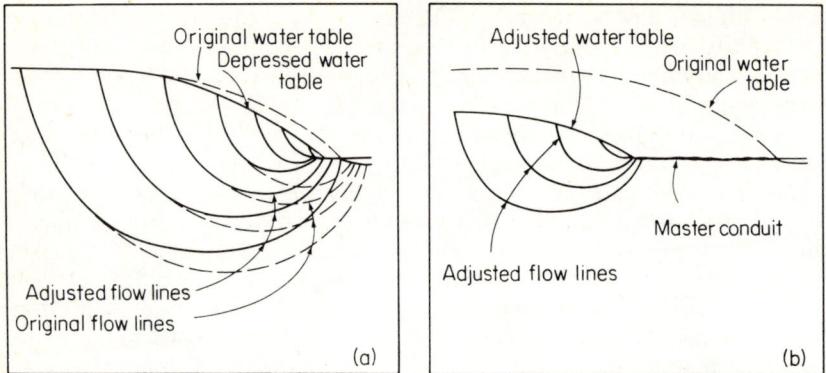

Figure 5.10 Development of a master conduit and corresponding adjustment of lines of groundwater flow in a compact limestone (from Rhoades and Sinacori[89], courtesy University of Chicago Press)

UK, but the complexity of the factors controlling the development of flow paths and the uniqueness of many individual flow systems has been emphasised[74].

The response of many springs in the Carboniferous Limestone to rainfall is relatively rapid. The permeability of the fissure systems is high and the amount of storage available for storing the winter recharge is small. From the point of view of water resource development such springs have the principal disadvantage of surface water sources that are not supported by storage, namely uneven temporal distribution. Water infiltrating karstic limestones can only be fully developed if the springs are supported by extensive storage systems that provide a more even flow or that exist below the spring outlet (as in confined systems) and can be pumped at times of low flow and refilled during subsequent periods of high infiltration. Some springs flow under pressure and these often have a more constant flow.

The limestones and sandstones of the Yoredale Series in the north of England are independent aquifers separated by shales. In such a sequence groundwater flow often tends to be concentrated along the bedding planes at the base of the aquifers. These rocks are rarely developed for water supply, typical reliable yields being no more than 1 l/s. When mineral deposits were mined extensively in these rocks 6 l/s was considered a large inflow of water[91], emphasising the low permeability of the aquifers but also the moderate yield that could be obtained from a well striking a good fissure system.

The groundwater conditions of the Yoredale sequence in Upper Teesdale were examined in some detail in connection with the site investigations prior to the construction of the Cow Green Reservoir[92]. Although perched water tables existed above the regional groundwater level, piezometric pressures below the regional level were consistent with a relatively uniform pattern of groundwater flow. The bulk hydraulic conductivity of the limestones decreased with depth from the relatively low values of about 1 m/d in the near-surface fissured layers to as little as 10^{-3} m/d at depths of about 30 m.

The Carboniferous sequence in Northumberland includes the Fell Sandstone, a friable fine- to medium-grained rock. The mean porosity is about 14 per cent and the intergranular hydraulic conductivity 0.3 m/d[93].

Secondary and minor aquifers

The less important aquifers include Pleistocene sands and gravels, Tertiary sands, the Upper Greensand, the Spilsby Sandstone and the sandstones of the Hastings Beds, the Coal Measures and the Millstone Grit. Excluding mine-drainage water they account for some 10–15 per cent of the total groundwater abstraction in England and Wales.

In contrast with the world in general, alluvial deposits are not important aquifers in the UK. Extensive spreads of alluvial gravels occur in the Trent[67,94] and Thames[95-97] valleys and to a lesser extent in other principal valleys[98], but the average thickness is only about 5 m. However, the hydraulic conductivity may range from 50 to 400 m/d and the specific yield from 15 to 25 per cent, and efficiently constructed large-diameter wells can yield 25–50 l/s. The Woking and District Water Company developed gravels in the Thames valley by means of three wells interconnected with 366 m of perforated collector pipes, to give a dry weather yield of 22 Ml/d.

The Pleistocene Crags[53,99] of East Anglia and the various Tertiary sands in the London[7] and Hampshire basins usually provide about 5–10 l/s, although in favourable situations large-diameter wells (say 600 mm in diameter) do yield up to 25 l/s.

The Upper Greensand is a poor aquifer where it crops out but yields are higher, with upper limits of 25–30 l/s, where it is overlain by the Chalk in the west of England.

In south-east England the Tunbridge Wells Sand yields 5–10 l/s, and up to 25 l/s from large-diameter wells. The sandstones of the Ashdown Beds are somewhat better, typically providing 10–25 l/s and in favourable circumstances up to 50 l/s.

The Spilsby Sandstone in Lincolnshire can yield as much as 40–50 l/s but 10–25 l/s is more typical. Further south in Norfolk, the Sandringham Sands provide about 5 l/s.

Yields from the sandstones of the Coal Measures are very variable but generally amount to no more than a few litres per second[9,67]. The sandstones of the Upper Coal Measures do provide somewhat larger yields, for example some shafts in the Midland coalfields yield 20–50 l/s, but borehole yields are much smaller[67]. Considerable volumes of water are pumped from the Coal Measures for mine-drainage. The coalfields of South Wales and Scotland are particularly wet, pumping about 8 and 10 m^3 of water respectively per tonne of coal mined. The average drainage-output ratio for mines in England and Wales in 1962 was 2 m^3 of water per tonne of coal mined. Groundwater flow is controlled by geological structure and water tends to collect in synclines, as for example in the East Midlands[67] and South Wales[100] coalfields, where collieries in synclines may pump as much as 13 Ml/d for drainage purposes. Groundwater flow through sandstones in the Coal Measures can be modified by fissures caused by mining subsidence. Studies in South Wales[101] have shown that subsidence produces irregular zones of tensional and compressional strains. The rate of groundwater flow in tensional zones in the area studied was about 570 m/d, implying very high fissure permeability compared with the low intergranular permeability of the sandstones of about 10^{-5} m/d. An appreciation of groundwater conditions is particularly important in mining areas where inflows of water from permeable rocks or from disused workings are a constant hazard[67,100].

The sandstones of the Millstone Grit commonly provide 15–20 l/s and exceptionally up to 50 l/s[9,67]. Springs have been developed for supply by means of collecting chambers and adits to give as much as 50 l/s. The construction of the Bowland Forest Tunnel, forming part of the aqueduct from Hawes Water to Manchester, afforded an opportunity to study the hydrogeology of the Millstone Grit[102]. High inflows (up to 100 l/s) were recorded from fissured zones, but the flow from individual fissures tended to decline with time and inflow to the tunnel was generally not significant when the depth of cover exceeded about 200 m.

Argillaceous deposits

The argillaceous deposits, which separate individual aquifers in the geological sequence, play a decisive role in controlling flow rates in regional flow systems. Clays, marls and mudstones are not completely impermeable and rarely comprise 100 per cent clay. Rowe[103] concluded that relatively permeable fabrics in argillaceous rocks are more common than is generally supposed. Sand and coarse silt commonly occur in thin vertical joints and layers. Many clays are fissured and although these may not be open they influence the permeability.

Weathered fissured clay near the surface may have a hydraulic conductivity of 10^{-2} m/d. The overall value for clays at relatively shallow depths is about 10^{-3} m/d, but at greater depths values as low as 10^{-6} to 10^{-5} m/d are typical for unfissured clay.

Some consequences of groundwater development

Under natural conditions the hydrological cycle is in a state of dynamic equilibrium. The abstraction of a quantity of groundwater, however small, disturbs this equilibrium and changes in storage and flow rates occur as the system adjusts to a new equilibrium. Because groundwater storage is large in relationship to abstraction from individual wells, the consequences of limited development are not evident, but the cumulative effects of abstraction from many individual wells eventually do become apparent.

The principal consequences are a decline of groundwater levels and a reduction of river flows; in coastal areas saline intrusion may occur and in some situations the ground surface may subside.

The consequences of groundwater abstraction are time-dependent, that is they are not felt simultaneously throughout the system. There is a time lag because of the large volume of water stored in the ground and its slow rate of flow. One of the main tasks of the groundwater hydrologist is to anticipate the consequences of development and ensure that undesirable effects do not occur. The principal responses to groundwater abstraction are discussed in the following paragraphs by referring to typical situations in the UK.

Decline of water levels in a confined aquifer

Groundwater abstracted from a well in a confined aquifer is drawn initially from storage in the vicinity of the well. As the hydrostatic pressure is reduced, the water is released by compression of the aquifer and by a slight expansion of the water. If pumping continues, the pressure is reduced over an increasing area until a recharge or discharge area is intersected, when a new equilibrium is established either by an increase in flow through the confined aquifer or a reduction in the discharge from the aquifer.

The rate of expansion of the cone of depression around a pumping well is inversely proportional to the coefficient of storage. In a confined aquifer this is small, of the order of 10^{-4}, and as a result the cone expands rapidly.

A classic example of this sequence developed in the London Basin[7], which is actually a syncline where the Chalk and Tertiary sands are confined in the centre of the syncline by the London Clay *(Figure 5.6)*. In the natural state groundwater flow was from the outcrop of the Chalk and Tertiary sands, on the north and south limbs of the syncline, through the confined aquifer to outlets in the Thames valley.

The basin has been developed for water supply since the eighteenth century. Abstraction from the London area steadily increased to some 227 Ml/d between 1925 and 1940. As a result groundwater levels declined by as much as 75 m and the Tertiary sands and Chalk were dewatered over 900 km². Ultimately natural discharge from the confined area ceased, and the only outlet became pumped wells. In addition to a reduction of the volume of groundwater held in storage, other consequences were:

1. a reduction in the flow of rivers draining the Chalk and Tertiary sands outcrop;
2. leakage of sulphate-rich water from the overlying London Clay into the Tertiary sands and Chalk;
3. intrusion of saline water from the Thames estuary into the Chalk and Tertiary sands;
4. slight subsidence of the ground surface due to settlement of the London Clay.

The total abstraction from the confined central part of the London Basin, over the 200 years since development commenced, is estimated to have been 5680 million m³. Reduction of groundwater storage has contributed about 18 per cent and increased flow through the confined aquifer, as a result of the change in the hydraulic gradient, about 80 per cent *(Table 5.3)*. The total flow through the aquifer has been 67 per cent greater over the 200 years than it would have been had natural conditions continued.

Since the early 1940s abstraction has declined to about 186 Ml/d, and an approximate steady-state situation now exists with flow through the aquifer from the recharge areas providing virtually the entire abstraction. The decline of water levels, indicating water was being taken from storage, was only a transient phase in the development of the aquifer. It could only be temporary, since storage is finite and the extent to which water levels can be lowered is controlled to a large extent by economic factors such as the increased pumping lift and decline of yields as the water level falls.

The Chalk and Tertiary sands in the central part of the London Basin have provided an average of 90 Ml/d in the period 1800–1965. To develop this

Table 5.3 GROUNDWATER BALANCE FOR THE CENTRAL CONFINED PART
OF THE LONDON BASIN, 1800–1965, IN MILLION m³
(AFTER WATER RESOURCES BOARD[7])

Abstraction from Chalk and Lower London Tertiaries		5682
Sources of replenishment:		
Leakages from London Clay	109	
Consolidation of London Clay	27	
Saline intrusion	29	
Change in storage	1032	
Underflow	5460	
Natural discharge to rivers		975
Total	6657	6657

large resource it was necessary to increase the hydraulic gradient and increase the flow from the outcrop. The considerable volume of rock dewatered (equivalent in water storage to seven times the total surface reservoir capacity available to provide London with water) is potentially available for replenishment by artificial recharge, and this is now under investigation.

Effect on river flow

Continuous abstraction of water from an unconfined aquifer that is discharging to a river sooner or later reduces the flow of springs from the aquifer and hence the flow of the river. Initially water is taken from storage in the aquifer, but as the cone of depression expands the hydraulic gradient to the river is reduced and groundwater discharge to the river is intercepted to an increasing extent. With continuous pumping a new equilibrium is established, and river flow is reduced at all times by the amount being pumped; the baseflow curve of the river with pumping is parallel to the natural baseflow curve, but below it by the amount being abstracted. However, this ideal state does not always occur. In some situations the fissure distributions controlling groundwater flow to spring outlets may be such that river flow is reduced in summer by less than the abstraction rate, and in winter by somewhat more. A new equilibrium is not established immediately. The time lag depends upon the geometry and hydraulic properties of the aquifer and the distance of the well from the river. The problem has been examined with a mathematical model of an idealised situation[104]. The effects of abstraction were related to different values of a term T/SL^2 (referred to as the aquifer response time), where T is transmissivity, S is the storage coefficient and L the distance from the river to the catchment divide parallel to the river.

The time that elapses before a new equilibrium is established is directly related to the response time. The higher the value and the nearer the well is to the river, the sooner a new equilibrium is attained. For very low response times a new equilibrium may not be reached for decades *(Figure 5.11)*.

The situation just described applies where a hydraulic gradient is

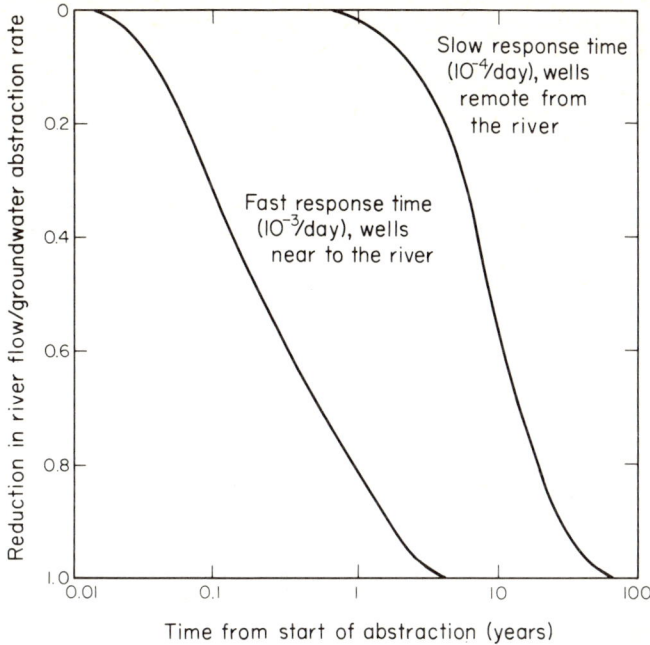

Figure 5.11   Reduction in groundwater flow to a river due to continuous abstraction from wells when the hydraulic gradient is maintained towards the river (from Downing et al.[105], courtesy Elsevier Scientific Publishing Co.)

maintained towards the river despite abstraction. However, in some circumstances, as for example with very heavy abstraction, a hydraulic gradient may develop in the aquifer from the river towards the abstraction centre. In this situation, if the river bed is permeable, water flows from the river into the aquifer; again the decline of river flow equals the amount pumped at all times once a new equilibrium is established, providing the flow of the river is greater than the abstraction rate.

If the river bed is impermeable or semi-permeable, so that groundwater can discharge from the aquifer to the river but the aquifer cannot be recharged readily by the river, the cone of depression extends beyond the river and groundwater levels in the aquifer fall below river level. Loss of water from the river in these circumstances is related to the hydraulic conductivity of the river bed deposits, their thickness, and the difference in head between the river and water in the aquifer[105]. The natural groundwater discharge to the river, from within the area in which groundwater levels are lowered, will be intercepted by wells and the balance between this quantity and the abstraction rate will be taken from storage in the aquifer. Natural infiltration during the winter replenishes the depleted storage preferentially, and the result is that the flow of the river is reduced to a greater extent in the winter than in the summer[106].

The relationship between aquifer properties, well location and the permeability of a river bed is exploited during regional groundwater development in England. Pumping regimes are designed to have minimal effect on river flows. This is discussed further in Chapter 8.

Surface settlement

Consolidation is the compression of soils and rocks under pressure due to the expulsion of water from the voids. It can occur, leading to settlement of the ground surface, when groundwater levels are lowered by pumping because this causes an increase in the effective overburden pressure. Settlement in clays and peats can be considerable and may be appreciable in sands, especially if they are loose and the water table fluctuates appreciably; dense sands and gravels are not usually affected, however, except in situations where pumping from wells results in the loss of fine material[107].

Drainage of the Fenlands has led to shrinkage of the peat deposits and oxidation of the organic matter. This, together with wind erosion, has resulted in considerable wastage of the peat in this region.

Land settlement also occurs when pressure in an artesian aquifer is reduced by pumping. Water drains from the confining layer, with a consequent reduction in its thickness due to consolidation. The settlement corresponds to the total volume of water drained from the confining layer. The total settlement is given by:

$$\text{Settlement} = m_v \times H \times \Delta\sigma \tag{5.5}$$

where m_v = the coefficient of volume compressibility
H = the original thickness of the confining layer
$\Delta\sigma$ = the change in effective stress

Considerable settlement occurs where the confining layer comprises relatively unconsolidated clays and silts, for example alluvial sediments. Under these conditions more than 6 m have been reported from parts of the San Joaquin valley in California where some 3–8 m decline of artesian pressure results in 0.3 m of settlement[108]. In other parts of the world (for example Mexico, Japan and the Po valley in northern Italy) similar cases have been reported.

Settlement has occurred in the London Basin[109,110] where artesian pressure in the Chalk and Tertiary sands has been lowered by as much as 75 m. The confining London Clay is an overconsolidated clay and the average settlement is only 0.06 m, although the maximum is about 0.5 m[7,110].

There is a time lag between the decline of water levels and the drainage of water from confining clays or silts, because of their low permeability[111]. The rate of settlement declines with time but continues at a reduced rate even if the decline of water levels is arrested. The time required for a particular percentage of total settlement to occur is given by:

$$t = \frac{T_v H^2}{c_v}$$

where t = time for a given percentage of total settlement to occur (referred to as the degree of consolidation)
T_v = a dimensionless time factor
H = the thickness of the confining layer
c_v = the coefficient of consolidation and is equal to $k/\gamma_w m_v$, where k is hydraulic conductivity and γ_w is the unit weight of water.

If drainage can occur from both the top and bottom of the confining layer, H^2 is replaced in the formula by $(H/2)^2$. There is a set of theoretical relationships between the degree of consolidation and the time factor for different distributions of consolidation pressure[111].

In the London Basin, the mean thickness of the confining layer, over the area where water levels have declined, is about 40 m, and for a mean decline of water level of 30 m 90 per cent of the consolidation (which amounts to 0.6 m) will occur in about 1200 years[7]. Thin interbedded fine-grained layers, such as occur in the Lower London Tertiaries, attain hydraulic equilibrium with the pressure reduction in the aquifer more rapidly than is the case with thick clays such as the London Clay. Studies in the San Joaquin valley, California, have shown that during periods of rapid water-level decline consolidation occurs at a maximum rate, but during periods of rising water levels the consolidation rate declines[108].

Land settlement due to declining artesian head causes a permanent decrease in volume, as there is a rearrangement of the grains of the deposit and plastic flow occurs with displacement of the rock particles.

When pressure is reduced in an artesian aquifer by pumping, water is derived from three sources:

1. expansion of the confined water;
2. consolidation of the aquifer;
3. consolidation of adjacent and included clay layers[112].

Of the three, the latter is usually the most significant. The change of storage in the central part of the London Basin due to the first two factors yielded 5.5 million m³, but consolidation of the confining layers has yielded 27 million m³ so far[7]. However, most of the water yielded by consolidation of clay layers can only be provided once because, if the load on the clay is reduced due to an increase in artesian pressure, the void ratio will eventually increase again, but not to the original value. Rock types that form aquifers, on the other hand, behave as elastic media and the storage is replaced if the water level recovers, there being virtually no permanent consolidation. An important point is that the coefficient of storage of an aquifer calculated from short-term pumping tests may represent only a small part of the water that can be removed from storage in the aquifer and associated clay layers in the long term[113].

Localised ground subsidence (creating sink-holes) has been reported from the United States and South Africa when groundwater levels have been lowered in cavernous limestones and dolomites that are overlain by thick superficial deposits[114,115]. It occurs where superficial deposits are of a non-uniform and heterogeneous character and the natural groundwater level is in the superficial deposits above the bed-rock surface. The cause is the drying out and consolidation of the superficial deposits following the lowering of the water level. The hydrostatic support is reduced and the sediments collapse into fissures or caverns in the limestone. The collapse may be encouraged by increased groundwater flow through the limestone due to pumping. The formation of such sink-holes is commonly associated with a very irregular buried limestone surface and may be triggered by unusual events such as heavy rainfall. Collapse of overlying deposits into buried sink-holes is known where the Chalk and Carboniferous Limestone outcrop.

If groundwater resources are to be developed on a large scale, an estimate of the amount and the consequences of any settlement should be made, if it is likely to be significant. It is necessary to measure relevant soil properties such as c_v and m_v.

Saline intrusion

In coastal aquifers a natural hydraulic gradient exists towards the coast, and fresh groundwater discharges into the sea thereby limiting the encroachment of salt water. Fresh water and sea water have different densities, sea water being 1.025 times heavier than fresh water. Because of this fresh water occurs above salt water in coastal areas, the boundary or interface between the two waters being in a state of dynamic equilibrium.

Ghyben and Herzberg independently studied this equilibrium in northern Europe, and the equation that bears their names approximately defines the position of the interface between the two fluids:

$$z = \frac{f}{s-f}h = \frac{1}{1.025-1}h = 40h \qquad (5.7)$$

where z = depth below sea level to the interface
f = density of fresh water (i.e. 1.0 g/cm³)
s = density of sea water (i.e. 1.025 g/cm³)
h = head of fresh water above sea level

The interface slopes inland forming the upper surface of a salt-water wedge *(Figure 5.12)*. This advances and recedes according to changes in the hydraulic gradient caused by natural seasonal changes, tidal variations or pumping patterns. The upper part of the interface only moves inland if fresh water no longer occurs above sea level in the vicinity of the coast. This arises when the natural hydraulic gradient is reversed by excessive pumping.

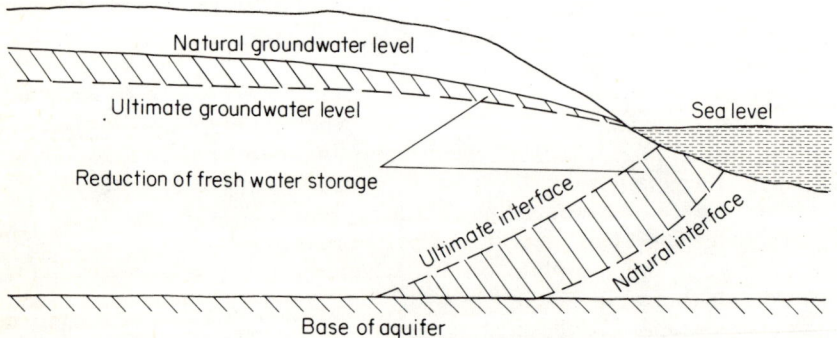

Figure 5.12   Reduction of freshwater storage following groundwater development in a coastal aquifer (after Santing[116])

The relationship defined by Ghyben and Herzberg assumes that ground-water is static. Nevertheless, in many situations where a natural equilibrium exists, the formula forecasts the approximate position of the saline interface.

Where the natural state has been disturbed by pumping or the aquifer is very fissured the formula is less perfect.

Because of the relative densities of fresh water and sea water, pumping from a well in an aquifer that contains a saline interface some distance below the well may raise the interface and draw saline water into the well. The tendency for this to occur is reduced if water-level drawdowns are limited and the required yield is obtained by many small-yielding wells rather than a few of large capacity.

As groundwater is constantly moving a dynamic rather than an hydrostatic equilibrium exists between fresh and salt water, and the depth to the interface is actually greater than that forecast by the Ghyben–Herzberg equation[90].

The interface between fresh and salt water is not sharp but is usually marked by a zone of diffusion. This is due to both mechanical dispersion and molecular diffusion, diffusion being the dominant process[117]. The width of the zone of dispersion is related to the porosity, the form of the pore spaces and the relative magnitude of the water level fluctuations due to the tide and to natural recharge and discharge. Cooper[117] suggested that sea water flows in a cycle from the sea floor into the zone of diffusion and back into the sea, the circulation being induced by dispersion produced by the reciprocative motion of the salt-water front due to the tidal and groundwater level fluctuations referred to above *(Figure 5.13)*. The flow of the salt water

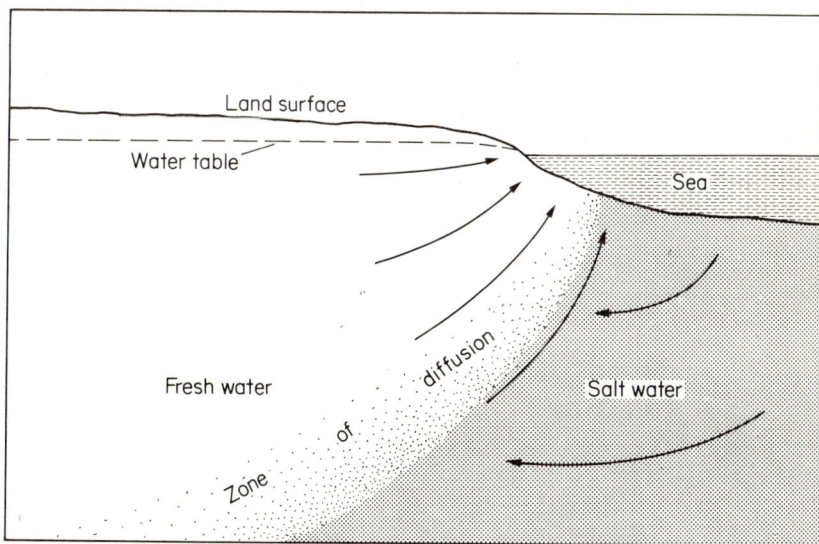

Figure 5.13   Circulation of salt water from the sea to the zone of diffusion and back to the sea (from Cooper[117], courtesy American Geophysical Union)

reduces head losses in the salt-water environment, which also reduces the extent to which salt water moves into the aquifer. An increase in the thickness of the zone of diffusion usually results from abstraction in a coastal area.

In Britain saline intrusion has occurred mainly where the Chalk and Triassic sandstones have been heavily exploited for water supply near coastlines or estuaries containing brackish water. The cause is generally industrial abstractors concentrated in coastal sectors of major towns. The wells causing the intrusion are often sited very close to the coast. In many cases the volume of water pumped in a particular locality is limited by the chloride value, and if the rate of increase in salinity cannot be controlled the wells eventually become disused. Therefore the problem is to some extent self-correcting. Heavy abstraction near the coast associated with low pumping levels often serves to protect inland areas from more extensive intrusion of sea water.

The principal areas where saline intrusion has occurred are along the estuaries of the Mersey, Humber, Orwell, Stour and Thames at the centres and in the areas shown in *Figure 5.14*. In addition it has occurred locally at other centres around the coast, for example at West Hartlepool and near

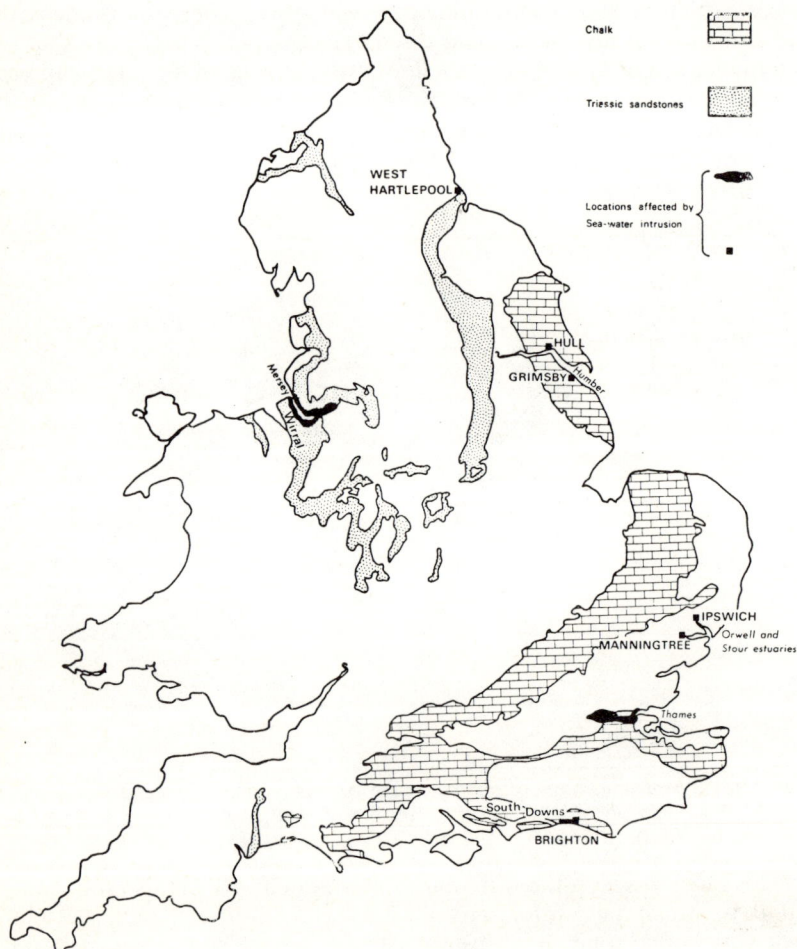

Figure 5.14   Main areas of sea water intrusion in England and Wales

Brighton. The aquifer involved at West Hartlepool, and probably also locally along the coast to the north for some 30 km, is the Magnesian Limestone of Permian age[72], otherwise the aquifers affected are the Chalk and Triassic sandstones.

In Grimsby groundwater in the Chalk has been developed since the middle of the nineteenth century and as a result groundwater levels are below sea level in and around the town, and over appreciable areas more than 5 m below sea level[118]. The salinity of the groundwater near the coast exceeds 250 mg/l in those areas where pumping is heaviest. Studies of the vertical salinity profile in wells in the town indicated that a wedge of saline water extended at least 2.5 km inland.

Along the Thames estuary to the east of London, the Chalk and Tertiary sands form the coastline. These aquifers have been developed for industrial use and saline intrusion has occurred along both banks of the estuary. The saline zone defined by the 150 mg/l isochlor is up to 4 km wide on the north bank. Near the river front extensive areas have values[7] in excess of 500 mg/l. In general terms the saline water is moving laterally and vertically from the river in response to both local and regional hydraulic gradients. Because of the fissured nature of the Chalk, saline water may be more concentrated at particular levels in the aquifer and fresh water can occur below saline layers.

The problem of saline intrusion along the Mersey was commented upon as long ago as the middle of the nineteenth century[119,120]. Since that time, with the steady increase in abstraction alongside the river, the extent of the intrusion has increased. The Triassic sandstones along both banks of the river are affected, and locally the chloride concentration exceeds 1000 mg/l—as for example in Birkenhead, Liverpool and Widnes[62]. Hibbert[121] in a study of the south bank of the river, in the Wirral Peninsula, pointed out that the Ghyben–Herzberg principle did not apply, the saline interface being considerably below the level that would be anticipated from an application of the equation. This also applies on the north bank of the river[62,66].

Saline intrusion has occurred locally where the Chalk of the South Downs forms the coastline in southern England *(Figure 5.14)*[122]. Chloride values of more than 1000 mg/l were recorded in one well near Brighton, situated 2650 m from the coast, at the end of the prolonged dry summer of 1949[123].

Over the last twenty years the abstraction from individual wells in this region has been related to seasonal variations in the amount of water stored naturally in the Chalk. This has involved pumping during the winter from wells situated near the coast or rivers, to reduce the natural loss of water from the aquifer when this is high, and pumping from inland wells during the summer while at the same time limiting abstraction from coastal wells. During the summer, abstraction from coastal wells is controlled by the chloride content[124]. This approach has increased the volume of water storage in the aquifer as a whole and considerably increased the annual volume of water that may be pumped from individual wells.

Abstraction of groundwater from a coastal aquifer reduces the freshwater head, and therefore the flow of freshwater to the sea, thereby causing an inland movement of the saline interface. Aquifer management is concerned with deciding upon the ultimate landward extent of the saline water and calculating the amount of natural discharge necessary to maintain the interface in this predetermined position. The difference between this discharge and the infiltration represents the amount of water that can be developed. The wells used for development are sited inland of the ultimate position of the toe of the interface.

As an aquifer is steadily developed groundwater levels decline and the average volume of fresh water in storage is permanently reduced. When this occurs in a coastal aquifer the decline of groundwater levels results in the inland movement of the saline interface, fresh water being displaced from the aquifer and replaced with saline water *(Figure 5.12)*. The fresh water displaced by the movement of the interface represents a considerable volume, which escapes to the sea unless intercepted by shallow collector wells or drains near the coast[116]. Investigations have indicated that part of the flow to the sea, which is necessary to maintain the position of the interface, can also be intercepted by coastal collector systems without extending the saline zone inland[125].

Saline intrusion in a fissured aquifer may be more complicated than in a homogeneous aquifer. For example, in the Chalk fissure density is greatest along the lines of narrow valleys, and these zones of high transmissivity are potential areas for saline intrusion. The extent of saline intrusion in individual fissure systems in contact with the sea depends on the shape of the fissure in vertical section and the relative heads of fresh water and sea water in the fissure system.

The various methods available for controlling sea-water intrusion have been summarised by Todd[126] as follows:

1. Reduce and/or rearrange the pumping pattern.
2. Recharge artificially to supplement natural resources.
3. Develop a pumping trough parallel to and adjacent to the coast.
4. Maintain a fresh-water ridge, above sea-level and parallel to the coast, by artificial recharge.
5. Construct an artificial sub-surface barrier.

As already mentioned the third method is generally the unintentional factor limiting more extensive intrusion in many situations in England.

REFERENCES

1. Water Resources Board, *Third annual report,* HMSO, 58 (1966)
2. Scottish Development Department, *A measure of plenty; water resources in Scotland—a general survey,* HMSO, 99 (1973)
3. Earp, J. R. and Eden, R. A., 'Amounts and distribution of underground water in Scotland,' *Water and Water Engng,* **65,** 255–259 (1961)
4. Manning, P. I., 'The development of the water resources of Northern Ireland, progress towards integration', *Q. J. Engng Geol.,* **4,** 335–352 (1971)
5. Land, D. H., 'Hydrogeology of the Bunter Sandstone in Nottinghamshire', *Hydrogeological Report No. 1,* Geol. Surv. Gt Brit., 38 (1966)
6. Downing, R. A. and Williams, B. P. J., 'Groundwater hydrology of the Lincolnshire Limestone', *Pub. No. 9,* Water Resources Board, Reading, 160 (1969)
7. *The hydrogeology of the London Basin,* Water Resources Board, Reading, 139 (1972)
8. Land, D. H., 'The hydrogeology of the Triassic sandstones in the Birmingham–Lichfield district', *Hydrogeological Report No 2,* Geol. Surv. Gt Brit., 30 (1966)
9. Gray, D. A., Allender, R. and Lovelock, P. E. R., 'The groundwater hydrology of the Yorkshire Ouse River Basin', *Hydrogeological Report No 4,* Inst. Geol. Sci., 40 (1969)
10. Ineson, J. and Downing, R. A., 'Some hydrogeological factors in permeable catchment studies', *J. Instn Water Engrs,* **19,** 59–80 (1965).
11. *First Periodical Survey,* Hampshire River Authority, Winchester, 45 (no date)

12. Skeat, W. O. and Dangerfield, B. J. (Ed.), *Manual of British water engineering practice,* Vol. II, Ch. 2 (1969)
13. Lewis, W. V., 'Some aspects of percolation in south-east England', *Proc. Geol. Assoc.,* **54,** 171–184 (1943)
14. Lapworth, C. F., 'Percolation in the Chalk', *J. Instn Water Engrs,* **2,** 97–108 (1948)
15. Slater, R. J., in discussion on 'Hydrological Measurements', *J. Instn Water Engrs,* **7,** 251–255 (1953)
16. Theis, C. V., 'The relation between the lowering of the piezometric surface and the rate and duration of discharge of a well using groundwater storage', *Trans. Am. Geophys. Union,* **16,** 519–524 (1935)
17. Walton, W. C., *Groundwater resource evaluation,* McGraw-Hill, London, 664 (1970)
18. Ineson, J., 'Yield-depression curves of discharging wells, with particular reference to Chalk wells, and their relationship to variations in transmissibility', *J. Instn Water Engrs,* **13,** 119–163 (1959)
19. Lennox, D. H., 'Analysis and application of step-drawdown test', *J. Hydraulics Div., Am. Soc. Civil Engrs,* **92,** 25–48 (1966)
20. Jacob, C. E., 'Drawdown test to determine effective radius of artesian well', *Trans. Am. Soc. Civil Engrs,* **112,** 1047–1070 (1947)
21. Rorabaugh, M. I., 'Graphical and theoretical analysis of step-drawdown test of artesian well', *Proc. Am. Soc. Civil Engrs, Sep No. 362,* **79,** 1–23 (1953)
22. *Great Ouse Groundwater Pilot Scheme—Final Report,* Great Ouse River Authority, Cambridge, 103 (1972)
23. Backshall, W. F., Downing, R. A. and Law, F. M., 'Great Ouse Groundwater Study', *Water and Water Engng,* **76,** 215–223 (1972)
24. *Groundwater and wells,* ed. J. Johnson Inc., St Paul, Minnesota, 440 (1966)
25. Monkhouse, R. A., 'The use and design of sand screens and filter packs for abstraction wells', *Water Services,* **78,** 160–163 (1974)
26. Gray, D. A. and Tate, T. K., 'Hydrogeological significance of borehole investigations', *Hydrology Group Discussion,* Instn Civil Engrs (1972)
27. Batt, L. S. and Brereton, N. R., 'The use of well-logging techniques in water resources development', *Trans. Second Ann. Symp. Soc. pour l'avancement de l'interpretation des diagraphires,* Paris (1973)
28. Gray, D. A., 'The stratigraphical significance of electrical resistivity marker bands in the Cretaceous strata of the Leatherhead (Fetcham Mill) borehole, Surrey', *Bull. Geol. Surv. Gt Brit.,* **23,** 65–114 (1965)
29. *Nuclear well logging in hydrology,* Int. Atomic Energy Agency, Vienna, 90 (1971)
30. Ineson, J. and Gray, D. A., 'Electrical investigations of borehole fluids', *J. Hydrology,* **1,** 204–218 (1963)
31. Downing, R. A. and Howitt, F., 'Saline groundwaters in the Carboniferous rocks of the English East Midlands in relation to the geology', *Q. J. Engng Geol.,* **1,** 241–269 (1969)
32. Bullerwell, W., in Worssam, B. C. and Ivimey-Cook, H. C., 'The stratigraphy of the Geological Survey borehole at Warlingham, Surrey', *Bull. Geol. Surv. Gt Brit. No 36* (1971)
33. Gray, D. A., 'Instrumentation in groundwater studies', *Water and Water Engng,* **68,** 185–188 (1964)
34. Patten, E. P. and Bennett, G. D., 'Methods of flow measurement in well bores', *US Geol. Surv., Water Supply Paper 1544C* (1962)
35. Tate, T. K., Robertson, A. S. and Gray, D. A., 'The hydrogeological investigation of fissure-flow by borehole logging techniques', *Q. J. Engng Geol.,* **2,** 195–215 (1970)
36. *Log interpretation: Vol. 1 Principles,* Schlumberger, New York, 112 (1972)
37. Gray, D. A., with appendix by Tate, T. K., 'The measurement of groundwater levels', *Water and Water Engng,* **65,** 431–437 (1961)
38. Manufactured by Sparling Envirotech Ltd, Burgess Hill, Sussex.
39. Toth, J., 'A theoretical analysis of groundwater flow in small drainage basins', *J. Geophys. Research,* **68,** 4795–4812 (1963)
40. Freeze, R. A. and Witherspoon, P. A., 'Theoretical analysis of regional groundwater flow: 1 Analytical and numerical solutions to the mathematical model', *Water Resources Research,* **2,** 641–656 (1956); '2 Effect of water table configuration and sub-surface permeability variation', *Water Resources Research,* **3,** 623–634 (1967)
41. Hitchon, B., 'Fluid flow in the Western Canada sedimentary basin', *Water Resources Research,* **5,** 186–195 and 460–469 (1969)

42. Black, M., 'The constitution of the Chalk', *Proc. Geol. Soc. Lond.* No. 1499, 81–86 (1953)

43. Young, B. R., in Gray, D. A., 'The stratigraphical significance of electrical resistivity marker bands in the Cretaceous strata of the Leatherhead (Fetcham Mill) borehole, Surrey', *Bull. Geol. Surv. Gt Brit.,* **23,** 65–114 (1965)

44. Higginbottom, I. E., 'The engineering geology of Chalk', *Proc. Symp. on Chalk in Earthworks and Foundations,* Instn Civil Engrs, 1–15 (1965)

45. Hancock, J. M., 'The hardness of the Irish Chalk', *Ir. Nat. J.,* **14,** 157–164 (1963)

46. Thomson, D. Halton, 'A 100 years record of rainfall and water levels in the Chalk at Chilgrove, West Sussex', *Trans. Instn Water Engrs,* **43,** 154–181 (1938); 'Hydrological conditions in the Chalk at Compton, West Sussex', *Trans. Instn Water Engrs,* **26,** 228–261 (1921); **36,** 176–206 (1931); *J. Instn Water Engrs,* **1,** 39–68 (1947) and **8,** 138–144 (1954)

47. Headworth, H. G., 'The analysis of natural groundwater level fluctuations in the Chalk of Hampshire', *J. Instn Water Engrs,* **26,** 107–124 (1972)

48. *Report on the Lambourn Valley pilot scheme (1967–1969),* Thames Conservancy, Reading, 172 (1972)

49. Binnie and Partners, *Water resources of the Great Ouse Basin,* Ministry of Housing and Local Govt, London, 167 (1965)

50. Wright, C. E., *Combined use of surface and groundwater in the Ely Ouse and Nar catchments,* Water Resources Board, Reading, 43 (1974)

51. Edmunds, W. M., Lovelock, P. E. R. and Gray, D. A., 'Interstitial water chemistry and aquifer properties in the Upper and Middle Chalk of Berkshire, England', *J. Hydrology,* **19,** 21–31 (1973)

52. Ineson, J., 'A hydrogeological study of the permeability of the Chalk', *J. Instn Water Engrs,* **16,** 449–463 (1962)

53. Woodland, A. W., 'Water supply from underground sources of Cambridge–Ipswich District', *Wartime Pamphlet No. 20 Pt 10,* Geol. Surv. Gt Brit. (1946)

54. Headworth, H. G., personal communication

55. *Great Ouse Pilot Scheme, Fourth Progress Report for 1971,* Great Ouse River Authority, Cambridge, 48 (1972)

56. Brown, E. O. F., 'Underground waters in the Kent coalfield', *Proc. Instn Civil Engrs,* **215,** 27–65 (1923)

57. Plumptre, J. H., 'Underground waters of the Kent coalfield', *Trans. Instn Min. Engrs,* **119,** 155–164 (1959)

58. Harold, C. H. H., *Thirty-second report on the results of the bacteriological, chemical and biological examination of the London waters for the year 1937,* Metropolitan Water Board, London, 104 (1938)

59. Williams, B. P. J., Downing, R. A. and Lovelock, P. E. R., 'Aquifer properties of the Bunter Sandstone in Nottinghamshire, England', *24th Int. Geol. Cong., Montreal,* Section 11, 169–176 (1972)

60. *Groundwater resources of the Vale of Clwyd,* Water Resources Board, Reading, 58 (1973)

61. Bow, C. J., Howell, F. T. and Thompson, P. J., 'Permeability and porosity of unfissured samples of Bunter and Keuper Sandstones of south Lancashire and north Cheshire', *Water and Water Engng,* **74,** 464–466 (1970)

62. *First Periodical Survey,* Mersey and Weaver River Authority, Warrington, 284 (1969)

63. Sherrell, F. W., 'Some aspects of the Triassic aquifer in east Devon and west Somerset', *Q. J. Engng Geol.,* **2,** 255–286 (1970)

64. Lovelock, P. E. R., *Aquifer properties of the Permo-Triassic sandstones of the United Kingdom,* PhD thesis, University of London (1972)

65. *Development of groundwater in the Triassic sandstones of the Fylde for combined use with surface resources,* Water Resources Board, Reading (in press)

66. Crook, J. M. and Howell, F. T., 'The characteristics and structure of the Permo-Triassic sandstone aquifer of the Liverpool and Manchester industrial region of north-west England', *International Symp. on Groundwater,* Palermo (1970)

67. Downing, R. A., Land, D. H., Allender, R., Lovelock, P. E. R. and Bridge, L. R., 'The hydrogeology of the Trent River Basin', *Hydrogeological Report No. 5,* Inst. Geol. Sci., 104 (1970)

68. Monkhouse, R. A., *An assessment of the groundwater resources of the Lower Greensand in the Cambridge–Bedford region,* Water Resources Board, Reading (1974)

69. Mather, J. D., Gray, D. A., Allen, R. A. and Smith, D. B., 'Groundwater recharge in the Lower Greensand of the London Basin—results of tritium and carbon-14 determinations', *Q. J. Engng Geol.,* **6,** 141–152 (1973)

70. Bennison, G. M. and Wright, A. E., *The geological history of the British Isles,* Edward Arnold, London, 406 (1969)

71. Burton, A. R., 'Hydrological study of the Latton groundwater source', *J. Instn Water Engrs,* **22,** 287–293 (1968)

72. Anderson, W., 'Water supply from underground sources of north-east England', *Wartime Pamphlet No. 19 Pt III,* Geol. Surv. Gt Brit., 17 (1945)

73. Cairney, T., 'Hydrological investigation of the Magnesian Limestone of south-east Durham', *J. Hydrology,* **16,** 323–340 (1972)

74. Sweeting, M. M., *Karst landforms,* Macmillan, London, 362 (1972)

75. Sweeting, M. M., 'Karst of Great Britain', in *Karst, important Karst regions of the Northern Hemisphere* (ed. M. Herak and V. T. Stringfield), Elsevier, Amsterdam, 565 (1972)

76. Eyre, J. and Ashmead, P., 'Lancaster Hole and the Ease Gill Caverns, Casterton Fell, Westmorland', *Trans. Cave Res. Group Gt Brit.,* **9,** 65–123 (1967)

77. Ford, D. C., 'The origin of limestone caverns: a model from the Central Mendip Hills, England', *Bull. Nat. Speleological Soc. America,* **27,** 109–132 (1965)

78. Smith, D. I., Atkinson, T. C. and Drew, D. P., *Hydrology of limestone terrains,* David & Charles, Newton Abbot (in press)

79. Carter, W. L. and Dwerryhouse, A. R., 'The underground waters of north-west Yorkshire', Pt I, *Proc. Yorkshire Geol. Polytech. Soc.,* **14,** 1–18 (1904); Pt II, *ibid,* **15,** 248–292 (1905)

80. Atkinson, T. C., Smith, D. L., Lavis, J. J. and Whitaker, R. J., 'Experiments in tracing underground waters in limestones', *J. Hydrology,* **19,** 323–349 (1973)

81. Drew, D. P., 'Limestone solution within the east Mendip area, Somerset', *Trans. Cave Res. Group of Gt Brit.,* **12,** 259–270 (1970)

82. Tratman, E. K., 'The hydrology of the Burrington Area, Somerset,' *Proc. Univ. Brist. Speleological Soc.,* **10,** 22–57 (1963)

83. Drew, D. P., Newson, M. D. and Smith, D. I., 'Mendip Karst hydrology research project, phase three', *Occ. Pub. Wessex Cave Club,* ser. 2, 28 (1968)

84. Atkinson, T. C., 'The dangers of pollution of limestone aquifers, with special reference to the Mendip Hills, Somerset', *Proc. Univ. Brist. Speleological Soc.,* **12,** 281–290 (1971)

85. Newson, M. D., 'A model of subterranean limestone solution in the British Isles', *Trans. Inst. Brit. Geographers,* **54,** 55–70 (1971)

86. Atkinson, T. C., Bradshaw, R. and Smith, D. I., *Quarrying in Somerset,* Somerset County Council, Taunton, 58 (1973)

87. Edmunds, W. M., 'Hydrogeochemistry of groundwaters in the Derbyshire Dome with special reference to trace constituents', *Rep. No. 71/7,* Inst. Geol. Sci., 52 (1971)

88. Drew, D. P., Newson, M. D. and Smith, D. I., 'Water-tracing of the Severn Tunnel Great Spring', *Proc. Univ. Bristol Speleological Soc.,* **12,** 203–212 (1970)

89. Rhoades, R. and Sinacori, M. N., 'Pattern of groundwater flow and solution', *J. Geol.,* **49,** 785–794 (1941)

90. Hubbert, M. K., 'The theory of groundwater motion', *J. Geol.* **48,** 785–944 (1940)

91. Dunham, K. C., 'Geology of the Northern Pennine Orefield, Vol I—Tyne to Stainmore', *Mem. Geol. Surv. UK,* 357 (1948)

92. Kennard, M. F., and Knill, J. L., 'Reservoirs on limestone, with particular reference to the Cow Green Scheme', *J. Instn Water Engrs,* **23,** 87–113 (1969)

93. Hodgson, A. V. and Gardiner, M. D., 'An investigation of the aquifer potential of the Fell Sandstone of Northumberland', *Q. J. Engng Geol.,* **4,** 91–109 (1971)

94. Broadhead, J. A. and Mackey, P. G., 'Use of Trent alluvial gravels', *Proc. Symp. on Advanced Techniques in River Basin Management: Trent Model Research Programme,* Instn Water Engrs (1972)

95. Glossop, R. and Collingridge, V. H., 'Notes on groundwater lowering by means of filter wells', *Proc. Second Int. Conf. Soil Mechanics and Foundation Engineering,* **II,** 320–322 (1948)

96. Naylor, J. A., *The groundwater resources of the river gravels of the Middle Thames Valley,* Water Resources Board, Reading (1974)

97. Gray, D. A. and Foster, S. S. D., 'Urban influences upon groundwater conditions in

Thames Flood Plain deposits of central London', *Phil. Trans. R. Soc. Lond., A,* **272,** 245–257 (1972)

98. Price, M. and Foster, S. S. D., 'Water supplies from Ulster valley gravels', *Proc. Instn Civil Engrs,* **57,** Pt 2, 451–466 (1974)

99. Downing, R. A., 'A note on the Crag in Norfolk', *Geol. Mag.,* **96,** 81–86 (1959)

100. Ineson, J., 'Groundwater conditions in the Coal Measures of the South Wales Coalfield', *Hydrogeological Report No. 3,* Geol. Surv. Gt Brit., 69 (1967)

101. Mather, J. D., Gray, D. A. and Jenkins, D. G., 'The use of tracers to investigate the relationship between mining subsidence and groundwater occurrence at Aberfan, South Wales', *J. Hydrology,* **9,** 136–154 (1969)

102. Earp, J. R., 'The geology of the Bowland Forest Tunnel, Lancashire', *Bull. Geol. Surv. Gt Brit.,* 1–12 (1955)

103. Rowe, P. W., 'The relevance of soil fabric to site investigation practice', *Geotechnique,* **22,** 195–300 (1972)

104. Oakes, D. B. and Wilkinson, W. B., 'Modelling of groundwater and surface water systems, I Theoretical relationships between groundwater abstraction and base flow', *Pub. No. 16,* Water Resources Board, Reading, 37 (1972)

105. Downing, R. A., Oakes, D. B., Wilkinson, W. B. and Wright, C. E., 'Regional development of groundwater resources in combination with surface water', *J. Hydrology,* **22,** 155–177 (1974)

106. Ineson, J. and Downing, R. A., 'The groundwater component of river discharge and its relationship to hydrogeology', *J. Instn Water Engrs,* **18,** 519–541 (1964)

107. Tomlinson, M. J., *Foundation design and construction,* Sir Isaac Pitman and Sons Ltd, 749 (1963)

108. Lofgren, B. E., 'Measurement of compaction of aquifer systems in areas of land subsidence', *US Geol. Surv. Prof. Paper 424-B,* 49–52 (1961)

109. Longfield, T. E., 'The subsidence of London', *Ordnance Surv. Prof. Paper, New Series, No. 14* (1932)

110. Wilson, G., and Grace, H., 'The settlement of London due to under-drainage of the London Clay', *J. Instn Civil Engrs,* **19,** 100–127 (1942)

111. Terzaghi, K., *Theoretical soil mechanics,* J. Wiley and Sons, London, 510 (1943)

112. Jacob, C. E., 'On the flow of water in an elastic artesian aquifer', *Trans. Am. Geophys. Union 21st Ann. Mtg,* Pt 2, 574–586 (1940)

113. Poland, J. F., 'The coefficient of storage in a region of major subsidence caused by compaction of an aquifer system', *US Geol. Surv. Prof. Paper 424-B,* 52–54 (1961)

114. Foose, R. M., 'Surface subsidence and collapse caused by groundwater withdrawal in carbonate rock areas', *23rd Int. Geol. Cong., Prague,* **12,** 155–166 (1968)

115. Jennings, J. E., Brink, A. B. A., Louw, A. and Gowan, G. D., 'Sinkholes and subsidences in the Transvaal dolomite of South Africa', *Proc. Sixth Int. Conf. on Soil Mechs and Foundation Engng,* 51–54 (1965)

116. Santing, G., 'The groundwater in the Coastal Plain as a source for water supply', *Pub. No. 23,* Water Planning for Israel Ltd, Tel Aviv (1957)

117. Cooper, H. H., 'A hypothesis concerning the dynamic balance of fresh water and salt water in a coastal aquifer', *J. Geophys. Research,* **64,** 461–467 (1959)

118. Gray, D. A., 'Groundwater conditions of the Chalk of the Grimsby area, Lincolnshire', *Research Report No. 1, Water Supply Papers,* Geol. Surv. Gt Brit. (1964)

119. Stephenson, R., *Report on the supply of water to the town of Liverpool* (1850)

120. Braithwaite, F., 'On the infiltration of salt water into the springs of wells under London and Liverpool', *Proc. Instn Civil Engrs,* **14,** 507–509 (1855)

121. Hibbert, E. S., 'The hydrogeology of the Wirral peninsula', *J. Instn Water Engrs,* **10,** 441–469 (1956)

122. Green, F. N., 'A method of water conservation and automation', *Proc. Soc. Water Treat. Exam.,* **13,** 4–6 (1964)

123. Warren, S. C., 'Some notes on an investigation into seawater infiltration', *Proc. Soc. Water Treat. Exam.,* **11,** 38–42 (1962)

124. Warren, S. C., 'Chemical aspects of controlled pumping and automation', *Proc. Soc. Water Treat. Exam.,* **13,** 7–11 (1964)

125. *Experimental coastal groundwater collectors—Israel,* FAO, United Nations, Rome, 128 (1968)

126. Todd, D. K., *Groundwater hydrology,* J. Wiley & Sons Inc., New York, 336 (1959)

Chapter 6

Surface Water

Measurement techniques

The earliest published measurements of the natural flow of a British river are probably those of J. F. La Trobe Bateman[1]. These measurements were made during the construction of Manchester Corporation's reservoirs in the Longdendale catchment in 1833 and the years following. Sharp-edged weir plates were used and several notable floods were recorded. In the years that followed increased understanding of hydraulics enabled more complex weirs and sluices to be rated and employed for river flow measurement. The longest extant record is that from 1851 at Feildes Weir on the River Lee[2]. This weir gauges a catchment of 1040 km² immediately to the north of London, but the intensive urbanisation and water supply development that have taken place during the last century have made the record non-homogeneous. Many long reservoir records exist that do not have this complication; they have been worked up from weir measurements of compensation discharges and flood spills, metered drawoffs and changes in storage[3-7]. At least eleven such records date back to before 1921 while many more records owe their origin to the stimulus provided by the 1933–34 drought.

Limited records of doubtful accuracy exist for the Thames[8] for various periods during the first part of the Victorian era. Then in the early 1880s the Teddington sluices were rated and since 1883 have provided a continuous record at the tidal limit[9]. The Severn is the other major river to have been gauged for many decades.

Early flow measurements were rarely published except in evidence to Parliament or in the occasional paper to the Institution of Civil Engineers. An unfortunate practice grew up whereby consulting engineers would keep their own measurements to themselves and it was left to a private individual, Captain W. N. McClean, to pioneer hydrometric publications. His organisation, River Flow Records, made available a widespread series of current-meter gaugings of Scottish rivers beginning in the 1920s[10,11]. Growing interest, encouraged by the British Association, led to the Government sponsoring the Inland Water Survey in 1935, while publication of the *Surface Water Yearbook of Great Britain* began at the same time (the year book is published by HMSO for the Department of the Environment and the Scottish Development Department). Some catchment boards, such as the Nene[12], took particular interest in gauging and the number of stations began to grow. Records for 28 stations were published in 1935, but the *Surface Water Year Book* for 1966–70 contains details of 782 stations for which records are available, reflecting the increasing responsibilities of the river boards, then the river authorities and now the regional water authorities. Taking the UK as a whole, the density of the river-gauging network is about one station per 300 km^2, considerable variations in density occurring from one area to another. A considerable growth in the network took place from 1965 onwards, largely as a result of the 'hydrometric schemes' that could be established under the terms of the 1963 Water Resources Act. *Table 6.1* shows the numbers of different types of gauging station that existed in 1966.

Table 6.1 SUMMARY OF BRITISH GAUGING STATIONS IN 1966

Type of structure	Number of stations	Average area of basin (km²)
Flat-V weir	44	120
Critical-depth flume	66	122
Crump weir (with or without crest tappings)	177	154
Sharp-edged weir	50	178
Broad-crested weir	49	412
Rated river section	286	675
Calibrated sluices	15	2160
Other types including compound structures	95	—

Increasingly records from the river gauging network have been processed by a central authority and published in the *Surface Water Year Book*. The practice up to 1965 was to provide a table of monthly flow data for each station, but subsequent year books are arranged differently. A list of all gauging stations for which records are held is presented, together with their characteristics. Tables and graphs are given for selected stations, and a list of the programs by which data may be retrieved from the computer-based archive. Now that most of the records are stored on magnetic tape, they can be processed by the computer in a variety of ways as the need arises. The responsibility for collecting, processing and storing river-flow records on a national basis belongs to the Water Data Unit. Some river authorities published their own hydrometric yearbooks and this is a practice taken up by the water authorities. Records from the Trent at Colwick, the Severn at Bewdley and the Thames at Teddington appear in the UNESCO/IHD publication *Discharge of Selected Rivers of the World*.

Records of some gauging stations are in private hands, or are held by an authority not normally considered to be connected with hydrometry. Thus it is necessary to remember local authorities, water companies, research establishments, the British Waterways Board and other bodies with some interest in water before all the possibilities for data location are exhausted.

Water level measurement

At early gauging stations water levels were often read only once a day on a staff gauge. Later float-operated water level recorders, normally with a weekly chart, were introduced[13], only to be superseded at many sites by the patent Lea flow recorder. This recorder gives a direct chart reading of flow for a structure with a known rating formula by the use of a spirally engraved drum mechanism. Many waterworks installed this type of recorder; they frequently had integrators attached to eliminate the chore of calculating total daily flow by hand from an ever-fluctuating hydrograph. However, this militated against revising the flow formula. The practice today is to install the type of instrument that records level on a chart with as open a time-scale and as wide a level-range as possible. Chart recorders are supplemented at many sites by the type of recorder that punches water level on paper tape at 15-minute intervals. The Fischer & Porter punched-tape water level recorders operate on 16-track tape, which has to be translated into 5- or 8-track paper tape for use in a computer. The Ott punched-tape water level recorder uses computer-compatible tape and this avoids the need for translation. Translation, however, provides an opportunity for editing tapes and correcting the faults and errors that occur. The Water Data Unit provides a translation service for punched tape recorders; between 400 and 500 tapes a month were being handled in 1976. Some authorities prefer to digitise levels recorded on their chart recorders and to feed this information into a computer to calculate discharges.

Both analogue and digital recorders are normally installed over a common stilling well in such a way that their floats can move over the range of expected river levels. Many early stilling wells were connected to the river by a single pipe of up to 150 mm diameter and, as a consequence, wave fluctuations sometimes appear on the record. These fluctuations are damped out by the use of a small-bore connection; to prevent this connection blocking with sediment[14], one or more additional pipes can be laid into the well at different levels. To check that the stilling well faithfully follows variations in river level, it is common practice to install an internal staff gauge, or to use an electrical zeroing device[15] to determine the depth to water from the recorder datum. The float is connected to the recorder gears by a graduated steel tape, which can also be read and compared with a staff set in the river or on the river bank and levelled in to the same datum.

Devices other than those relying on the movement of a float have been employed[16] to record water level at certain sites where there are problems in installing the conventional system, for example the ultrasonic gauge, the electrical resistance gauge, the purge-bubble gauge and the load cell.

Although opinion favours the recording of river level as frequently as possible, one report[17] has shown that for assessing monthly and annual flows negligible errors would arise from recording water level at eight-hourly intervals. Some two years of 15-minute records from thirteen contrasting catchments were employed in this study, which also demonstrated that if all stations had been sampled at 30-minute intervals and the majority at one-hour intervals, the error would only amount to 0.25 per cent of the daily mean discharge. This could be considered as unnecessarily accurate when the larger errors from other sources are taken into account.

Structures

Probably no other country has such a variety of structures employed for river flow measurement, a variety that results in part from the many different agencies involved in their design and construction. Structures are suitable for the smaller rivers on economic and practical grounds. They are particularly suited to the many reservoir sites where flows are required to be recorded by law. Existing structures may be calibrated by hydraulic model tests: where the design of a new structure is well-founded in theory the calibration may also be obtained from that theory. *Table 6.2* lists the formulae for determining flow for various types of gauging structure.

Recurrent problems at structures are silting of weir pools (creating high, asymmetrical approach velocities), drowning-out at high flows (especially

Table 6.2 GAUGING STRUCTURES

Type	Reference	Formula for flow, Q	Variants
Sharp-edged weirs			
(a) V-notch	18	$\frac{8}{15} \cdot (2g)^{\frac{1}{2}} \cdot C_d \cdot \tan\frac{\theta}{2} \cdot H^{5/2}$	
(b) rectangular notch	19	$\frac{2}{3} \cdot (2g)^{\frac{1}{2}} \cdot C_d \cdot b \cdot h_g^{3/2}$	With side contractions[20]
Broad-crested weirs	21	$\frac{2}{3} \cdot \left(\frac{2g}{3}\right)^{\frac{1}{2}} \cdot (B_t - 2\delta_*)$ $\cdot (H - \delta_*)^{5/2}$	Square-nosed[22]
Triangular-profile weirs			
(a) modular Crump	23, 24	$1.96g^{\frac{1}{2}} \cdot b \cdot H^{3/2}$	Truncated weirs[25]
(b) non-modular Crump	25	As above, times reduction factor for submergence	
(c) flat-V	26	$\frac{4}{5} \cdot C_{De} \cdot g^{\frac{1}{2}} \cdot n \cdot H^{5/2}$	1 : 2 u/s and d/s slopes
Villemonte weirs	27	Varies with throat, sill, and culvert entrance control	
Critical-depth flumes			
(a) triangular	28	$C_s \cdot C_v \cdot \frac{2}{3} \cdot \left(\frac{2g}{3}\right)^{\frac{1}{2}} \cdot b_e \cdot h_e^{3/2}$	Glass-fibre moulded[29]
(b) trapezoidal	28		Double or Butterfly[30]
(c) venturi: bottom contracted	31		Side contracted, Parshall[32] and/or non-modular
(d) Plynlimon drop type	33	$\frac{2}{3} \cdot \left(\frac{2g}{3}\right)^{\frac{1}{2}} \cdot B_t -$ $2\delta_*\left[\left(H - \frac{6t}{4m_t}\right) - \delta_*\left(\frac{(1+m_t)^2}{m_t} - \frac{1}{2m_t}\right)\right]^{\frac{3}{2}}$	

where θ = angle included between sides of notch, e.g. 90°
 b = notch width
 h_g = gauged head +0.0012 m
 n = crest cross slope (1 vertical: n horizontal)
 C_d or C_{De} = discharge coefficients (see reference 34)
 C_s = throat-shape coefficient
 C_v = coefficient of velocity of approach
 h_e = effective measured head relative to invert
 b_e = bed width
 g = gravitational acceleration
 H = total head of water
 B_t = water surface width in throat
 δ_* = boundary layer displacement thickness
 m_t = side slope of triangular floor of throat

with channel weed growth) and non-standard construction. Different situations have produced very different solutions, as the following examples show.

1. A Villemonte weir[27] has been used in a road embankment box culvert on the Gallica stream near Yeovil. The control passes from the weir sills to the culvert entrance in floods. It causes no loss of culvert capacity and allows debris to wash through.
2. A variety of 'butterfly' flumes[30] have been used in Essex to ensure sensitivity over a wide flow range.
3. A modified flat-V weir[35] has been modelled and built in the Dunsop valley to record up to the extreme of a 'catastrophic' flood. The flow 'hurdles' the structure in such an event.
4. Drop flumes[33] have been installed at six sites at Plynlimon, central Wales, to cope with high Froude numbers and prevent sediment from depositing.

Rectangular notch weirs, made of steel plate and with a machined crest, are still employed for temporary low-flow gauging. Their main use has been in the gauging of compensation flows and spills below dams, because at such sites there is no sediment deposition and it is possible to ensure that the notch is not drowned. Early hydraulic model tests[20] meant that weir coefficients were accurately known.

In recent years the Crump weir[23] has been widely employed because of its stable discharge coefficient right up to 75 per cent submergence (i.e. downstream head above the crest divided by upstream head above the crest = 0.75). To improve the sensitivity of a Crump weir the crest can be compounded by constructing a central low-flow section. Where sensitivity is paramount, or migratory fish must pass the gauging site without difficulty, the flat-V weir is normally adopted; it is more difficult, however, to build and requires a special rating above the 'notch-full' level.

Flumes of various types are used where head-drop across the structure must be minimised and sediment load is high *(Figure 6.1)*. However, flume capacities are usually small *(Table 6.1)* because of the expense involved in constructing the larger ones. Harrison[34] has discussed the criteria of

Figure 6.1   Critical depth flume, Stonebridge, Norfolk

accuracy, head loss and cost that are so important in choosing the best structure, cost often being paramount. He showed that, for a rectangular gauging section, the volume of a structure is a function of the Froude number and the discharge coefficient. Because the Froude number rises as the stream slope steepens, so accurate gauging becomes more difficult and expensive.

Velocity-area stations

The first current-meter gauging appears to have been carried out on the Severn at Worcester, and at Vyrnwy at the site of the present dam. Deacon, who was responsible for these gaugings (which were commenced in 1878 and were carried on for about eleven years), made a thorough study of the velocity profile at many verticals across the channel at each site[36] to establish the mean velocity. He also made use of an electrically operated level recorder in a well connected by a pipe to a sump in the river bed. When he required, at Vyrnwy, more sensitive readings at low flows than a current meter could give, he installed a temporary sharp-edged weir.

Since Captain McClean's privately financed work on the rivers Garry and Ness[37] and Dixon's gauging of the Severn in 1921[38], most velocity-area stations, at least in England and Wales, have been operated by catchment boards and their successors. A number of velocity-area stations had been established on the Wye and on the Nene by 1939[39,40], but the majority of stations of this type were installed in the late 1950s and early 1960s. Both the method of using the equipment and the analysis of the observations have largely followed American practice. Cup and propeller meters are used, either on wading rods or suspended from a cableway across the channel. One British development has been a low-cost current meter, the Braystoke, embodying a plastic, instead of stainless steel, impeller. Current meter calibration tanks exist at the Hydraulics Research Station, Wallingford, and at the British Hovercraft Corporation.

Sites chosen for the use of the velocity area method have generally been on the lower reaches of rivers draining catchments of 250–2500 km² in area, where extensive flood plains exist. Occasionally the river has been constrained between flood banks to stop the station being outflanked, but normally access difficulties have prevented the proving of stage–discharge relationships above bank-full level. The uncertainty of extrapolating measured relationships to greater depths of water can lead to repeated revisions of flow records and a consequent lack of confidence in such records. Thus, wherever a structure was a feasible alternative to a velocity-area station, a structure has usually been installed. A study carried out at Southampton University[41] on gauging costs showed that a velocity-area station, at 1970 prices, would require an expenditure of about £4000 on capital works almost regardless of the size of the river. Maintenance costs were put at £50 per annum, chart changing at £100 per annum and current metering at £100 per annum. No costs were put to the analytical time spent on reviewing the ratings at stations with a shifting bed, and of course this is an important problem at a number of sites.

Although at one time mean velocity in the cross-section would be computed from readings at 0.2 and 0.8 times river depth, there seems to have been a swing towards making observations at a single depth. This single-point method is based on the assumption that a reading at 0.6 depth, or at 0.5 depth with the application of a correction factor, will give a measurement of mean velocity with a standard error that is acceptable for most purposes. Lambert[42], in a study of observations from three rivers, derived standard errors for single observations (as a measurement of mean velocity in a single vertical) as follows:

> 0.2/0.8 depth : 2.7%
> 0.6 depth : 4.2%
> 0.5 depth : 3.3% (using measured velocity × 0.95)

He concluded that, provided a reasonable number of verticals were taken across the width of the cross-section, the standard error derived from the use of single-point observations would be small compared with other errors, and that the 0.5 depth method was to be preferred to the 0.6 depth because of its smaller standard error and its simplicity of calculation in the field.

The Yorkshire River Authority[43] has carried out simultaneous metering at several positions in a gauging section and it was discovered that in irregularly shaped channels distinct pulses of higher velocity occurred. The conclusion was that current metering should be carried out over a sufficiently long period to smooth out any pulsations that might occur. British Standards have been published on many aspects of current metering procedure[44].

Ultrasonic gauging

This method depends upon the measurement of the velocity of flow at a certain depth by simultaneously transmitting sound pulses through the water from transducers sited on either side of the channel. The transducers, which both transmit and receive, are located diagonally opposite one another, so that the pulses travel with the flow in one direction and against it in the other, usually at angles of about 30° and 60° respectively. The difference between the travel time of the pulses moving upstream and that of those moving downstream is directly related to the average velocity at the depth of the transducers. This velocity can be related to the average velocity of flow in the whole cross-section by current metering at a large number of points. Alternatively, the transducers can be moved in a vertical plane so that the pulses are transmitted over a number of different horizontal paths and a relationship established between the velocity at one chosen level and the mean velocity for the whole cross-section.

Herschy[45] described the feasibility study of this method undertaken by the Water Resources Board and the ultrasonic gauging site established on the Thames at Sutton Courtenay by the WRB, the Atomic Energy Research Establishment, Harwell, and the Thames Conservancy. The cross-profile of the bed was determined from echo soundings checked by manual soundings, while water level was measured continuously by a resistance gauge. This information, and that derived from the ultrasonic gauging, was fed into processing and recording equipment[46] which output discharge digitally on paper tape and on chart. Measurements of velocity were made over three-minute periods every 15 minutes. The accuracy of a single measurement of discharge was estimated as 4 per cent at the 95 per cent confidence level, but when bank-full conditions are approached in the Thames recorded velocities have to be increased by 11 per cent to correspond to the mean velocity over the larger cross-section.

The ultrasonic method of gauging possesses a number of advantages that conventional approaches lack. The spatial integral of velocity provides information that could only be obtained by a large number of point measurements made simultaneously, and because the measurement is made over a very short time it can function in rapidly changing conditions of flow. It causes no obstruction in the channel and it can function whether flow is upstream or downstream in direction. Such a system is eminently suited to measurements in deep rivers with regular stable cross-sections. A number of sites have been selected where ultrasonic gauges could be installed, mainly in the lower reaches of rivers at sites where discharge has not been recorded hitherto. At some sites, where levels fluctuate over a wide range, multi-path rather than single-path gauging systems are preferred.

Electromagnetic gauging

Current meters and pipe-flow meters, whose operation is based on Faraday's Law of electromagnetic induction, have been developed in a number of countries, but the UK is one of the few where the same principle has been applied to measuring flow in rivers. For this purpose a coil, extending across the width of the channel, is located in the bed of the river and a probe is built into each bank but electrically isolated from it. The coil is shaped to produce a magnetic field of uniform strength. The interaction of the field with the water in motion sets up a potential between the two probes; the magnitude of this potential is a function of the velocity.

Tests of an electromagnetic (EM) gauge were undertaken by the Plessey Laboratory for the Water Resources Board[47], and later a prototype electromagnetic station was installed on the River Rother at Petersfield immediately below a Crump weir[48]. The configuration of both coil and probes was given considerable attention, and in addition to measurements of water level it was found that changes in the resistivity of the bed and the water itself required these factors to be measured. The discharge *(Q)* is a function of:

$$\frac{\text{probe voltage} \times \text{stage factor} \times \text{water resistivity}}{\text{coil current} \times \text{bed resistivity}}$$

Measurements of these parameters are fed into an on-site processor, and after 'noise' is compensated for, the measured discharge is output at 15-minute intervals on punched paper tape. Results for the initial trial period indicate that the EM gauge produces flow measurements that closely match those of the Crump weir[49].

Indirect flow measurements

Under this general heading artificial controls, natural controls, the slope method and reservoir records are considered.

Artificial controls

A control in a river causes a unique streamflow–water-level relationship upstream of its site; a gauging structure is specifically designed to make this relationship sensitive and regular. Many types of artificial control exist that can be utilised for flow measurement—for example, mill weirs for diverting flows into mill races, and fishing weirs. Culverts, particularly those under large railway embankments, can prove useful, as can bridge openings where the channel invert is lined with concrete, brick or masonry.

Such sites are normally insensitive at low and medium flows but they can yield valuable information in times of flood[50] when an observer has recorded the highest stage. Many British rivers have a substantial and historic weir at their tidal limit which cannot be replaced by a standard structure for financial or amenity reasons. Nevertheless, with upstream and downstream recorders adequate records can be obtained, particularly if rating by current meter is possible. Examples of this type of site include Chester weir, and Constantine weir on the Gipping at Ipswich. Reversal of flow across such a weir at high spring tides requires an open-scale record of level and an adjustment to the usual method of obtaining mean daily flows.

Natural controls

Water levels in many reaches are controlled by a submerged rock sill or a stable gravel riffle. A flood may leave a clear trash mark on the river banks upstream of the control, making it possible to estimate the peak flow. A survey of the River Lyn after the Lynmouth flood disaster in 1952 provided sufficient information[51] to build a 1:48 scale hydraulic model. The peak flow was estimated by direct simulation. In simple cases it may be possible to use hydraulic formulae as in the Hydraulics Research Station study of the Forest of Bowland flooding in 1967[52]. Straight channel sections are required for this approach, otherwise super-elevation effects on bends may cause misleading results.

The slope method

Inferring the flow in a channel from the slope of the water surface was the method employed in 1908 for the study of the Medway at Allington, Maidstone, part of the Royal Geographical Society's *Investigation of Rivers*[36]. Two Palatine weekly recording level gauges with a 1:6 scale reduction were installed in stilling wells 3048 m (10 000 feet) apart. Flow was computed by the Chezy–Kutter equation:

$$\text{Velocity} = C\sqrt{mi}$$

where m = the hydraulic radius (or mean depth) = $\dfrac{\text{cross section}}{\text{wetted perimeter}}$

N = Kutter's coefficient of rugosity

i = the water surface slope

$$C = \frac{[41.6 + (1.811/N) + (0.00281/i)]\sqrt{m}}{\sqrt{m} + N[41.6 + (0.0028\,1/i)]}$$

The cross-profile of the river was measured by soundings from a boat, while the coefficient N was determined by surface float discharge calculations as 0.03. The gauges were levelled to an accuracy of 3 mm (0.01 ft) because the majority of slope readings were 25 mm (1 in) or less in 3048 m (10 000 ft). Difficulties occurred in synchronising the recorder clocks and the information obtained was limited to the period from September 1908 to August 1910. Reverse slopes were occasionally seen due either to wind set-up or the sudden closing of sluices downstream.

Engineers have made great use of Manning's equation for uniform open-channel flow ever since it was propounded in 1890[53]:

$$V = \frac{1.486}{n} m^{2/3} i^{1/2} \text{ ft/s (retaining the original units)} \qquad (6.2)$$

where V = velocity
        m = the hydraulic mean depth (feet)
        n = a roughness constant

The appropriate British Standard suggests values for Manning's n in channels of varying type[54], subject to the proviso that the method should be avoided in channels of high curvature or of very flat slope with high sediment concentration. *Table 6.3* shows some typical values of n. The most commonly adopted figure for n for well maintained reaches and the design of new channels in lowland England is 0.025.

Table 6.3   VALUES OF MANNING'S n FOR VARIOUS CHANNELS

Earth	very uniform/good condition	{ 0.018 on completion { 0.023 after weathering
	ordinary condition, little weed	0.028
	poor condition, stone and weeds	0.030 to 0.035
Gravel	fine, well-rammed	0.019
	coarse, well-rammed	0.024
Masonry	well-laid brickwork and ashlar	0.013
	rubble masonry in cement	0.017
	brickwork/stonework in poor condition	0.017
Flood plains	short pasture	0.030
	mature crops	0.040

A detailed comparison of Manning's formula and weir measurements was made on the canalised River Bain by the Lincolnshire River Authority. In a reach 1342 m long the level at each end could be read to 0.3 mm (0.001 ft) and the depth indicators were levelled in to ±0.9 mm (0.003 ft). Measurements taken weekly since 1967 have revealed that marked rises in values of n results from weed growth. Winter values of 0.04 at 3 m³/s fall to 0.021 at channel capacity (39 m³/s), but rise to values as high as 0.6 or more in July. More representative late summer values are 0.04 to 0.08, following channel maintenance, but the fall back to winter conditions commences in August or September. Other measured values of n are[39,55]:

Towy at Tycastell	0.029
Irfon at Abernant	0.023
Wye at Cadora	0.033 ±0.003

Wye at Erwood   0.029
Wye at Rhyader   0.033 rising to 0.14 for low flows
Nar, Ter and Wallop Brook   0.035 generally near bank-full

Reservoirs

Many of the longest and most consistent records of runoff come from instrumented dam sites[4-6]; valuable information, particularly for monthly flows, can be gained from a lake water balance. The records of reservoir drawoff, releases for compensation or regulation, scour discharges and spills are adjusted for the change in storage over the time period concerned to obtain 'natural' inflow. Such inflow figures include the effects of rainfall on to, and evaporation from, the lake as well as runoff to the lake perimeter.

A few reservoirs are equipped with punched-tape water-level recorders and these records allow inflows to be calculated. Reservoir discharges and overflows are almost always measured at a gauging structure at the toe of the dam; overflow recorders exist but are less useful and frequently have too coarse a scale for accurate interpretation. This comes from the need to have long flat weir crests for flood control. Reservoir drawoff is normally measured by venturi tube on the supply main close to the dam, although more recently electromagnetic meters have been installed to do the same task. For example, the meters on the Cow Green reservoir operate by the flowing water deflecting an electromagnetic field imposed from the pipe walls[56]. These meters can only be used under pipe-full conditions and are not suitable at low velocities.

Reservoir records are still being discovered[7] and converted into records of

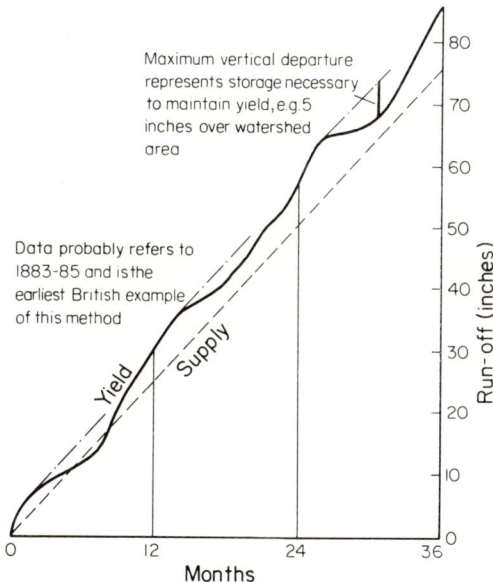

Figure 6.2   Mass flow curve for Llyn Bran, Denbighshire (after Turner[57])

flow; such records depend upon the dedication of reservoir keepers maintaining a reliable record over many decades. Occasional records, like that for Llyn Bran above Brenig reservoir, seem to get lost save for a tantalising published digest; *Figure 6.2* provides such an example[57] for a rare dry period in the nineteenth century.

Dilution gauging

Dosing flowing water with a tracer of known concentration and sampling the resulting mixture for the diluted concentration is a method of determining flow that was discovered by Schloesing[58] in France in 1863. This method reached England sometime after[59] but it has flourished for practical purposes only since 1958[60,61]. First used for measuring pipe flows, it was later applied to river-flow measurement by universities[62] and the Water Research Association[63]. The method rapidly established itself as the most practical means of calibrating gauging stations in turbulent and well mixed rivers where catchment areas did not greatly exceed 100 km[2]. The Water Resources Board sponsored a mobile team[64] for carrying out such work to consistent standards; research into the use of dilution gauging now continues at the Institute of Hydrology, particularly for applications other than calibration of gauging stations, such as assessing where water may be leaking from the bed of a stream in a dry valley.

Dilution gauging is primarily used for spot checks and it is doubtful if a complete station rating has ever been achieved by this technique. Some development of a continuous sampling device has proceeded[65] at the Institute of Hydrology in an effort to develop an unattended station based solely on the dilution principle. At most sites interpretation of dilution gaugings is aided by continuous water level recordings and a postulated stage-discharge formula for the river section or gauging structure concerned.

A wide variety of tracers[63] has been tried, including sodium chloride, sodium dichromate, lithium chloride, fluorescein, Rhodamine B, bacteria and radioactive isotopes.

The subsequent analysis of river water samples has involved conductivity measurement, titration, flame photometry, plate counts or radioactivity counts. The main factors affecting choice of tracer have been cost, the type of equipment available in the organisation, the volume of the material required for any one gauging and its acceptability to riparian interests. Dichromate is normally favoured and has been tested successfully on some slow-moving lowland rivers as well as in turbulent mountain streams. Two basic methods of analysis are practised, both depending on the fact that mixing a soluble tracer, of known amount and concentration, with the flow of a stream yields a concentration in the stream inversely proportional to the discharge.

The gulp method

A small known volume of tracer is injected instantaneously into the river and the concentration time curve is recorded at the first point downstream where a complete mix can be achieved. By following the mass of the tracer as it passes down the reach, this equation can be applied:

$$C_1V = Q \int_{t_1}^{t_2} (C_2 - C_0)\,dt \tag{6.3}$$

where
C_1 = the tracer concentration at injection (mg/l or µg/l)
C_0 = the background concentration of trace chemical (mg/l or µg/l)
V = the injection volume (l)
Q = stream discharge (l/s)
C_2 = the downstream concentration as the pulse passes (mg/l or µg/l)

Sampling continues until the background concentration is re-established, and integration of the $(C_2 - C_0)$ curve follows to solve for Q.

If a salt tracer is used, and the gulp is followed by conductivity meter measurements, it is possible to make quite rapid on-site flow computations. For accurate results, however, it is necessary to correct any uncompensated meter readings by river water temperature[66]. Normally other tracer samples are subjected to automatic colorimetric analysis under laboratory conditions. Although the gulp method is simple and economical, a rigorous sampling procedure is required and generally less accurate results are obtained than with the constant-rate method.

The constant rate method

With this method a tracer is injected into the river for a long enough period to create a steady plateau of concentration at a downstream point. The Conservation of Mass principle gives:

$$QC_0 + qC_1 = (Q + q)C_2 \tag{6.4}$$

where q = rate of injection (l/s)

This relationship is normally simplified because, if Q is much greater than q, C_1 will be much greater than C_2. Thus

$$Q = \frac{q\,C_1}{C_2 - C_0} \tag{6.5}$$

Steady injection is crucial and it is necessary to use a Marriote vessel or similar device to achieve this *(Figure 6.3)*. When this method is used it is recommended that the reach is evaluated to ensure adequate mixing will take place and to check that no chemical adsorption is taking place. However, it is more expensive and time-consuming than the gulp method, and for this reason is less suitable for flood gauging[67] when flows vary rapidly. Nevertheless, it has the advantage that only a few samples are needed to determine the plateau concentration, samples that may be collected at any time during this period. On the other hand there is the requirement for a Marriote vessel and a larger amount of tracer than is usually used in the gulp method.

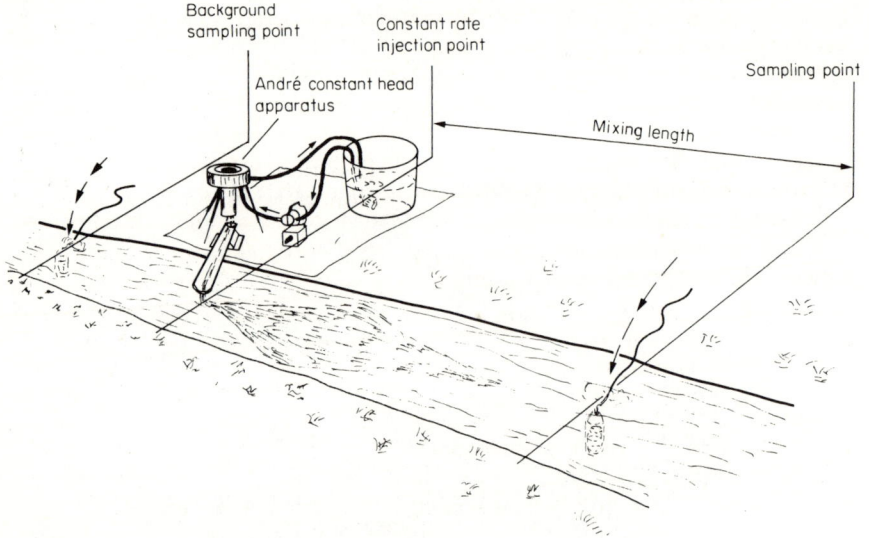

Figure 6.3   Dilution gauging: constant rate injection

When used with care, dilution techniques can give a measure of discharge to an accuracy better than ±2 per cent. Although the standard methods[68] require a stream to have a low constant background concentration (if any) of the tracer chemical, ingenious methods have been used in heavily polluted rivers. Harmless bacteria can be dosed and recaught in lower densities. When subsequently grown in a culture medium the numbers of colonies indicate the concentration of individual bacteria in the injected and diluted samples. Similarly, radioactive isotopes can be employed[69] but there are stringent Government regulations that inhibit their regular use. For this reason isotope gauging is normally carried out by AERE Harwell. In an early study[70] flows were checked in the Aylburton Brook in Gloucestershire and in the Usway Burn and Alwin Rivers in Northumberland. Sodium-24, bromine-82 and phosphorus-32 were used as tracers, the first two in the forms of sodium bicarbonate and ammonium bromide, and the last in solution in dilute hydrochloric acid, with potassium hydrogen phosphate as a carrier to prevent adsorption on the river sides. Tritium and iodine-131 have also been used successfully. Tritium only emits low energy particles and therefore cannot be detected in the field; samples must be collected for analysis in the laboratory.

Averages and departures from average

The discharge hydrograph, expressing the sequence of relationships that exist between runoff and the other components of a basin's water balance, takes into account their adjustments to the physical characteristics of the basin. By summing the area under the hydrograph, discharge may be expressed as the total flow for any chosen period, such as a month or a year.

By dividing the total flow by the appropriate time interval, the mean discharge can be obtained.

Records of flow in the form of averages and means are more readily available than instantaneous discharges. For the latter it is usually necessary to consult the authority maintaining the record. It is worthwhile noting that when annual flows are quoted they frequently refer to the water year (October to September) rather than to the calendar year.

Average annual flows broadly reflect the distribution of average rainfall amounts across the country. There is generally a greater variability in annual flows in the south and east than in the north and west *(Figure 6.11)*.

Average flow patterns

Average monthly volumes of flow exhibit a distinct seasonal pattern when

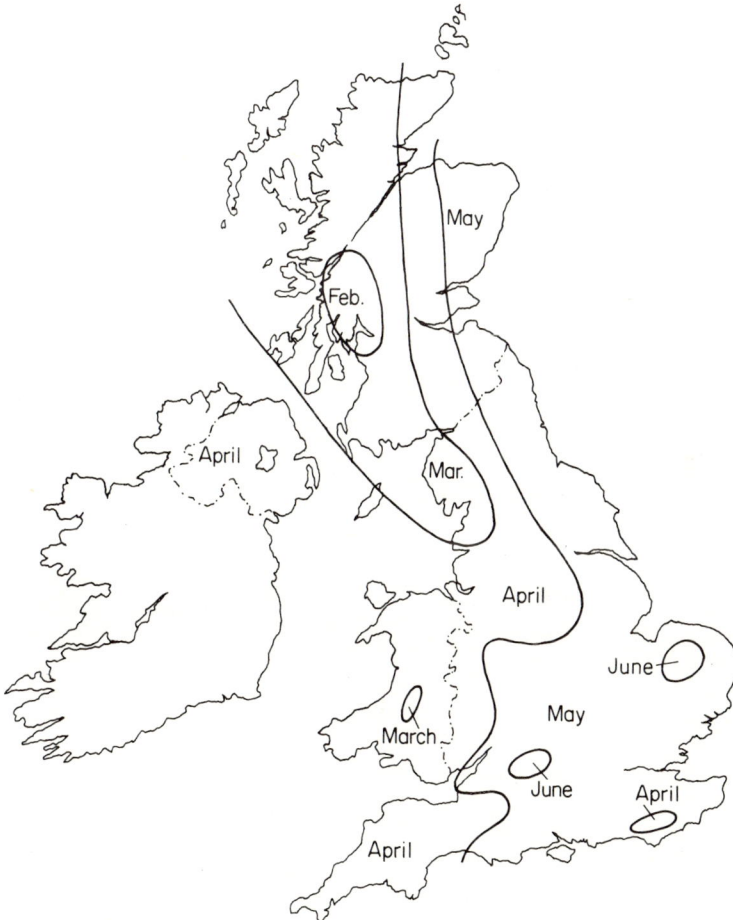

Figure 6.4   Summer flow season map, indicating the first month of the lowest six months of flow (based on long average conditions)

Table 6.4 DIMENSIONLESS MONTHLY FLOWS

Catchment	Period of record	Monthly flow as percentage of annual flow volume											
		J	F	M	A	M	J	J	A	S	O	N	D
Wissey, Norfolk	1956-71	14.1	12.8	12.4	10.1	8.0	5.5	4.9	4.1	4.1	4.8	7.6	11.6
Coln (Bibury)	1964-70	13.0	12.4	12.6	9.4	8.1	7.1	5.8	4.6	3.8	4.5	7.5	11.2
Winsford Brook, Som.	1934-65	14.9	13.0	10.9	7.9	5.6	4.4	3.6	3.4	4.5	6.3	12.1	13.4
Avon (Bath)	1939-65	15.2	13.9	11.2	7.3	5.0	4.2	3.2	3.3	4.3	6.2	11.9	14.1
Thames (Teddington)	1883-1968	15.2	13.5	12.6	9.0	6.9	4.9	3.7	3.4	3.6	5.5	9.1	12.6
Ely Ouse (Denver)	1926-64	15.5	14.8	14.3	10.3	7.6	4.9	4.1	3.5	3.4	4.5	8.2	10.9
Bedford Ouse (Bedford)	1933-66	16.8	16.3	14.6	9.2	5.3	3.3	2.6	2.1	2.2	3.9	10.2	13.5
Middle Level, Cambs.	1948-66	18.2	17.3	14.9	9.8	4.5	3.4	1.9	2.2	2.3	3.4	8.6	13.5
Nene (Orton)	1940-63	17.8	17.2	15.0	8.3	4.8	3.4	2.5	2.3	2.2	3.5	10.7	12.3
Bain (Fulsby Lock)	1962-70	13.6	12.0	12.6	10.8	7.4	3.8	3.4	3.2	4.4	4.3	10.1	14.4
Minningsby Beck, Lincs	1927-70	19.5	16.2	13.5	8.0	5.1	2.3	2.2	1.8	2.5	3.8	11.4	13.7
Blithe, Staffs.	1938-64	15.6	13.0	10.6	6.9	4.7	4.0	3.5	4.5	5.5	6.5	12.5	12.7
Scottish Dee (Woodend)	1930-65	10.7	8.8	9.2	10.0	7.5	4.8	4.5	5.6	6.2	9.2	11.5	12.0
Swale	1956-66	14.4	11.7	10.6	8.1	5.3	3.8	4.4	5.6	5.4	7.2	10.2	13.3
Tees (Broken Scar)	1956-66	13.1	10.9	10.0	8.3	5.1	3.3	4.5	6.5	6.3	8.0	10.3	13.7
Fernworthy Res., Dartmoor	1943-66	14.9	12.0	9.9	7.0	5.3	3.6	2.9	3.9	5.6	7.9	12.4	14.6
Cheddar	1916-60	13.5	11.6	9.4	7.3	5.6	4.4	4.2	5.2	6.1	8.3	11.8	12.6
Wallers Haven	1952-70	17.6	13.3	10.7	7.0	4.4	2.8	2.0	2.6	3.4	7.1	13.1	16.5
Elan Res.	1908-65	13.0	9.8	8.1	6.2	4.4	3.6	4.7	6.4	7.2	10.0	12.5	14.1
Taf Fechan	1913-65	13.5	9.8	8.1	6.5	5.0	3.9	4.9	6.1	6.9	9.9	12.0	13.4
Towy	1958-70	13.7	9.5	7.3	7.4	6.3	3.5	3.5	4.8	7.1	10.2	12.3	14.4
Severn (Bewdley)	1937-63	14.9	13.5	9.6	6.8	4.7	3.8	3.2	4.2	5.9	7.5	12.1	13.8
Vyrnwy	1910-64	13.5	10.5	8.0	6.1	4.7	3.6	4.5	6.3	6.7	10.1	12.6	13.4
Dee (Erbistock)	1937-68	14.1	11.0	8.4	6.3	4.8	3.6	3.9	4.9	7.0	9.3	12.3	14.4
Brenig	1923-55	14.0	11.3	8.2	5.8	4.5	3.7	2.9	4.6	6.5	11.1	14.5	12.9
Yorkshire Bridge	1906-23, 30-66	13.1	10.5	9.3	7.0	5.0	4.3	5.5	6.1	6.3	8.4	11.3	13.2
Deep Hayes	1915-64	14.3	11.6	8.9	7.3	5.0	3.7	4.3	5.8	5.8	8.1	12.4	12.8
Erne, Fermanagh	1900-47	16.2	11.6	9.1	5.9	4.5	3.3	3.2	4.9	5.7	8.9	12.2	14.5
Stocks Res.	1927-72	12.6	10.3	6.4	6.0	4.0	3.5	4.8	7.0-	7.8	11.7	12.9	13.0
Rivington Res.	1932-68	11.9	9.2	6.8	5.7	4.9	3.9	5.9	7.7	8.4	11.0	11.8	12.8
Leven (Windermere)	1939-73	11.5	9.2	7.4	7.1	5.2	4.3	5.1	6.8	8.9	10.2	11.6	12.7
Lune	1945-49, 53-69	11.8	8.5	7.1	6.7	5.1	4.0	5.5	7.4	9.5	10.5	11.1	12.8
Haweswater	1939-73	11.7	8.4	7.5	6.9	5.2	5.2	4.8	7.1	9.2	9.8	11.9	12.8
Allt Uaine, Argyll	1950-68	10.4	7.0	8.6	6.5	5.5	5.1	6.6	7.7	9.4	10.6	9.6	13.0

Note. Figures in italic refer to the summer (6-month) low flow season they are expressed in dimensionless form such as a percentage of the average annual value. This pattern often emerges with less than ten years of records and is probably equally as stable as the rainfall that is the prime cause. *Table 6.4* shows a wide selection of monthly percentages; it is laid out so that stations with similar patterns are together. Points to note are:

1. Catchments with large outcrops of aquifers exhibit the least range and the latest dates for extremes.
2. Catchments with low rainfalls have the widest extremes.
3. Snowmelt in Scotland causes a subsidiary peak in early spring.
4. The driest six months (the 'summer') occur earlier further west and further north (*Figure 6.4* shows 'summer' beginning in months varying from February to June).

This last characteristic is due to the fact that average monthly rainfalls decline early in the year in the mountainous west, particularly towards the north of Britain. Thus the six months with the lowest amount of precipitation in west Scotland are normally February to July; this period does not coincide with the season of highest evaporation so the amount of runoff is increased.

The consistency of such patterns makes it possible to infer data for ungauged areas and to check data obtained from a gauging site of uncertain quality. The check, although coarse, is particularly useful for ensuring homogeneity[7] or consistency with other gauges on the same river. In a very similar approach Ward[71] defined monthly coefficients of flow as (monthly average flow rate)/(annual average flow rate) and these coefficients were graphed to see how well they fitted Parde's[72] early work on river flow classification. The regional variations in the highest and lowest runoff months of the year are shown in *Figure 6.5;* the addition of further stations, often of widely varying catchment size, has not upset Ward's basic pattern.

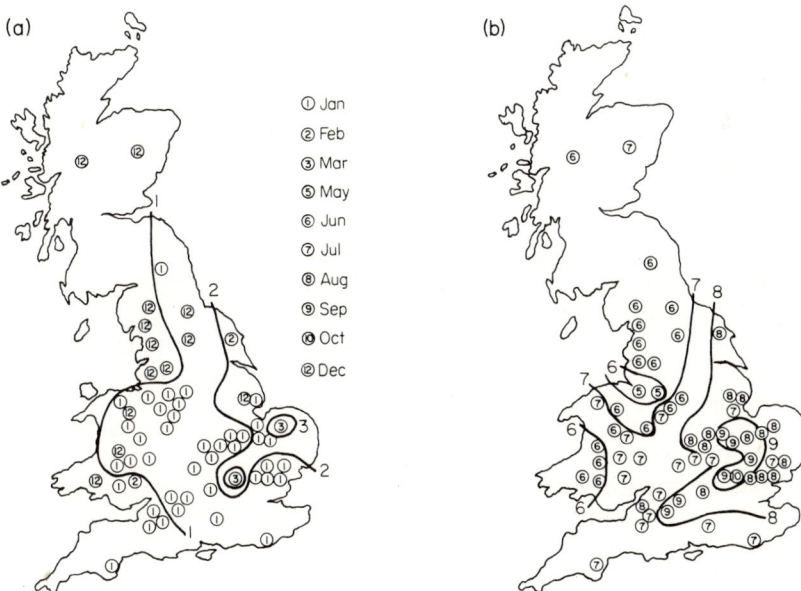

Figure 6.5   *Months of (a) maximum and (b) minimum flows (after Ward[71])*

Menzies[73] pointed out that one third of the volume of annual flow occurs in summer: consequently the average winter runoff is twice that of summer, the consistency of this simple rule being demonstrated in *Table 6.4*.

Flow variability

Many of a river's flow characteristics can be best appreciated from a flow-duration curve, i.e. a curve showing the percentage of time that any stated flow is equalled or exceeded *(Figure 6.6)*. Normal practice[74] is to prepare flow-duration curves both for individual water years and for long periods. Occasionally separate curves have been constructed for wet and dry seasons as a basis for yield calculations; this is because of the clear way in which these seasonal curves show the cycle of surplus and deficiency[78,79]. As the curve is obtained by reordering the basic data (usually in groups) from the smallest to largest, such methods are liable to overestimate the stored water needed to balance the natural hydrograph. They also involve the arbitrary assumption of the length of dry season that will be critical for the reservoir.

Frequently duration curves are plotted on semi-logarithmic graph paper in order to contain the wide range of flows that occur on all but spring-fed streams. A long-period curve will often approximate to a straight line; departures from linearity at low flows can be due to abstractions or effluent returns, with departures at the other extreme resulting from overbank storage.

A great deal can be learned from making duration curves dimensionless: the units on the flow scale are divided by the average flow rate for the period concerned. As *Figure 6.6* shows, Britain's rivers can be represented by a small family of curves. It is noticeable that the average daily flow is exceeded for about 30 per cent of time, that the median flow is about 70 per cent of the average flow and that the flow most likely to occur (the modal flow) is exceeded about 70 per cent of time. Hall[80] mapped the ratio (30 per cent time flow)/(70 per cent time flow) for many Devon gauging stations and demonstrated the similarity between the main features of his map and those of the geological map of the area. The dimensionless duration curve for an ungauged site can be obtained with reasonable precision by choosing a line from *Figure 6.6* for an area comparable in terms of topography, geology and raininess. The correct units can be obtained by multiplying the dimensionless units by an estimate of average daily flow.

Flow duration curves have been used extensively for the design of gravity catchwaters ('leats')[81,82] and pumped transfers. A catchwater is a small contour canal or piped aqueduct that conveys water from intakes on streams in catchments adjacent to the reservoired catchment, the water being conveyed to the reservoir to augment its yield. By summing the quantities under successively higher flow levels on a curve like *Figure 6.6* it is possible to construct a graph that indicates the diminishing returns for increasing catchwater capacity[83].

Although it is a standard technique to obtain mean sediment load by multiplying a sediment rating curve by the flow-duration curve at a measurement station[84], published British examples of sediment loads are

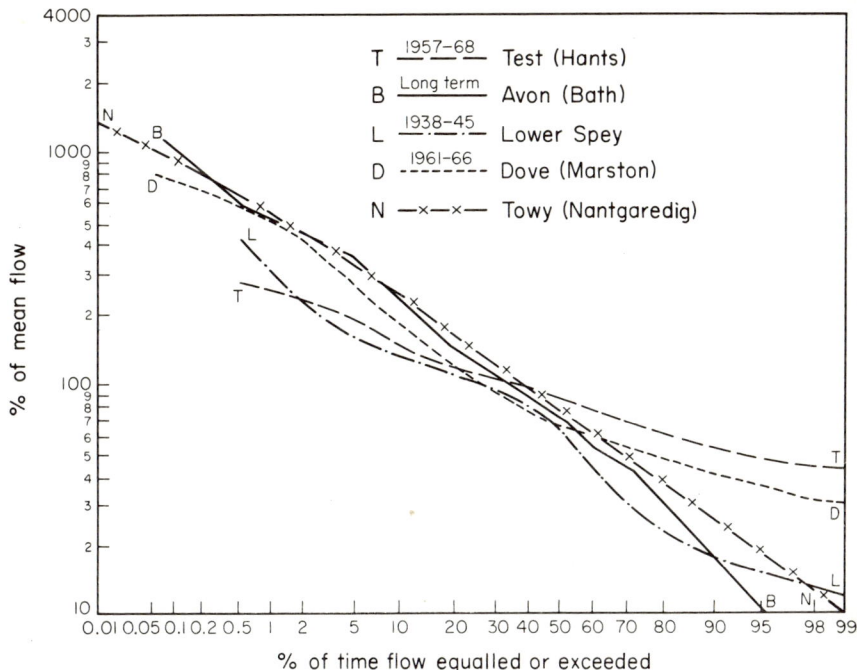

Figure 6.6   Dimensionless flow duration curves. Sources of additional curves:
(a) river authority first periodical surveys 1970–74, under Section 14 of Water Resources Act 1963
(b) Surface Water Yearbooks prior to 1965
(c) River Shin (Lairg) 1949–56, and Allt Uaine (Sloy) 1950–60: see Aitken[75]
(d) four large British rivers, five hilly catchments: see Cochrane[76]
(e) fifteen rivers, mainly in Scotland: see Baxter[77] (includes photographs of rivers at ⅛, ¼, ½ and full average daily flow)
(f) eight rivers in N. Wales and NW. England: see Hoyle[78]

few[85]. This is probably due to the apparently small quantities of suspended and bed load material in rivers in the UK compared with the massive amounts conveyed downstream in other parts of the world; the result is that such work has had a 'Cinderella' status.

Regional variations in flows

Although the UK is small in area, the diversity of climate and topography leads to important variations in the local severity of wet and dry periods. These variations are demonstrated in *Table 6.5* which contains the rankings of annual flows for a representative sample of gauging stations, with a French station included for comparison. While there is some consistency between stations for the wettest and driest years, groups of wet and dry years show little similarity from station to station. Such a table can be used to re-rank a short record within a region so that a probability interpretation can be made more intelligently.

Table 6.5 RANKED ANNUAL FLOWS, 1932–1965

Year	Thirlmere	Stocks Res.	Ladybower Res.	Deep Hayes, Staffs.	Alwen Res.	Vyrnwy Res.	Severn (Bewdley)	Elan (Caban Coch)
1932	20	18	17	27	19	15	24	19
1933	1	1	1	6	1	1	1	1
1934	27	12	6	1	7	8	2	11
1935	22	26	27	15	20	20	21	28
1936	25	9	25	19	32	21	27	21
1937	6	3	14	23=	6	3	18	9
1938	33	33	18	11	22	27	11	25
1939	20	14	24	12	27	25	28	24
1940	4	7	12	21	8=	16	25	12
1941	3	4	23	17	8=	5	17	8
1942	26	11	9	7	5	7	4	14
1943	30	22	8	4	17	23	10	20
1944	21	27	28	25	26	24	15	16
1945	13	16	7	8	11	18	8	2
1946	28	28	29	29	28	32	30	32
1947	10	8	10	14	13	12	22	6
1948	31	23	15	20	23	30	20	27
1949	14	20	11	16	16	14	9	7
1950	24	32	16	23=	24	28	23	22
1951	32	31	32	30	30	31	32	29
1952	8	17	22	18	12	10	16	12
1953	15	13	4	3	3	9	3	3
1954	34	34	34	34	34	34	33	34
1955	2	2	3	10	2	2	14	5
1956	5	10	26	22	10	6	6	15
1957	23	24	21	28	29	29	26	31
1958	7	21	30	33	25	22	29	26
1959	12	5	2	2	14	11	19	17
1960	19	29	31	31	33	33	34	33
1961	16	30	19	26	21	17	12	18
1962	17	19	20	13	13	19	13	23
1963	18	15	13	5	15	13	5	10
1964	9	6	5	9	34	4	7	4
1965	11	25	33	32	31	26	31	30

Note. Rank 1 = year with lowest total volume of flow; Rank 34 = year with highest total volume of flow

Floods

'. . . it wasn't much good having anything exciting like floods if you couldn't share them with somebody . . .' (Winnie the Pooh)

Britain's worst recorded floods

Nearly every year one or more parts of the country experience a flood that is, by general consensus, catastrophic. Considerable disruption occurs, widespread and costly damage and, sometimes, loss of life. Lists of notable floods recorded by diarists, historians and natural scientists can be awe-inspiring because the centuries have brought their extremes to nearly every community[86]. Both the variety and severity of floods are demonstrated in the

Taf (Fechan Res.)	Minningsby Beck, Lincs.	Ouse (Bedford)	Ely Ouse (Denver)	Thames (Teddington)	Scotland: Dee (Woodend)	Ireland: Erne	France: Loire (Blois)
20	16	20	19	24	28	15	28
1	10	8	6	13	4	2	16
12	1	2	1	1	31	13=	20
13	19	22	5	18	21	19	33
16	24	34	29	30	9	28	29
6	31	33	34	33	25	13=	24
19	12	13	9	5	18	32	6
29	32	29	31	31	26	17	31
10	20	32	26	27	6	20	32
7	29	28	30	26	24	6	34
11	4	10	18	12	14	22	18
21	8	4	20	10	11	25	12
18	21	1	3	2	15	10	25
22	11	6	10	3	22	11	8
32	26	14	16	22	16	26	3
9	25	27	28	28	19	31	4
27	5	3	4	4	30	24	7
5	2	9	2	7	3	18	1
30	23	23	14	17	13	33	9
31	33	30	32	34	32	16	27
14	7	19	23	21	5	4	30
4	15	15	13	6	7	1	2
33	31	16	24	23	33	34	14
2	22	24	21	15	2	8	19
8	13	5	8	8	20	7	17
26	14	18	11	16	17	21	11
28	28	26	33	29	27	27	23
24	6	25	22	19	12	3	13
34	34	21	27	32	34	29	22
25	17	30	25	25	8	23	10
15	9	12	15	14	23	12	15
17	18	16	17	20	29	5	21
3	3	10	7	11	1	9	5
23	27	7	12	9	10	30	26

Annex to this chapter (pages 276–282), which summarises the main features of some notable floods for which published descriptions are available. These floods usually occurred where there was no orthodox means of gauging; consequently quoted runoff rates are generally estimates with a considerable error. The same applies to the estimated hydrographs and the timings (sometimes British Summer Time, sometimes Greenwich Mean Time) by witnesses at the different sites. Checks have been made to ensure that the volume of flood runoff is not unrealistically large.

As can be seen from the listed source material in the Annex, flood information can be found in diverse publications. Britton's collation and critical review of manuscripts[86] before 1450 is particularly valuable. Not only does it list in chronological order the original quotations on climate, and therefore flooding, but it also shows how errors can creep into flood lists. Whenever assessing flood severity from verbal descriptions, allowance has

Figure 6.7 July 1968 flood at Cheddar Gorge (courtesy Dr. M. D. Newson)

to be made for the exaggerated (or under-played) style of the times. A less satisfactory source that brings English agricultural climatology up to date is Stratton's *Agricultural Records AD 220–1968*[87]. Much more detail of the flood sequence on the major rivers of England and Wales, the Thames[88], Severn[89] and Trent[90] is found in individual papers. A broad summary was given by Brooks and Glasspoole in their book *British Floods and Droughts*[91]. Volume 4 of the *Flood Studies Report*[92] tabulates several historic flood levels and flow series stretching back a century or more. However, some caution is required in interpreting records where man-made influences have affected river conditions. One example is the Thames, where the change from 'flash' weirs, originally used for barge navigation, to modern sluice structures has raised bank-full flows.

Table 6.6   RELATIVE SEVERITY OF MEASURED FLOODS[92]

Gauging station	Max. flow (×AMAF)	Date
Lud, at Louth	28.5	29 May 1920
West Lyn, at Lynmouth	17.2	15 Aug 1952
Darent, at Hawley	15.4	15 Sep 1968
Whitewater, at Lodge Farm	10.0	21 Mar 1947
Bourne, at Hadlow	8.9	15 Sep 1968
Dunsop, at Footholme	8.9	8 Aug 1967
Chew, at Compton Dando	8.3	11 Jul 1968
Eden, at Vexour Bridge	7.2	15 Sep 1968
Wandle, at Connolly's Mill	7.1	15 Sep 1968
Findhorn, at Forres	6.2	18 Aug 1970
Burbage Brook	6.0	1 Jul 1958
Tas, at Shotesham	5.5	17 Sep 1968
Stour, at West Mill	5.5	17 Sep 1968
Mole, at Ifield Weir	5.5	15 Sept 1968
Stour, at Kidderminster	5.2	27 Mar 1955
Tweed, at Peebles	5.0	12 Aug 1948
Kislingbury Branch, at Upton	4.6	17 Mar 1947
Otter, at Dotton	4.6	10 Jul 1968
Wye, at Cefn Brwyn	4.0	6 Aug 1973

Table 6.6 shows the relative severity of measured flood peaks in terms of multiples of the mean annual flood (AMAF) at the site concerned. It is noticeable how extreme a severe event can be, relative to even a 70-year (once-in-a-lifetime) flood. Clearly villages and towns that lie near a river or stream must expect that some day their turn for a flood emergency will come. For many towns in eastern and southern England that event was in March 1947[93,94] when a sudden thaw coupled with repeated rain produced a flood of immense volume.

Major dams are not only threatened by floods; they can initiate them too if the design, siting or construction is wrong. The phrase 'once in a Sheffield flood' (meaning a rare event) preserves the memory of the dreadful events of 11 March 1864 when the Bradfield 'Dale Dyke' earth dam failed when it was first filled[95]. A total of 245 people were killed and a public inquiry followed[96]. In 1930 legislation was passed for England, Scotland and Wales to ensure that reservoirs impounding more than 5 million gallons (23 Ml) were regularly inspected to ensure their safety. The Institution of Civil Engineers followed this up by forming a committee to study floods in relation to reservoir practice. Its report[97], reissued in 1960 with additional flood data, contains an enveloping curve of recorded floods on small upland catchments. This 'normal maximum curve', multiplied as necessary to represent catastrophic events, was for about forty years the basis for British spillway design. It has now been replaced by a more comprehensive guide to floods and reservoir safety[98].

The publication of Volume 4 of the UK *Flood Studies Report* makes available the bulk of flood-flow measurements for the country. Original records can be readily traced from this volume, thus making it possible to locate a flow record in almost every part of Britain and Ireland. The most common practice is for the local water authority to keep tabulations of mean daily and peak flows for all stations in its area, a copy also being held by the Water Data Unit. Other sources of descriptive material on flooding include the volumes of *British Rainfall* and the British Museum newspaper library at Colindale.

Determining flood flows

The problems of predicting the discharge characteristics of ungauged catchments are considerable. Yet assessment of the peak flows, time to peak, the distribution of flow and the relationship between peak discharge and return period are required for engineering design and other purposes. As a result, over the years, a considerable amount of attention has been given to these problems and this has led recently to the simulation of the complete hydrograph by both deterministic and stochastic modelling. Nash[99] has provided an account of the development of methods for determining runoff from rainfall, tracing the origins of the Rational method, often referred to as the Lloyd-Davies method[100], to the work of Mulvaney[101] in 1850. The basis of the Rational method is the assumption that for every catchment there is a period, known as the time of concentration (T_c), required for a particle of water to flow from the most distant point in the catchment to the gauging station. The peak discharge (Q) occurs at the time T_c after the start of rain, and:

$$Q = CAp \qquad (6.6)$$

where C = the coefficient of runoff
A = area
p = mean intensity of rainfall in the time T_c

Frequency of flood peak flows

Flood-flow probability techniques only became widely used in the UK in the early 1960s. The first paper on a basin scale seems to have been that of Dury[102] for the Nene and Ouse. His work concerned annual maximum mean daily flows, and he adopted an Arithmetic–Gumbel probability distribution for their extrapolation to ten-year return period values.

The Ministry of Agriculture, Fisheries and Food, because of its land drainage responsibilities, encouraged 'worthwhileness tests' for new flood protection schemes and these embryo cost-benefit studies involved flood designs of stated probability.

Employing the concept of regional analysis for estimating flood peaks, Cole[103] first computed the *arithmetic mean annual flood* (AMAF) and then applied a multiplying factor to estimate any rarer event up to about a 100-year return period. The AMAF is obtained from the series of highest annual flood peaks in the record; it has a return period of 2.33 years on Cole's assumption that the annual flood series follows a Gumbel extreme-value distribution[104]. Records from 56 gauging stations with 731 years of records were employed in this study and the results were grouped into six regions covering England and Wales. Between the extremes of region 1, which covered most of the areas of chalk in southern England, and region 6, the mountainous parts of Wales and the Lake District, the AMAF increased about eighty times. On the other hand there was little difference between the scaling factor for estimating rarer flood peaks between the east and west of the country.

This technique has been widely used and extended to Scotland[105] and Ireland[106] and it is usually preferred to the techniques of calculating flood peaks from catchment characteristics and relating flood frequency to rainfall frequency[107,108].

A major step forward was taken when the Government accepted the recommendation of the Institution of Civil Engineers[109] that a comprehensive study of floods in the UK should be undertaken. A Flood Studies Team was formed under Sutcliffe in 1969 at the Institute of Hydrology, and a small group set up at the Meteorological Office to undertake the meteorological studies. Cooperation with the Irish authorities has enabled these studies to cover all the British Isles. The Team's report was published in 1975 in four volumes:

Volume 1   'Hydrological studies'
Volume 2   'Meteorological studies'
Volume 3   'Flood routing studies'
Volume 4   'Hydrological data'

As a result an instantaneous flood peak of any given probability up to 1000 years can be found by an improved version of Cole's method. Where a catchment is gauged the AMAF can be directly computed and then entered at the appropriate point in *Figure 6.8.* Adjustments can be made to short records to get better AMAF estimates where other local information is available. If nothing at all is known of the flood regime, regression equations can be used to predict AMAF from mapped catchment variables. The relationships vary regionally but follow the form[92]:

$$\text{AMAF} = a \text{ AREA}^b \text{STMFRQ}^c \text{S1085}^d$$
$$\times \text{ RSMD}^e \text{SOIL}^f (1 + \text{LAKE})^g (1 + \text{URB})^h \qquad (6.7)$$

where AREA $=$ catchment size (km²)

STMFRQ $=$ the number of stream confluences/km² as shown on a 1:25 000 OS map

S1085 $=$ main stream slope between two points 10 per cent and 85 per cent of the stream length from the site (m/km)

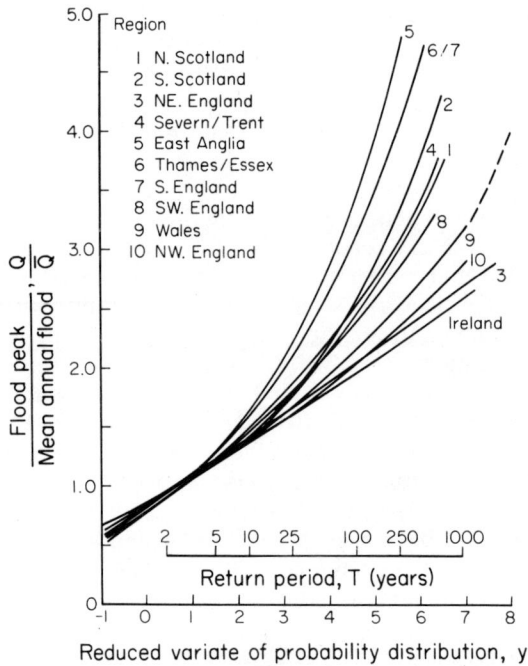

Figure 6.8 Flood peak probability related to mean annual flood Q (from Flood Studies Report[92], *Vol. 1, courtesy Institute of Hydrology)*

RSMD = the five-year return period one-day rainfall minus the 'effective soil moisture deficit' (which has a complex definition but is conveniently calculated from a map; it varies from 2 mm in mountain areas to 12 mm or more in south-east England); typical RSMD values are shown in *Figure 6.9*

SOIL = an index of the infiltration capacity of catchment soils (using *Table 4.3*) = 0.15(% Cl. 1 area) + 0.3(% Cl. 2) + 0.4(% Cl. 3) + 0.45(% Cl. 4) + 0.5(% Cl. 5); the index is said to be a measure of runoff fraction in a typical storm, and varies from 0.15 in the most pervious areas to 0.5 in the most impervious

URB = the fraction of the area covered by urban development, found by summing 'grey areas' on a 1:625 000 OS map

LAKE = the fraction of the total area to the site that drains into major lakes and reservoirs, found by examining a 1:250 000 scale map; a major lake is defined as having a surface area exceeding 1 per cent of its catchment

Table 6.7 gives the regional values for the above equation. The data used in the study exceeded 6000 station years at 533 sites.

It was found that 68 per cent of floods lay between +47 per cent and −32 per cent of the *prediction* by these equations; the error expressed as a

Figure 6.9   RSMD (mm), the net one-day rainfall of 5-year return period (courtesy Institute
of Hydrology)

percentage of the *actual peak flow* can be very much larger. However, the equations meet many needs despite the possibility of anomalous results from certain catchments. Nevertheless caution must be exercised in the use of such formulae. Wherever an important decision has to be made it is advisable to gauge the river concerned; this is on the basis that even a single year of discharge records provides a better estimate of the mean annual flood than is given by most methods of prediction[107].

Table 6.7 REGRESSION COEFFICIENTS FOR MEAN ANNUAL FLOOD PREDICTION

Region	a	b	c	d	e	f	g	h
N. Scotland	0.0186	0.94	0.27	0.16	1.23	1.03	−0.85	0
Central area	0.0213	0.94	0.27	0.16	1.23	1.03	−0.85	0
East Anglia	0.0530	0.94	0.27	0.16	1.23	1.03	−0.85	0
Thames/Lee/Essex	0.0373	0.70	0.52	0	0	0	0	2.5
S. England	0.0234	0.94	0.27	0.16	1.23	1.03	−0.85	0
SW. England	0.0315	0.94	0.27	0.16	1.23	1.03	−0.85	0
Ireland	0.0172	0.94	0.27	0.16	1.23	1.03	−0.85	0

Flood volumes

Probably most attention has been given to peak flow, while methods of determining volumes of water in floods longer than a few hours' duration have been rather neglected. Yet without a knowledge of flood volumes, flood forecasting and protection methods must be inadequate. The potential for flood plain inundation is determined by the way in which the flood volume locally exceeds the river channel's ability to carry it away. Again, if storage area on the flood plain is progressively taken up by houses and factories the flood volume will only be accommodated by its taking up a greater depth. Consequently 'flood parks' are becoming an accepted solution in new towns; spates are temporarily but deliberately stored on recreation land behind control structures until the downstream channel can accept the water.

Table 6.8 REGRESSION COEFFICIENTS FOR MEAN ANNUAL FLOOD VOLUME PREDICTION

$$CALMAF = a\ AREA^{0.9475}\ STMFRQ^{0.4068}\ RSMD^{0.628}\ SOIL^{0.7102} \qquad (6.8)$$

Region	Coefficient 'a'
North Scotland	0.0395
South Scotland	0.0417
North-east England	0.0410
Severn/Trent	0.0360
East Anglia	0.0279
Thames/Essex	0.0250
Southern England	0.0428
South-west England	0.0585
Wales	0.0539
North-west England	0.0459

Note. The units are m³/s and an average flow rate over the day is implied

In the *Flood Studies Report*[92] the calendar day flood volumes recorded at 236 stations in ten regions were examined. Multiple regression was used to relate catchment characteristics to the *mean annual calendar day flood* (CALMAF) experienced at each station, where CALMAF is the arithmetic mean of the series of highest daily flood volumes in each year. *Table 6.8* shows the best equations using the same catchment characteristics that were used to estimate peak runoffs. The factorial standard error of estimate at 1.39 suggests that the equations should be used only where there are few or no measurements.

Flood volumes for other durations up to ten days are obtained by a dimensionless technique based on the way in which the average flood flow rate decreases with time. A curve for an ungauged site can be predicted simply by measuring its catchment slope (S1085) and using its logarithm (LS1085) in the following equations:

$$\text{Reduction factor at 3 days} = \text{antilog}(-0.101 - 0.081\,\text{LS}1085) \tag{6.9}$$

$$\text{Reduction factor at 10 days} = \text{antilog}(-0.269 - 0.127\,\text{LS}1085) \tag{6.10}$$

The curve is then sketched in through the three known points. Care is needed when applying this technique to events of a longer return period than the values used in deriving it.

For such areas as the Fens a great deal of experience of dealing with volume floods has been built up by internal drainage boards. For instance the Middle Level Commissioners with their St. Germans pumping station are responsible for pumping the entire flow from 692 km² up into the Great Ouse river above Kings Lynn. The catchment's average annual rainfall is about 550 mm. Since 1947 the entire flow has been pumped and the station capacity has been gradually raised from 4 mm/day to 6 mm/day. Records of pumping hours show that whilst 50 mm has been pumped in a month on six occasions in 22 years, there have been seven years when pumping did not exceed 25 mm in any month.

Several regions of a character similar to the Fens, lying at or below sea level, have areas designated as 'washlands', which are grasslands where floods can be preferentially spilled to protect more vital areas. Semi-natural areas include the Somerset 'moors'[110], the Ouse 'washes' between the Old Bedford River and the Hundred Foot River[111], and areas by the lower Rother in Kent[112]. A recent scheme to protect Doncaster adopts the same principle[113] of off-channel storage on land lying below river flood levels.

Provision of flood control storage in impounding reservoirs is much rarer in the UK. The operating rules for Llyn Clywedog[114] above Llanidloes on the Severn and Llyn Celyn[115] above Bala on the Dee both have a seasonally varying capacity for flood control. These capacities are equivalent to runoffs from the respective catchments of up to 18 mm and 16 mm in early winter, but they reduce to nil and 5 mm in summer. Proposals to control floods by storage on the Medway were made public by Kent River Authority in 1969 and 1971[116,117]. Low embankment dams across the river above Edenbridge and above Tonbridge would not normally impound any water until flood flows exceeded a critical level approaching the channel capacity through these towns. Thereafter it would be possible to store 23 mm and 6 mm respectively from their total catchment areas of 270 km² and 900 km².

A system of weirs and sluices at the outlet from Bala Lake enables floods on the Dee to be mitigated[118] by raising natural lake levels. However the consequent prolongation of bank-full discharges on the lower Dee during flood recession is reputed to cause waterlogging of farmland and land drainage problems due to overlong submergence of tile drain outlets. River regulation by hydro-power reservoirs is varied in flood periods by the North of Scotland Hydro-board to give a measure of downstream protection. Their reservoirs include most of the examples of gated spillways in the UK[119].

The unit hydrograph

The unit hydrograph, the hydrograph resulting from an excess rainfall of unit volume and duration, is frequently criticised as an over-simple linear concept. Nevertheless it is a concept that is valuable in both flood forecasting and flood control. Nash[120] has been the leading exponent of the unit hydrograph approach, but few examples of unit hydrographs for British rivers had been published[121-127] prior to 1975, when the 'Data' volume of the UK *Flood Studies Report* gave values for 139 catchments. Nash[128] and O'Donnell[129] explored the form of the *instantaneous unit hydrograph* (IUH), which is the limit to which the unit hydrograph tends as the period of effective rainfall is diminished indefinitely, one definition of the IUH being:

$$u = \frac{1}{K\Gamma(n)} (t/K)^{n-1} \exp(-t/K) \qquad (6.11)$$

where u = the ordinate at time t
       K = a time constant in a first order linear system
       Γ = a gamma function
       n = a numerical parameter

Nash[130] also related the moments (m_1, m_2) of the IUH to the topographical characteristics of 90 catchments and devised a method for predicting the lag-time and the approximate unit hydrograph of an ungauged catchment from measures of its topography. The relationships were of the form:

$$m_1 = 27.6A^{0.3}(\text{OLS})^{-0.3} \qquad (6.12)$$
$$m_1 = 20L^{0.3}(\text{EA})^{-0.33} \qquad (6.13)$$
$$m_2 = 1.0m_1^{-0.2}(\text{OLS})^{-0.2} \qquad (6.14)$$
$$m_2 = 0.41L^{-0.1} \qquad (6.15)$$

where m_1 and m_2 = lag times (hours)
             A = area (square miles)
             L = length from gauging site to most distant watershed (miles)
       EA = a measure of main channel slope (parts per 10 000)
    OLS = a measure of overland slope (parts per 10 000)

As a demonstration of the practical value of the IUH approach, Nash[131] was able to show how the response of the River Wandle in South London was likely to be altered by drainage improvement works. The post works unit hydrograph was predicted to have a higher peak and a shorter lag-time than the actual hydrograph.

One criticism of these unit hydrograph studies was that not all the conditions usually considered necessary for the construction of a unit hydrograph were adhered to. For example, some very small storms and some non-uniform storms were employed in constructing unit hydrographs. The Flood Studies Team[92] attempted to avoid these pitfalls by specifying that a 10 mm/hour unit hydrograph should only be obtained from floods with peaks greater than half the mean annual value. In addition attention was restricted to:

1. areas less than 500 km²
2. rainstorms that were relatively homogeneous in space
3. catchments with autographic rainfall records
4. rivers having short-term response to heavy rain

About 1500 floods were analysed to provide unit hydrographs for 139 gauging stations. Regression analysis of the factors expected to influence unit hydrographs were not entirely successful. The recommended equation is:

$$T_p = 46.6 \text{MSL}^{0.14} \text{S}1085^{-0.38} \text{URBT}^{-1.99} \text{RSMD}^{-0.4} \text{ hours} \quad (6.16)$$

where
T_p = time to peak, of 1 hour UH
MSL = length of the main stream (km), using dividers in 0.1 km steps on a 1:25 000 map showing the stream
URBT = 1 + URBAN, where URBAN is the urban fraction of the catchment defined by grey areas on the 1:63 360 map

S1085 and RSMD are defined on pages 247–248

However if direct observation has been made of the lag time, i.e. the time from the centroid of storm rainfall to peak flow, then this should be used in preference to find T_p from:

$$T_p = 0.9 \text{ lag time} \quad (6.17)$$

Only a simple triangular unit graph was found to be warranted; the dimensions of this graph are peak flow $Q_p = 2.2/T_p$ m³ s⁻¹ km⁻² and time base $T_b = 2.52T_p$ hours, giving a runoff of 10 mm after one hour's effective rainfall. Should an equivalent unit graph be required for a time period T hours rather than one hour, it is necessary first to compute:

$$\text{new } T_p = \text{old } T_p + (T-1)/2 \quad (6.18)$$

The unit graph is still applied but with the new T_p value.

Difficulty has been found in simulating floods of 'catastrophic' magnitude[92] by unit hydrograph methods. It is clear that a linear technique of this sort cannot quite cope with the accelerating severity of flooding when overland flow is initiated. This is particularly so in hilly areas with large outcrops of aquifers where lag time shortens noticeably as depths of rain increase. The current recommendation of the Floods Team is that the time to peak for creating a 'probable maximum flood' should be shortened to 0.67 T_p as an empirical estimate of the added severity. As an increasing number of studies are performed with synthetic unit hydrographs[132], the accumulated experience should permit an element of non-linearity with storm magnitude to be introduced.

Probable maximum floods

The *probable maximum flood* (PMF) is defined as the flood hydrograph resulting from *probable maximum precipitation* (PMP)[133], including snow-melt where applicable, coupled with the worst flood producing catchment conditions that can be realistically expected in the prevailing meteorological conditions. (See definition of PMP in Chapter 2.) The steps in the computation of PMF for a catchment are as follows:

1. Compute the PMP over the catchment for each time increment.
2. Estimate the additional equivalent rainfall due to any concurrent snowmelt.
3. Reduce the rainfall to allow for storm losses.
4. Distribute 'rainfall excess' symmetrically in time so that the most intense effective rainfall occurs in the heart of the storm.
5. Convert (4) by means of a unit hydrograph into a flood hydrograph.
6. Calculate antecedent flow and add it as a baseflow to the quick response hydrograph of step (5) to obtain PMF.

These steps involve considerable computation and judgement[134] (and potential controversy on the validity of the philosophical assumptions that must be introduced). Reference should be made to Chapter 6 of Volume 2 of the *Flood Studies Report*[92], and in particular to the worked example.

The effects of urbanisation

Most authorities would agree that the discharge from an urban catchment is far more flashy than from an area of similar size used for agriculture. The extension of the storm-water drainage system and the increase in imperme-able surfaces appear to be the two main causes of this difference, but in areas of clay or chalk the changes may not be so great.

Hollis[135] studied Canon's Brook in Essex, a 21.4 km² clay catchment where between 1953 and 1968 the paved area increased from nil to 17 per cent. He found that by comparison with a nearby rural catchment the summer runoff of the brook appears to have increased. An examination of the seasonal floods in classes also revealed this change, although there was much less of a change for the rarer storms. The progressive attenuation of the unit hydrographs for the brook was pronounced *(Figure 6.10)*.

Crawley New Town at the head of the River Mole in Sussex is a difficult site for drainage, and five gauging stations were installed between 1952 and 1961 to monitor flood peaks[136,137]. In an area of very slight relief and impermeable clay soils there have been some hydrograph changes, but they seem more subtle and less marked than those at Canon's Brook. The Institute of Hydrology is investigating the change in urban runoff processes as the new city of Milton Keynes develops[138]. A raingauge network, and a flow-measuring station on the main drain, have been completed and experiments are being conducted on the rainfall runoff translation of a variety of impervious surfaces, including roads, roofs and gullies. Analysis will proceed using the exact solution of the continuity and momentum equations, and simplified derivatives of these. The city's Development Corporation is tackling the problem of increased flooding[139] by creating

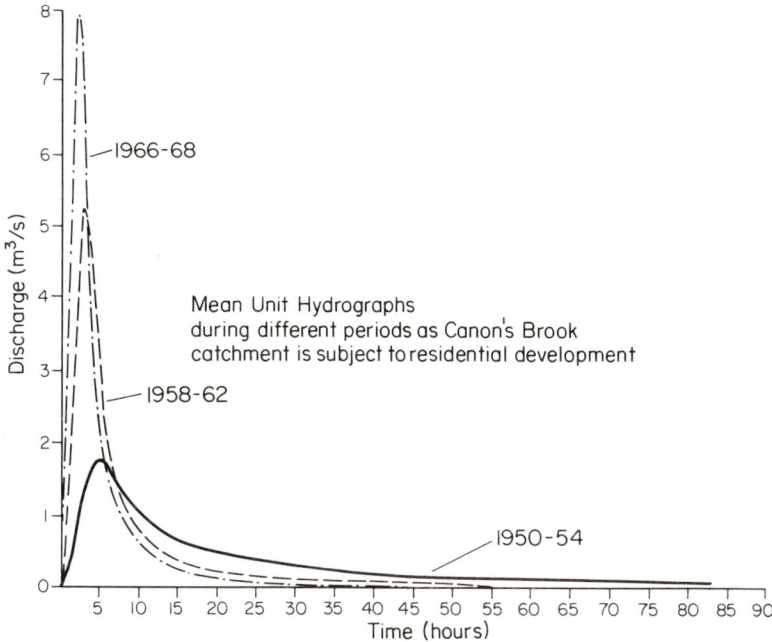

Figure 6.10   Accentuation of unit hydrograph due to urbanisation, Harlow (from Hollis[135], courtesy Institute of British Geographers)

temporary storage on areas otherwise used for recreation. Similar schemes are becoming widespread; an early successful example, using gravel pit storage, is on the River Cray[140] in the Kent suburbs of London.

Increasing storm-water drainage imposes a pollution load on a stream. In Birmingham, experiments by the city Rivers and Sewers Department[141] on a 2.4 km² area with a combined sewerage system led to the conclusion that it would be best to mitigate the quality problems caused by polluted storm water overflows by the provision of retention tanks to catch the worst first flush through the drains in each storm. The vast urban complex of Birmingham at the head of the River Tame effectively killed the Trent as a potable water source, and this led to the detailed study of the river (Chapter 8). One result was the passing of the Trent River Authority Act in 1973, which permits the Severn-Trent Water Authority to build purification lakes directly on the river at Kingsbury. By utilising old gravel pits for river water retention, and an old quarry for storage of dried sludge extracted from these lakes, it is hoped that the scheme will lead to an improvement of the river[142]. It is intended that the lowest of the series of lakes will become an amenity area, thus turning a liability into an asset. Chemical, sediment and hydraulic studies have been combined in the lake design.

The primary technique for the design of modern storm water drainage systems in urban areas is the Road Research Laboratory (RRL) hydrograph method[143,144]. This is primarily aimed at the economic sizing of the pipe network by allowing for the routing effect of the storage volume in the pipes. However, as a by-product it becomes possible to predict the flood hydrograph from an existing drained urban area for all but those rare events in which the system surcharges and significant overland flow is initiated. Standard computer programs exist for performing the necessary time–area method calculations once the drainage system has been surveyed and coded in the requisite manner. Varying rainfall profiles and development densities can be examined; the proportion of paved or impervious area normally averages about 35 per cent in a town, but may reach 90 per cent in city centres and fall to 15 per cent in modern suburbs.

Droughts

Historic extremes

Low flows are not readily characterised, but one approach is to classify well-known droughts that covered much of the country in terms of their impact on water users. *Table 6.9* lists some of the major events of the past century (see also Chapter 2); it illustrates that no one event is dominant everywhere or for all sources. There are regional variations of drought duration and intensity and *Table 6.5* summarises the cross-country changes in the importance of annual droughts by ranking flows since 1932.

Table 6.9　HISTORIC BRITISH DROUGHTS SINCE 1874

Drought/duration	Type of source affected	Drought year(s)	Region
Single summer	River intakes, small reservoirs	1959	England, W. Wales, E. Scotland
		1949	Central England and Wales, Kent
		1921	South and Central England
		1887	E. Ireland, Central England
Two summers and intervening winter	Major reservoirs	1955–56	Cumbria
		1943–44	S. England
		1933–34	England, Wales
		1887–88	Wales
Three/four years	Major aquifers	1971–74	E. Scotland, E. England
		1962–64	E. England
		1901–03	S. England, Yorkshire

Dry weather flow (DWF) records, collected mainly from current-meter gauging surveys, have been published in summary form by the Water Resources Board[145]. The countrywide variation in unit DWF values is related to catchment geology rather than rainfall. Wright[146] has studied this phenomenon by classifying the response of different types of geological formation in the form of a geological index, and using it to predict lowest monthly flow (Q) in the summer/autumn period. By regression analysis of 64 records with an average length of five years, he was able to reduce the factorial standard error of estimate of the discharge (Q) to 1.27.

$$Q = \frac{9.84}{10^8} \text{AREA}^{1.05} \text{AAR}^{0.683} \text{GI}^{0.59} \text{WR}^{0.411} \text{SR}^{0.658} \qquad (6.19)$$

where AREA   is expressed in km² and Q in l/s
       AAR = average annual rainfall (mm)
         SR = summer rainfall (mm), i.e. April–September precipitation in the year in which the low flow occurs
      WR = winter rainfall (mm), i.e. precipitation in the six months prior to the summer rainfall
       GI = geological index, values for which are as follows:

Chalk	125 000
Sand and gravel on Chalk	,,
Chalky Boulder Clay	31 620
Great and Inferior Oolite Series	,,
Upper and Lower Greensand	,,
Brickearth and Crags (East Anglia)	,,
Chalky Jurassic Boulder Clay	15 850
Clay with flints	,,
Boulder Clay (undifferentiated)	,,
London Clay	10 000
Gault	,,
Stiff Boulder Clay	,,
Chalk Marl (in Lower Chalk)	,,
Oxford Clay	,,
Alluvium and Loam	,,

Dry-weather flow for a single day is readily affected by milling operations on the river, irrigation abstraction peaks and the pattern of industrial water use in the working week. The idea has been broached[147] that such flows can be standardised by using the arithmetic mean of a series formed from the lowest weekly flow rate per annum. This parameter, termed the Seven-Day Minimum Flow, probably stabilises with less than ten years of information. *Table 6.10* gives some examples from Yorkshire and South Wales[148].

Table 6.10 SEVEN-DAY MINIMUM FLOWS

River and station	Period of record	Minimum SDMF for period of record m³/s	year	Mean SDMF m³/s	% time exceeded	Min SDMF / Mean SDMF %
Hodge Beck (Bransdale)	1956–72	0.028	1959	0.072	91.8	42.4
Nidd (Hunsingore)	1955–72	0.78	1959	1.73	91.7	45.1
Swale (Leckby)	1956–72	1.79	1959	3.68	93.7	48.6
Ure (Westwick)	1958–69	1.16	1959	2.56	96.4	45.3
West Beck (Wansford)	1953–72	0.33	1965	0.72	83.7	45.8
Hull (Hempholme)	1962–72	0.58	1965	1.29	81.9	44.9
Wharfe (Ilkley)	1960–72	0.94	1972	1.87	92.1	50.3
Lwyd (Ponthir)	1967–72	0.59	1972	0.66	98.2	89.4
Honddu (Brecon)	1964–72	0.11	1964	0.21	97.0	52.4
Usk (Trallong)	1964–72	0.85	1964	1.06	96.0	80.2
Ebbw (Rhiwderin)	1958–72	1.15	1962	1.69	95.0	68.0
Usk (Chain Bridge)	1958–72	2.35	1959	4.72	93.0	49.8

Drought flow probability

Early runoff probability studies were carried out along lines laid down by
Hazen in the United States, but it was the impetus of a paper by
Thompson[149] in 1950 that stimulated British work in this field. He
demonstrated with data from the Derbyshire Derwent at Yorkshire Bridge
that runoff over any fixed period between 14 and 20 months was normally
distributed. Subsequent work shows that this holds over that part of the
country where the coefficient of variation of runoff (for the period
concerned) is less than 0.25[83]. This generally holds for durations greater
than six months in areas with an average annual rainfall exceeding 1250 mm.
Figure 6.11 indicates the range of the coefficient of variation of annual
events.

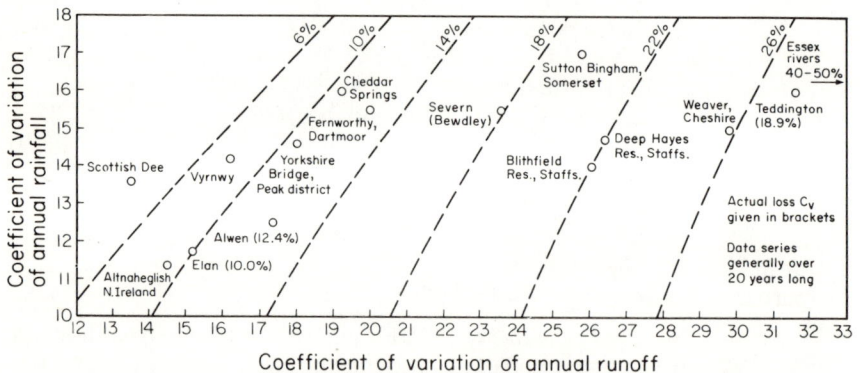

*Figure 6.11   Variability of annual rainfall, runoff and loss; the dashed lines indicate theoretical
C_v of loss, i.e. rainfall – runoff (assuming these factors are independent events; this is less true
where C_v is high in permeable areas)*

In the Ministry of Housing and Local Government's Hydrological
Surveys[150] for various regions of England, and in other papers[151], probability
was attributed to estimates of reservoir yield by modifying Thompson's
method, and using it to predict the lowest flow to be expected once in 100
years. The results were presented in the form of minimum runoff diagrams[19],
each point showing total flow for an interval of between 1 and 36 months.
More recently many water undertakings have adopted a design standard of
once in 50 years for the provision of drought storage. *Figure 6.12* shows the
extent to which such flows fall below average in the various parts of the UK.
It can be seen that the higher the average rainfall the lower is the
proportionate variability of discharge. Other factors that have a noticeable
effect in varying the size of the departure of drought conditions from the
normal are:

1. the extent to which aquifers sustain baseflow;
2. the magnitude of the coefficient of variation of annual rainfall (see
   Chapter 2);
3. the average timing of the season of low runoff: the later in the year that
   this season occurs, the worse the 2 per cent drought becomes
   relatively.

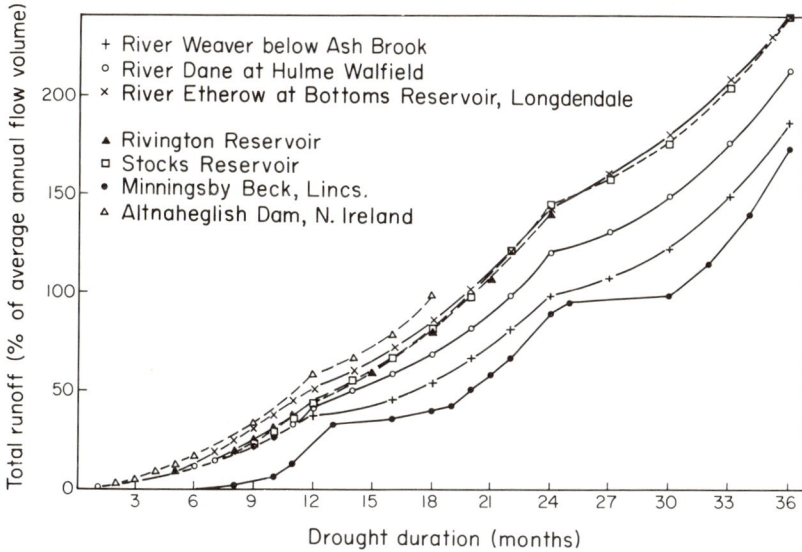

Figure 6.12   Dimensionless minimum runoff diagram for a fifty-year drought

The strong family pattern that is exhibited by dimensionless minimum runoff diagrams is demonstrated if other return periods are examined, and this pattern is useful for estimating drought potential at a site with little or no data.

Groundwater outflows

Spring flows, and their decay with time in the absence of recharge, have been described by a variety of equations. The most popular is:

$$Q_t = Q_0 \exp(-kt) \qquad (6.20)$$

where Q_t = the springflow t days after it was Q_0
  k = a recession constant (units: day^{-1})

This plots as a linear recession on 'log flow against time' plots. Recession constants fall in the range shown on Table 6.11. However, such recession constants do not hold at the lowest extremes; in perennial rivers they reduce as low ground water levels are reached. The reason for this is uncertain; it may be that at low discharges the secondary porewater drainage from previously saturated strata becomes relatively more important. Alternatively, it may be that the recession curve is actually a composite curve representing drainage from different 'zones' in the aquifer with different properties. The 'zones' could be both laterally distributed at increasing distances from the river and vertically distributed, that is at increasing depth. The departure from the theoretical drainage curve may also be influenced by evaporation from the riparian zone.

  Rather than separate total groundwater discharge (i.e. baseflow) from surface runoff in hydrograph analysis by an arbitrary under-envelope of the

Table 6.11　AQUIFER RECESSION CONSTANTS

River	Gauging station	Aquifer	k per day	Reference
Un-named streamlet	Southampton University	Bracklesham Beds	0.090	Webber[152]
Test	Broadlands	Upper Chalk	0.005	Headworth[153]
Itchen	Allbrook	Upper Chalk	0.005	Hydrological survey[154]
Meon	Mislingford	Chalk	0.011	Webber[152]
		Peat	0.152	Wright[155]
Lothian rivers		Sand and Gravel	0.080	
		Boulder Clay	0.061	
		Sandstone	0.034	
Aln	Hawkhill	Fell Sandstone, Limestones	0.007	Petrie[156]
Coquet	Morwick	Fell Sandstone, Limestones	0.016	Petrie[156]
Nar	Marham	Chalk	0.014	Binnie and Partners[157]
Wissey	Northwold	Chalk (drift cover)	0.012	Binnie and Partners[157]
Cam	Dernford	Chalk (drift cover)	0.012	Binnie and Partners[157]
Lark	Temple Weir	Boulder Clay over Chalk	0.028	Binnie and Partners[157]
Rhee	Ashwell	Chalk spring	0.008	Binnie and Partners[158]
Lambourn	Shaw	Chalk	0.004 (low flows)	Thames Conservancy[159]
Winterbourne	Bagnor	Chalk	0.005 (low flows)	Thames Conservancy[159]
Coln	Bibury	Jurassic limestones	0.010	Binnie and Partners[160]
Avon	Fordingbridge	Chalk	0.014 (low flows)	Avon and Dorset RA[161]
Nadder	Wilton	Chalk	0.006 (low flows)	Avon and Dorset RA[161]
Wylye	Wilton	Chalk	0.011 (low flows)	Avon and Dorset RA[161]
Frome	East Stoke	Chalk	0.009 (low flows)	Avon and Dorset RA[161]

obvious low points, preference is often given to graphical well-level–baseflow relationships. It is important that the relationship of the well level with baseflow is defined in absolutely dry conditions when tile drains have ceased to run. In general terms about a fortnight without rain on the catchment is a necessary minimum before it can be said that natural river flow is really baseflow.

Lakes and reservoirs

The largest freshwater lake in the UK is Lough Neagh[162] in Northern Ireland, with a surface area of 397 km². Natural lakes are most abundant in North Wales[163], the Lake District, the Highlands of Scotland[164] and Western Ireland. Loch Ness is the longest (37 km) and Loch Morar is the deepest (310 m) body of fresh water.

Defining a major lake as one with a circumference of at least 30 km limits the number of English lakes in this category to only two and Irish loughs to ten. There are no Welsh llyns on this scale, but at least eleven Scottish freshwater lochs reach this size, all lying north of Glasgow. Extensive bathymetrical surveys have been conducted[165-168] and good estimates of the volumes of many lakes are available. Windermere[169], for instance, contains about 310000 Ml at typical water levels; it has a maximum top-water variation of about 2.2 m which is similar to that of many other British lakes.

Some natural lake outlets have been dammed to enlarge the storage range and provide major water supplies. Thirlmere[170] and Hawes Water[171] were successfully developed in this way and bring Lake District water to Manchester; additional intakes on Windermere and Ullswater have raised the yield to 679 Ml/d in recent years. Similar works have been carried out at Loch Katrine for Glasgow[172]. In 1970–71 a barrage was built at the outlet of Loch Lomond[173] to give control of 1.2 m of storage on a lake area of 71 km^2 so permitting a supply of up to 455 Ml/d to be abstracted for consumers in the central lowlands. A similar but smaller scheme was built on the Dee to control the levels of Bala Lake[174]. Besides flood control, additional regulation of the flow of the river was achieved, enabling increased abstraction above Chester. The new Bala sluices replaced those of Telford, which were constructed in the mid-nineteenth century to regulate the river for canal use. An Irish example is provided by the Beleek Sluices below Lough Erne in Fermanagh[175].

The earliest reservoirs were either built by men like Capability Brown to beautify the estates of large landowners[176] or by canal engineers. Many of the earth dams containing canal reservoirs were large by any standard, but were completed well before the laws of soil mechanics became known. The register[177] of all dams greater than 15 m in height kept by the International Commission on Large Dams shows Coombs Dam as the first entry for Great Britain; it was completed in 1797. Three more dams were started in that century, 148 in the nineteenth and a further 200 by 1974. Capacities have steadily increased[178], and twenty-five of these reservoirs store more than 46000 Ml (see Appendix 6).

The Brenig Reservoir, due for completion to its Stage 1 height in 1975, will impound 60000 Ml initially, and 170000 Ml ultimately. Powers have been granted for Kielder Reservoir, on a tributary of the Tyne, which will also regulate the Wear and Tees from its storage of 200000 Ml; north-south transfers between these rivers will be made largely by tunnel aqueducts. If the proposals for the enlargement of Craig Goch are accepted and a dam 110 m high is built, some 550000 Ml will be impounded (see Chapter 8).

Sediment transport and hydraulic geometry

Although there has been a great deal of laboratory research in the UK into sediment processes and river regimes, this research was largely in support of the design of major barrages and irrigation canals in other countries. The absence of any serious problems of economic significance in the field of freshwater sedimentation in the UK has discouraged any coherent network of measurement stations. Consequently published data are, in the main, from research into erosion processes made as a part of catchment studies. One significant exception is the study of estuarine conditions, undertaken initially as a part of pollution control[179–181], and more recently for flood control barrage[182] and water supply storage investigations. With increasing attention being given to water-borne pollutants, there is an added incentive for research on the factors controlling their transport and deposition.

Regime conditions

There has been a continuous debate for many years about the validity of the regime equations used with success to design the irrigation canal systems of India and Pakistan. By 'regime' it is meant that a river is in a state of equilibrium with neither accretion nor deposition of transported sediment occurring. Initially these equations were derived empirically, but latter attempts have been made to find a satisfactory theory to explain naturally occurring channel dimensions, bank-full discharge, bed slope and meander patterns.

Significant early British contributions to the debate came from Lindley[183], Lacey[184] and Inglis[185]. Their work concentrated on readily measured factors such as water surface width (W), hydraulic mean depth (A/W), energy gradient (S) and velocity (V), and the relation of these factors to rugosity (N_a), Manning's roughness coefficient (n), silt factor (f) and bed material size (d).

Lacey's equations in their developed form[186] are not dimensionless and in their original ft/s units are:

$$f = 1.60 \sqrt{d} \qquad \text{where } d \text{ is in mm} \tag{6.21}$$
$$V = (Qf^2/4)^{1/6} \qquad \text{with area } A = \text{flow } Q \div V \tag{6.22}$$
$$W = 2.67 \sqrt{Q} \tag{6.23}$$
$$S = 5.47 \, 10^{-4} \, (f^{5/3})/(Q^{1/6}) \tag{6.24}$$
$$N_a = 0.0225 \, f^{1/4} \tag{6.25}$$

Subsequently world-wide progress on the physics of sediment transportation on a laboratory scale has enabled workers at the Hydraulics Research Station, notably Ackers and Charlton[187] to extend regime theory.

The importance of the bank-full discharge to river engineers stimulated Nixon[188] to attempt a synthesis of British conditions. He defined the bank-full discharge as 'the maximum discharge reached by a bank-full flood at any point in the course of the river', adding that 'small pockets of low-lying land adjoining the river may be inundated before the river reaches a bank-full state, and there will be sections of the river with high banks which will not be overtopped except in a major flood having a discharge far exceeding that of bank-full'.

Table 6.12 shows his results but it probably poses more questions than it answers. This is primarily because regime theory cannot be expected to cope other than with a river flowing through a homogeneous material like sand or silt. Rock intrusions, artificial constrictions, subsidence features and many other factors upset any simple approach. However, Nixon concluded that, for an average river in England or Wales, the bank-full flow was that exceeded 0.6 per cent of the time (see pages 240–241).

Table 6.12 BANK-FULL RIVER CHARACTERISTICS

River	Station	Width (m)	Depth (m)	Area (m²)	Bank-full discharge (m³/s)	Catchment area (km²)	Unit Bank-full discharge (m³ s⁻¹ km⁻²)	% of time exceeded	% of mean flow
Thames	Teddington	78	3.3	260	316	9870	0.032	1.1	470
Thames	Day's Weir	47	2.8	130	136	3440	0.040	1.6	505
Wey	Tilford	11	1.3	14	21	396	0.053	1.2	580
Blackwater	Swallowfield	11	0.9	10	13	355	0.037	0.9	530
Severn	Bewdley	52	3.9	200	285	4330	0.066	2.0	455
Severn	Montford	36	4.2	150	200	2030	0.099	—	500
Avon	Evesham	30	2.7	80	100	2210	0.045	1.0	745
Stour	Kidderminster	10	1.8	18	14	324	0.043	0.4	515
Avon	Bath	30	1.7	50	119	1600	0.075	0.5	610
Avon	Melksham	15	4.3	64	42	666	0.063	0.7	600
Semington Bk.	Semington	10	1.5	15	13	158	0.082	0.2	1080
Mersey	Irlam	26	2.6	68	140	679	0.207	0.2	1055
Wye	Cadora	64	5.0	320	512	4040	0.127	0.4	730
Wye	Belmont	48	4.0	195	308	1900	0.162	0.9	685
Wye	Erwood	61	3.6	200	442	1280	0.345	0.1	1260
Wye	Rhayader	23	1.4	33	62	167	0.371	—	955
Weaver	Ashbrook	18	2.4	42	70	609	0.115	0.1	1280
Tyne	Barrasford	64	3.0	195	415	2180	0.190	0.2	925
Trent	Nottingham	55	4.0	220	334	7490	0.045	2.0	420
Dove	Rocester	18	1.6	29	40	399	0.100	0.6	560
Ouse	Bedford	26	2.5	65	75	1460	0.051	0.6	800
Cam	Bottisham	23	2.5	57	35	811	0.043	—	1000
Dee	Erbistock	51	2.0	105	170	1040	0.164	1.1	550

Note. Gauging station locations can be found in the *Surface Water Year Book Supplement*. Data taken from Ref. 188 and metricated, except final two columns reworked with mean flows for longer record; additional data from river authority periodical surveys in accordance with Section 14 of Water Resources Act 1963

The movement of stream channel position within a valley is a process of very slow evolution that is occasionally hastened by a man-made diversion or an extreme flood[189]. It has often been commented that more sediment is transported in a few hours by a raging torrent than in years of normal flow.

Suspended load

The material being carried seaward by a river is partly in solution, the 'dissolved load', but mainly temporarily in suspension, the 'suspended load'. The former is a strictly chemical phenomena[190] and is dealt with in Chapter 7. Here information is presented on the erosion of catchments and the potential for deposition downstream in stiller waters.

No rigorous standard has emerged for the measurement of sediment concentrations in the UK. Techniques range from dipping a bottle in a stream just off the bank to the standards adopted by the United States

Geological Survey[191]. Perhaps the most common approach is to take depth-integrated samples by lowering a special bottle through a fixed vertical on a known river cross-section, often at a current-meter station with a cableway[192]. From such spot sampling, a sediment rating curve can be constructed showing the mean measured concentration against river stage. As *Figure 6.13* illustrates, this will normally show higher concentrations on the rising limb of a flood than on the recession. This follows from the 'washing down' of a catchment in the early rain and the picking up of bed and bank materials as velocities rise towards a peak.

Development of continuous turbidity recorders in the 1960s has added a new dimension to research on suspended sediment. These recorders measure the back-scattering of light as the river water passes between two parallel plates. A record of the variation of light intensity and a prior calibration of the instrument permits a measure of mean suspended load to be obtained *at the sampling point.* Alternatively some recorders function by

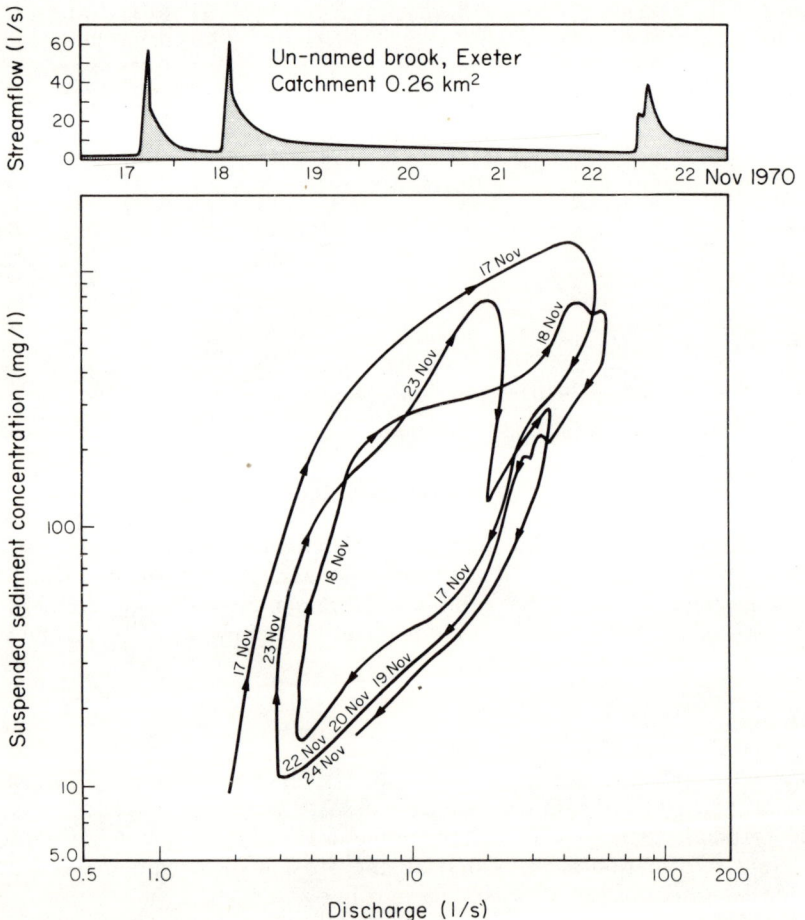

Figure 6.13   Sediment rating curve exhibiting hysteresis effects (from Walling[85], courtesy Institute of British Geographers)

comparing the intensity of two beams of light, one beam having an uninterrupted passage, the other passing through the river water. The sampling point will be either the small-bore pumped intake to the instrument housing or, in the case of the instrument devised by the Water Pöllution Research Laboratory[193], will be at the position at which the plates are fixed on the river bank. In this case the results are not strictly comparable with depth-integrated samples. Most automatic water-quality monitoring stations are equipped with a sediment meter.

Perhaps the first measurements of river sediment-carrying properties were those made by the Royal Geographical Society during the first decade of this century[36], the measurements being made in the Severn at Worcester, the Exe at Brampford Speke and the Medway at Maidstone. Subsequently extensive studies have been carried out on the Tyne[194] and the Clyde[195]. Results are frequently expressed as tonne/year and then related to catchment area; alternatively they may be expressed as the mean erosion rate of the catchment surface, in mm/year. Fleming[196] has drawn together some of the data and compared it with the much higher sediment movements of America and the Far East. *Figure 6.14* shows how scattered are the results before factors like slope, catchment geology and vegetation cover are taken into account. Except in special circumstances there is little scope for rivers to

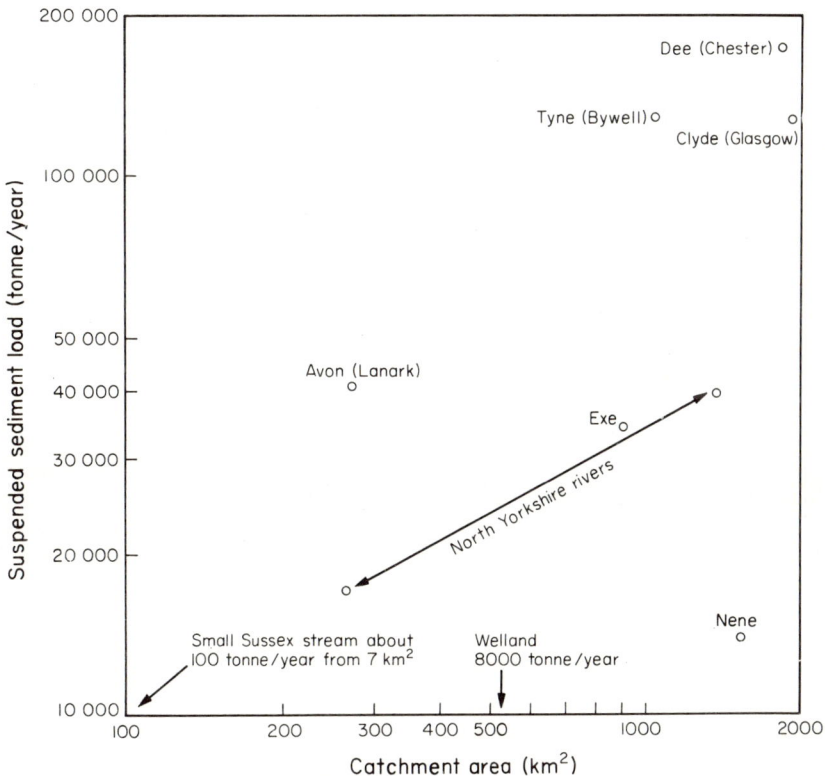

Figure 6.14   British suspended sediment load

carry large loads in suspension, because rainfall intensities are low, catchments are covered by vegetation for much of the year and most channels are not steep and are usually lined with materials that do not erode readily. One exception is basins where a land use change is under way, for example where building development is taking place. Research on a basin in this category is being carried out from Exeter University[197]. Similarly when catchment soils are stripped of their protective surface cover by locally intense surface runoff, erosion can continue for many months. Riverbank erosion is a main source of sediment, and at Bradgate Brook, Leicestershire, observations are being kept on a typical meander to account for its temporal change[198]; the measured annual rate of recession of the associated Keuper Marl cliff is 25 mm.

On the Plynlimon catchments the Institute of Hydrology has, since 1973, observed the difference in sediment production of areas that differ only in vegetative cover[199]. Suspended sediment from an area of moorland and another of coniferous forest has been sampled at 8-hour intervals with a vacuum sampler, supplemented by flood event sampling using a Tait-Binckley sampler. The latter is used to take perhaps five samples in 15 minutes or ten samples in 30 minutes in a search for an optimum. Already there is evidence of higher sediment yields created by the driving of forest roads.

Bed load

Bed load is exceedingly difficult to measure, but bed-load transport is an obvious feature of channels with sand and gravel beds. When the water level has subsided after a spate, fresh material can be seen lying upstream of any obstruction to flow such as a weir. Shoals come and go at any given point as conditions change and bed load moves.

Some idea of the quantities involved can be gained from the gravel traps operated at Plynlimon[199] and by the experiments conducted by the Catchment Board in the Lake District[200]. These studies suggest annual bed loads that vary from 15 to 60 tonne/km^2 on steep catchments. On the lower Clyde, where lesser slopes favour the deposition of bed load, the river gravels are commercially worked taking advantage of a much higher load of 95 tonne/km^2 per year[195].

Muir[201] tried to carry out measurements on the Tyne with a pressure-difference bed-load trap and with an acoustic device, but he had to resort to theoretical and empirical formulae. To apply these formulae required twenty-three sections on a relatively straight kilometre reach of river that was gauged and rated. During flood events water surface elevations at the ends of the reach were measured to the nearest 3.05 mm (0.01 ft) to find the energy surface slope. Bulk and areal samples of bed material were taken at six locations to determine particle size and shape, size grading being taken from a composite bulk sample weighing 0.67 tonne. Muir concluded that the Meyer-Peter and Muller equation[202] was most reliable for determining the bed load of gravel-bed rivers like the Tyne, and this equation produced an estimated average annual bed load of 15 100 tonnes for the 1956–66 period. The estimated bed-load yield of the Tyne catchment was 6.9 tonne/km^2 and

Table 6.13 LAKE SEDIMENT DEPOSITION RATES

River	Reservoir	Original volume (Ml)	Subsequent volume (Ml)	Deposition period (years)	Catchment area (km²)	Average inflow (m³/s)	Sediment deposit* (tonne/km² per year)	(mg/l)	Reference
Bradgate Brook	Cropston	2527	2323†	95	17.8	0.14	30	48.5	Cummins & Potter[206]
Churnet	Deep Hayes	455	435	116	9.8	0.14	4.4	3.9	Binnie & Partners[207]
Wyre	Abbeystead	831	455	85	47.3	1.70	23	8.3	Armstrong[208]
Loxley	Strines	2060	1953	87	10.9	0.27	28	13.5	Young[209]
Dodder (Eire)	Bohernabreena	709	500	66	28.0	0.84	28	4.8	

*Assumes a lake mud that is 90 per cent water/10 per cent sediment and of specific gravity 2.5
†Indicates a volume that would have occurred but for upstream settling ponds

this was higher than estimates derived from eight other methods. The bed load in the Tyne amounts to 11 per cent of the average annual suspended sediment load; the comparable figure for the Clyde[195] is 120 per cent. Thus, while it can be assumed that the suspended load in most British rivers is the main component of the sediment transport, this assumption is not always valid.

The composition of river-bed materials is now receiving attention from ecologists because of its importance in harbouring life that is important to fish. This involves sediment studies on a micro scale with especial attention being paid to the detritus and fine sands[203]. The high value placed on salmon and trout rivers has given an incentive to defining the flow and gravel bed conditions that are suitable for spawning[204].

Deposition in lakes and reservoirs

Lakes provide an ideal opportunity for the deposition of sediment, and where reliable bed surveys are carried out at widely spaced intervals of time a good estimate of total sediment load can be obtained. The most accurate method is to survey a reservoir's bottom prior to first impoundment and then to carry out the survey again in some major drought when the bed is re-exposed[205]. Such surveys are primarily concerned with computing the lost storage capacity and the results are usually expressed in terms of volume *(Table 6.13)*.

Comparison of the lake sediment deposit rates with the suspended and bed-load rates quoted earlier suggests that trap efficiency is not particularly high at typical small British reservoirs. However, some ornamental lakes built about 250 years ago have now almost completely silted up. Thoresby Lake[210] in the Dukeries of Nottinghamshire and Sherborne Lake[211] in Somerset provide good examples.

REFERENCES

1. Bateman, J. F. la Trobe, *History and description of Manchester Waterworks* (1884)
2. Medrington, N., 'The hydrology of a highly developed river basin—River Lee', *Informal discussion note,* Hydrological Group, Institution of Civil Engineers (1964)
3. Dixon, S. M. and Macaulay, F. W., 'Measurements of discharge over a rock-faced dam', *Mins Proc. Instn Civil Engrs,* **220,** 110–123 (1925)
4. Risbridger, C. A. and Godfrey, W. H., 'Rainfall, runoff and storage: Elan and Claerwen gathering grounds', *Proc. Instn Civil Engrs,* **3(3),** 345–388 (1954)
5. Lewis, W. K., 'Investigation of rainfall, runoff and yield on the Alwen and Brenig catchments', *Proc. Instn Civil Engrs,* **8,** 17–51 (1957)
6. Thompson, R. W. S. and Saxton, K. J. H., 'Rainfall and runoff in the Derwent Valley to Yorkshire Bridge, Derbyshire', *Proc. Instn Civil Engrs,* **25,** 147–163 (1963)
7. Law, F. M. and Macdonald, D. E., 'The discovery of a 50-year flow record at Deep Hayes, Staffordshire', *J. Instn Water Engrs,* **27,** 319–322 (1973)
8. Taylor, J., Discussion of 'On the floods in England and Wales during 1875', *Mins Proc. Instn Civil Engrs,* **45,** 101–102 (1876)
9. Andrews, F. M., 'Some aspects of the hydrology of the Thames Basin', *Proc. Instn Civil Engrs,* **21,** 55–90 (1962)
10. McClean, W. N., 'Rainfall and runoff, River Garry, Inverness-shire', *Trans. Instn Water Engrs,* **32,** 110–146 (1927)

11. McClean, W. N., 'River Dee (Aberdeenshire) at Woodend and Cairnton: fifteen years' record of water level, flow and rainfall', *River Flow Records,* London (1946)
12. Wileman, R. F. and Clark, H. W., 'The measurement of the discharges of the river basins of the White Nile (Sudan) and Nene (Great Britain)', *J. Instn Civil Engrs,* **26,** 267–295 (1946)
13. Chapman, S. C., 'Gauging and recording the flow of streams', *Trans. Instn Water Engrs,* **15,** 147–178 (1910)
14. Burgess, J. S., 'Stream-gauging in upland waters', Conference on Surface Water Resources and the Drought of 1959, *J. Instn Water Engrs,* **15,** 168–177 (1961)
15. British Standards Institution, 'The measurement of liquid level (stage)', Part 7 of *Methods of measurement of liquid flow in open channels,* BS 3680 (1971)
    See also Part 9, 'Water level instruments', and Part 9A, 'Specification for the installation and performance of pressure-actuated liquid level measuring equipment'
16. Charlton, F. G., *A review of the cost of installation, maintenance and operation of systems for the measurement of liquid flow with a free water surface,* CIRIA report (in preparation)
17. Herbertson, P. W., Douglas, J. W. and Hill, A., 'River level sampling periods', *Institute of Hydrology Report No. 9* (May 1971)
18. British Standards Institution, 'Thin plate weirs and venturi flumes', Part 4A of *Methods of measurement of liquid flow in open channels,* BS 3680 (1965)
19. Skeat, W. O. (ed.), *Manual of British water engineering practice,* 4th edn, Instn Water Engrs (1969)
20. Dowden, R. R., *Fluid flow measurement: a bibliography,* British Hydromechanics Research Association, Cranfield (1972)
21. Harrison, A. J. M., 'The streamlined broad-crested weir', *Proc. Instn Civil Engrs,* **38,** 657–678 (1967)
22. Harrison, A. J. M., 'Some comments on the square-edged broad-crested weir', *Water and Water Engng,* **68,** 445–448 (1964)
23. Crump, E. S., 'A new method of gauging stream flow with little afflux by means of a submerged weir of triangular profile', *Proc. Instn Civil Engrs,* **1,** 223–242 (1952)
24. Burgess, J. S. and White, W. R., 'The triangular profile (Crump) weir: two-dimensional study of discharge characteristics', *Report No. INT 52,* Hydraulics Research Station, Wallingford (1966)
25. Water Resources Board, 'Crump weir design', *Technical Note No. 8* (revised), Reading (1970)
26. White, W. R., 'The performance of two-dimensional and flat-V triangular-profile weirs', *Proc. Supplement Instn Civil Engrs,* Paper 7350S, 21–48 (1971)
27. Ree, W. O. and Crow, F. R., 'Measuring runoff rates with rectangular highway culverts', *Technical Bulletin T–51,* Oklahoma Agricultural Experiment Station (1954)
28. Ackers, P., 'Comprehensive formulae for critical-depth flumes', *Water and Water Engng,* **65,** 296–306 (1961)
29. Wright, C. E., 'Glass-fibre trapezoidal flumes', *Surveyor,* 4 Feb., 23–27 (1967)
30. Dutton, C., unpublished lecture course note, Chelmsford Technical College (1965)
31. Engel, F. V. A., 'The venturi flume', *Engineer (London),* **156,** 104–107 and 131–133 (1934)
32. Henderson, F. M., *Open channel flow,* Macmillan, New York, 213–214 (1960)
33. Hydraulics Research Station, 'Plynlimon experimental catchments: model investigation of a structure for flow measurement in steep streams', *Report No. EX 335,* Wallingford (1966)
34. Harrison, A. J. M., 'Factors governing the choice of a hydraulic structure for flow measurements', *Proc. Seventh Congress, Int. Commission on Irrigation and Drainage,* Mexico City, Paper 24, 117–137 (1969)
35. Hydraulics Research Station, 'Investigation of proposed gauging structure, at Footholme in the Forest of Bowland, Yorkshire', *Report No. EX 424,* Wallingford (1968)
36. Royal Geographical Society, *Final Report on the Investigation of Rivers* (1912)
37. McClean, W. N., 'Practical river flow measurement and its place in the Inland Water survey as exemplified on the Ness (Scotland) Basin', *Trans. Instn Water Engrs,* **38,** 233–261 (1933)
38. Dixon, S. M., Fitzgibbon, G. and Hogan, M. A., 'The flow of the River Severn 1921–36', *J. Instn Civil Engrs,* **6,** 81–109 (1937)
39. Croker, G. N., 'Records of flow in the River Wye system . . .', *J. Instn Water Engrs,* **5,** 39–76 (1951)

40. Clark, H. W., *Memorandum on river gauging, with special reference to the work carried out in the River Nene catchment area*, River Nene Catchment Board, Oundle (1950)

41. Bates, A. V. and Helliwell, P. R., 'An investigation into the costs of streamflow data production', *Report to the Water Resources Board*, Reading (1971)

42. Lambert, A., *Current metering in open channels*, unpublished dissertation in part qualification for Diploma in Hydrology, University of Newcastle-upon-Tyne (1964)

43. Johnson, D. and Tattersall, K. H., 'Hydrological data analysis for water resources problems', *J. Instn Water Engrs*, **25**, 181–198 (1971)

44. British Standards Institution, *Methods of measurement of liquid flow in open channels*, BS 3680 (1964)
    Part 3, 'Velocity area methods'
    Part 8, 'Current meters'
    Part 8A, 'Current meters incorporating a rotating element'
    Part 8B, 'Current meters: suspension equipment'
    Part 8C, 'Current meters: rating procedure' (in press)

45. Herschy, R. W., 'The ultrasonic method of river gauging', *Water Services*, **78**, 198–200 (1974)

46. Herschy, R. W. and Loosemore, W. R., 'The ultrasonic method of river flow measurement', *Proc. WRC/WDU Symp. on river gauging by ultrasonic and electromagnetic methods*, Reading (1974)

47. Water Resources Board, *Eighth Annual Report of the Water Resources Board for the year ending 30 September 1971*, HMSO (1972)

48. Water Resources Board, *Ninth Annual Report of the Water Resources Board for the year ending 30 September 1972*, HMSO (1972)

49. Herschy, R. W. and Newman, J. D., 'Electromagnetic river gauging', *Proc. WDU/WRC Symp. on river gauging by ultrasonic and electromagnetic methods*, Reading (1974)

50. Benson, M. A., 'Measurement of peak discharge by indirect methods', *World Meteorological Organisation Technical Note No. 90* (1968)

51. Dobbie, C. H. and Wolf, P. O., 'The Lynmouth flood of August 1952', *Proc. Instn Civil Engrs*, **2**, 522–588 (1953)

52. Hydraulics Research Station, 'Bowland Forest and Pendle floods of August 1967', *Report No. EX 382*, Wallingford (1968)

53. Manning, R., 'Flow of water in open channels and pipes', *Mins Proc. Instn Civil Engrs of Ireland*, **20**, 161–207 (1890)

54. British Standards Institution, 'Slope area method of estimation', Part 5 of *Methods of measurement of liquid flow in open channels*, BS 3680 (1970)

55. Harvey, A. M., 'Channel capacity and the adjustment of streams to hydrologic regime', *J. Hydrology*, **8**, 82–98 (1969)

56. White, W. R., 'A field comparison of the calibrations of a flat-V weir and a set of electromagnetic flowmeters', *Water and Water Engng*, **75**, 340–344 (1971)

57. Turner, J. H. T., in discussion of: A. W. Brightmore, 'Methods of collecting water for supply', *Trans. Liverpool Engng Soc.*, **13**, 51–53 (1892)

58. Schloesing, T., 'Water measurement by the chemical method, (in French), *Comptes rendus du deuxième semestre de l'Académie des Sciences*, Paris (1863)

59. Stromeyer, C. E., 'The gauging of streams by chemical means', *Proc. Instn Civil Engrs*, **160**, 349–363 (1905)

60. Spencer, E. A. and Tudhope, J. S., 'A literature survey of the salt-dilution method of flow measurement', *J. Instn Water Engrs*, **12**, 127–138 (1958)

61. Hutton, S. P. and Spencer, E. A., 'Gauging water flow by the salt dilution method', *Proc. Instn Civil Engrs*, **16**, 395–418 (1960)

62. Collinge, V. K. and Simpson, J. R. (editors), 'Dilution techniques for flow measurement', *Bull. Civil Engng Dept*, Newcastle University, 105 (1964)

63. Water Research Association, 'River flow measurement by dilution gauging', *a manual produced under contract to Water Resources Board*, WRA, Medmenham (1970)

64. Water Resources Board, *Sixth Annual Report of the Water Resources Board for the year ending 30 September 1969*, HMSO (1970)

65. Institute of Hydrology, *Research 1971–72* (1972)

66. Bottani, E., 'Teoria, errori ed approssimazione del metodo chimico-elettrico per la misura delle portate', *L'Elettrotecnica* (15 Feb 1926)

67. Mansell-Moullin, M., in discussion on: J. A. Cole, 'Dilution gauging by inorganic tracers', *Proc. Symp. on River Flow Measurement*, Institution of Water Engineers, 63–80 (1970)

68. British Standards Institution, *Methods of measurement of liquid flow in open channels,* BS 3680 (1964)
    Part 2, 'Dilution methods'
    Part 2A, 'Constant rate injection'
    Part 2B, 'Sudden injection' (in press)
    Part 2C, 'Radio-isotope techniques'
69. Pilgrim, D. H. and Summersby, V. J., 'Less conventional methods of stream gauging', *Civil Engng Trans.,* April 1–11, Institution of Engineers of Australia (1966)
70. Clayton, C. G. and Smith, D. B., 'A comparison of radioisotope methods for river flow measurement', *Proc. IAEA Conf. on radioisotopes in hydrology,* Vienna, 1–24 (1963)
71. Ward, R. C., 'Some runoff characteristics of British rivers', *J. Hydrology,* **6,** 358–372 (1968)
72. Parde, M., 'Hydrologie fluviale des Iles Britanniques', *Annales de Géographie,* **48,** 369–384 (1939)
73. Menzies, W. J. M., discussion on 'River utilisation and the preservation of migratory fish life', *Proc. Instn Civil Engrs,* **21,** 896 (1962)
74. Robertson, A. I. G. S., 'Preparation of discharge frequency diagrams, surface water survey', *Technical Note No. 13,* Ministry of Housing and Local Govt (1962)
75. Aitken, P. L., discussion on 'River utilisation and the preservation of migratory fish life', *Proc. Instn Civil Engrs,* **21,** 894 (1962)
76. Cochrane, N. J., 'An engineering calculation of risk in the provision for the passage of floods during the construction of dams', *Proc. 9th Congress ICOLD,* Vol. V, 325–341
77. Baxter, G., 'River utilisation and the preservation of migratory fish life', *Proc. Instn Civil Engrs,* **18,** 225–244 (1961)
78. Hoyle, N., 'Frequency curves and applications', *J. Instn Water Engrs,* **17,** 499–509 (1963)
79. Cole, J., 'Final report of Research Panel No. 5: Recession curves and frequency diagrams', *J. Instn Water Engrs,* **20,** 231–250 (1966)
80. Hall, D. G., 'The assessment of water resources in Devon, England, using limited hydrometric data', *Proc. IASH General Assembly, Berne 1967,* Pub. No. 76, 110–120 (1968)
81. Binnie, W. J. E. 'Water power problems', *Trans. Liverpool Engng Soc.,* **52,** 99 (1920)
82. Mansell-Moullin, M. M., 'Application of flow frequency curves to the design of catchwaters', *J. Instn Water Engrs,* **20,** 409–424 (1966)
83. Twort, A. C., Hoather, R. C. and Law, F. M., *Water supply* (2nd edn) Edward Arnold, London (1974)
84. United States Geological Survey, *Technique of water resources investigations of the USGS,* Book C, Ch. C1 to C3 on Fluvial Sediments, US Govt Printing Office, Washington (1970)
85. Walling, D. E., 'Suspended sediment and solute yields from a small catchment prior to urbanisation', *Fluvial processes in instrumented watersheds,* Inst. Brit. Geographers Special Publication No. 6, 169–192 (1974)
86. Britton, C. E., 'A meteorological chronology to A.D. 1450', *Geophysical Memoirs No. 70* (First Number, Vol. III), Meteorological Office (1937)
87. Stratton, J. M., *Agricultural records AD 220–1968,* John Baker, London (1969)
88. Symons, G. J. and Chatterton, G., 'The November floods of 1894 in the Thames Valley', *Q. J. R. Met. Soc.,* **21,** 28–89 (1895)
89. Rhodes, L. A., *Engineer's report on problems in connection with drainage and flooding within the area of the Board . . .,* Severn River Board (1950)
90. Potter, H. R., 'Introduction to the history of the floods and droughts of the Trent Basin', *unpublished typescript,* Trent River Authority (undated)
91. Brooks, C. E. P. and Glasspoole, J., *British floods and droughts,* Ernest Benn, London (1928)
92. *Flood Studies Report,* Natural Environment Research Council (1975)
93. Johnson, E. A. G., 'Flooding in the Fens and remedial measures taken', *Maritime Paper No. 10,* Instn Civil Engrs, 3–31 (1948)
94. Howorth, B., Mowbray, N. A., Haile, W. H. and Crowther, G. C., 'The spring floods of 1947', *J. Instn Water Engrs,* **2,** 12–46 (1948)
95. Harrison, S., *A complete history of the great flood at Sheffield on March 11 and 12,* Harrison, Sheffield (1864)
96. Rawlinson, R. and Beardmore, N., *'Report on the failure of the Dale Dyke reservoir',* 20 May 1864 (1864)

97. Institution of Civil Engineers, *Interim Report of the Committee on floods in relation to reservoir practice* (1933)
98. Institution of Civil Engineers Floods Working Party, *Reservoir Flood Standards Discussion Paper* (1975)
99. Nash, J. E., 'Determining runoff from rainfall', *Proc. Instn Civil Engrs,* **10,** 163–184 (1958)
100. Lloyd-Davies, D. E., 'The elimination of storm water from sewerage systems', *Mins Proc. Instn Civil Engrs,* **164,** 41 (1906)
101. Mulvaney, T. J., 'On the use of self-registering rain and flood gauges in making observations of the relations of rainfall and of flood discharges in a given catchment', *Trans. Instn Civil Engrs of Ireland,* **4 (2),** 18–33 (1850–1851)
102. Dury, G. H., 'Analysis of regional flood frequency on the Nene and Great Ouse', *Geographical Journal,* **125,** 223–229 (1959)
103. Cole, G., 'An application of the regional analysis of flood flows', *Proc. Symp. on river flood hydrology,* Instn Civil Engrs, 39–58 (1966)
104. Gumbel, E. J., 'Statistical theory of floods and droughts', *J. Instn Water Engrs,* **12,** 157–184 (1958)
105. Biswas, A. K. and Fleming, G., 'Regional flood frequency analyses for Scotland', *Proc. Instn Civil Engrs,* **25,** 313–315 (1966)
106. Lynn, M. A., 'Flood estimation for ungauged catchments', *Irish Engineer, J. Inst. Engrs of Ireland,* **24,** 7–24 (1971)
107. Nash, J. E. and Shaw, B. L., 'Flood frequency as a function of catchment characteristics', *Proc. Symp. on river flood hydrology,* Instn Civil Engrs, 115–136 (1966)
108. Rodda, J. C., 'The significance of characteristics of basin rainfall and morphometry in a study of floods in the United Kingdom', *Proc. IAHS/UNESCO/WMO Symposium on floods and their computation,* IAHS Pub. No. 85, 834–845 (1967)
109. Institution of Civil Engineers, *Flood studies for the United Kingdom,* Report of the Committee on Floods in the United Kingdom (1967)
110. Somerset River Authority Annual Reports
111. Binnie, A. M., 'Flooding of the Hundred-Foot Washes, River Great Ouse', *J. Instn Water Engrs,* 21, 432–434 (1967)
112. Midmer, F. N., 'Rother area drainage improvement scheme', *Association of River Authorities Year Book for 1969* (1969)
113. Nixon, M., 'Flood regulation and river training', Ch. 18 of *River engineering and water conservation works* (ed. R. B. Thorn), Butterworths, London (1966)
114. Fordham, A. E., Cochrane, N. J., Kretschmer, J. M. and Baxter, R. S., 'The Clywedog Reservoir project', *J. Instn Water Engrs,* **24,** 17–48 (1970)
115. Crann, H. H., 'The design and construction of Llyn Celyn', *J. Instn Water Engrs,* **22,** 13–42 (1968)
116. Macdonald, Sir M. and Partners, *Flood control in the Medway Valley,* Kent River Authority, Maidstone (1969)
117. Macdonald, Sir M. and Partners, *River Eden flood control,* Kent River Authority, Maidstone (1971)
118. Wright, G. A., 'Comprehensive scheme to utilise the resources of the River Dee', *J. Instn Water Engrs,* **9,** 229–244 (1955)
119. Fulton, A. A. and Dickerson, L. M., 'Design and constructional features of hydroelectric dams built in Scotland since 1945', *Proc. Instn Civil Engrs,* **29,** 713–742 (1964)
120. Nash, J. E., 'Determining runoff from rainfall', *Proc. Instn Civil Engrs,* **10,** 163–184 (1958)
121. O'Kelly, J. J., 'The employment of unit hydrographs to determine the flows of Irish arterial drainage channels', *Proc. Instn Civil Engrs,* **4,** 365–412 (1955)
122. Linsley, R. K., 'River forecasting', *Conference on forecasting river floods,* Imperial College, 21 May 1958 (Typescript paper in which Figure 11 is a 1-day UH for Day's Weir, R. Thames)
123. Lambert, A. O., 'An investigation into infiltration and interception rates during storm rainfalls, and their application to flood prediction', *J. Instn Water Engrs,* **21,** 529 (1967) (Sankey Brook, Lancashire)
124. Devon River Authority, *First report on hydrology* (1964) (Figure 14: Exe at Thorverton, Devon)
125. Andrews, F. M., 'Some aspects of the hydrology of the Thames Basin', *Proc. Instn Civil Engrs,* **21,** 82 (1962) (Figure 12 shows 2, 4, 6, 8 and 10-day 1-inch UHs for R. Thames at Teddington)

126. Lothians River Purification Board, *Annual Report for 1966* (1967) (Figure 4)
127. Wilson, E. M., *Engineering hydrology,* 2nd edn, Macmillan, 146 (1974) (UH for River Rother, Yorkshire)
128. Nash, J. E., 'The form of the instantaneous unit hydrograph', *Proc. IASH General Assembly, Toronto,* Pub. No. 45, 114–121 (1957)
129. O'Donnell, T., 'Instantaneous unit hydrograph derivation by harmonic analysis', *Proc. IASH General Assembly, Helsinki,* Pub. No. 51, 546–557 (1960)
130. Nash, J. E., 'A unit hydrograph study, with particular reference to British catchments', *Proc. Instn Civil Engrs,* **17,** 249–282 (1960)
131. Nash, J. E., 'The effect of flood-elimination works on the flood frequency of the River Wandle', *Proc. Instn Civil Engrs,* **13,** 317–338 (1959)
132. Bleek, J. M., *Estimation of storm runoff from urban catchments by synthetic unit hydrograph procedures,* unpublished dissertation, Imperial College, London (1972)
133. Paulhus, J., 'Manual for estimation of probable maximum precipitation', *Operational Hydrology Report No. 1,* World Meteorological Organisation, Geneva (1974)
134. British National Committee on Large Dams, *Proc. Symp. on the inspection, operation and improvement of existing dams,* Newcastle University (1975)
135. Hollis, G. E., 'The effect of urbanisation on floods in the Canon's Brook, Harlow, Essex', *Fluvial processes in instrumented watersheds, Inst. Brit. Geographers Special Publication No. 6,* 123–140 (1974)
136. Greenfield, D. J., 'Hydrologic effects of urbanisation at Crawley, Sussex', report in partial fulfilment of requirements for MSc, Imperial College, London (1970)
137. Hall, M. J., 'Synthetic unit hydrograph technique for the design of flood alleviation works in urban areas', *Proc. UNESCO/WMO/IAHS Symp. on design of water resources projects with inadequate data,* Madrid, **1,** 145–161 (1973)
138. Institute of Hydrology, *Research 1973–74,* 26 (1974)
139. Davis, L. H., 'Problems posed by New Town development with particular reference to Milton Keynes', *CIRIA Symp. on rainfall, runoff and surface water drainage of urban catchments, Bristol 1973,* 2-1 to 2-6 (1974)
140. Thompson, G., 'The use of balancing reservoirs and flow-regulating reservoirs in dealing with runoffs from urban areas', Ch. 9 of *River engineering and water conservation works* (ed. R. B. Thorn), Butterworths (1966)
141. King, M. V. and Hedley, G., 'Suggested correlation between storm sewage characteristics and storm overflow performance', *Proc. Instn Civil Engrs,* **48,** 399–411 (1971)
142. Lester, W. F., Woodward, G. M. and Rowan, T. W., 'River purification lakes', *Vol. 6, Trent Research Programme, Trent Steering Committee,* Water Resources Board, Reading (1972)
143. Road Research Laboratory, 'A guide for engineers to the design of storm sewer systems', *Road Note No. 35,* 20 (1963)
144. Watkins, L. H., 'The design of urban sewer systems', *Road Research Technical Paper No. 55,* HMSO (1962)
145. Water Resources Board, 'Dry weather flows', *Technical Note 12* (1970)
146. Wright, C. E., 'The influence of catchment characteristics upon low flows in south-east England', *Water Services,* **78,** 227–230 (1974)
147. Hindley, D. R., 'The definition of dry-weather flow in river flow measurement', *J. Instn Water Engrs,* **27,** 438–440 (1973)
148. Thoms, M. J. D. and Wain, A. S., communication on 'The definition of dry-weather flow in river flow measurement', *J. Instn Water Engrs,* **28,** 125 (1974)
149. Thompson, R. W. S., 'The application of statistical methods in the determination of the yield of a catchment from runoff data', *J. Instn Water Engrs,* **4,** 394–419 (1950)
150. Ministry of Housing and Local Government, *North Lancashire rivers hydrological survey,* HMSO (1964)
151. Bannerman, R. B. W., 'River flow statistics', Ch. 12 of *River engineering and water conservation works* (ed. R. B. Thorn), Butterworths (1966)
152. Webber, N. B., 'The baseflow recession curve: its derivation and application', *J. Instn Water Engrs,* **15,** 368–377 (1961)
153. Headworth, H. G., 'The analysis of natural groundwater level fluctuations in the Chalk of Hampshire', *J. Instn Water Engrs,* **26,** 107–124 (1972)
154. Ministry of Housing and Local Government, *Wessex rivers hydrological survey,* HMSO, 88 (1967)
155. Wright, C. E., 'Catchment characteristics influencing low flows', *Water and Water Engng,* **74,** 468–471 (1970)

156. Petrie, J. L., *A hydrogeological report on the Aln and Coquet catchments, Northumberland'*, unpublished MSc dissertation, University College, London (1972)
157. Binnie and Partners, *Water resources of the Great Ouse Basin,* Ministry of Housing and Local Government, Figure A4/2 (1965)
158. Binnie and Partners, *Report on Ashwell Springs recharge study* to Great Ouse River Authority (1974)
159. Thames Conservancy, *Report on the Lambourn Valley pilot scheme 1967–69,* Reading, Diagram 7.1 (1971)
160. Binnie and Partners, *Report on River Coln hydrological survey and the Buscot Source* to Oxfordshire and District Water Board (1972)
161. Avon and Dorset River Authority, *First Periodical Survey* (1970)
162. Gresswell, R. K. and Huxley, A., *Standard encyclopedia of the world's rivers and lakes,* Weidenfeld and Nicholson (1965)
163. Ward, F., *The lakes of Wales,* H. Jenkins (1931)
164. Calderwood, W. L., *The salmon rivers and lochs of Scotland,* E. Arnold (1909)
165. Murray, Sir J. and Pullar, L., *Bathymetrical survey of the freshwater lochs of Scotland,* reprinted from *Geographical Journal,* 1900–08, and published by the Royal Geographical Society (1908)
166. Kirby, R. P., 'The bathymetrical resurvey of Loch Leven, Kinross', *Geographical Journal,* **137,** 372–378 (1971)
167. Jehu, T. J., 'A bathymetrical and geological study of the lakes of Snowdonia and eastern Caernarvonshire', *Trans, R. Soc. Edinburgh,* **40 (2),** 419–467 (1902)
168. Ferrar, A. M., 'The depth of some lakes in Snowdonia', *Geographical J.,* **139 (3),** 516–519 (1973)
169. McLean, W. N., 'Windermere Basin: rainfall, runoff and storage', *Q. J. R. Met. Soc.,* **66,** 337–356 (1940)
170. Hill, G. H., 'The Thirlmere works for the water-supply of Manchester', *Mins Proc. Instn Civil Engrs,* **126 (IV),** 2–23 (1896)
171. Taylor, G. E., 'The Haweswater Reservoir', *J. Instn Water Engrs,* **5,** 355–380 (1951)
172. 'Glasgow's £797 000 Glen Finglas Reservoir opened', *Surveyor,* **135,** 17–18 (1965)
173. Cormie, W. M., 'The Loch Lomond water scheme', *J. Instn Water Engrs,* **24,** 291–318 (1970)
174. 'The Bala Lake scheme of the Dee and Clywd River Board', *Water and Water Engng,* **60,** 331–335 (1956)
175. O'Riordan, J. A., 'Runoff of the River Liffey and the Lower River Erne', *Instn Engineers of Ireland* (1949)
176. Kennard, M. F., 'Examples of the internal condition of some old earth dams', *J. Instn Water Engrs,* **26,** 135–147 (1972)
177. International Commission on Large Dams, *World register of dams* (1973)
178. Hetherington, R. le G., Presidential Address, *Proc. Instn Civil Engrs,* **54(1),** 1–10 (1972)
179. Inglis, Sir C. C. and Allen, F. H., 'The regime of the Thames as affected by currents, salinities and river flow', *Proc. Instn Civil Engrs,* **7,** 827–878 (1957)
180. Price, W. A. and Kendrick, M. P., 'Field and model investigation into the reasons for siltation in the Mersey estuary', *Proc. Instn Civil Engrs,* **24,** 473–518 (1963)
181. Hall, D. G., 'The pattern of sediment movement in the River Tyne', *Proc. IASH General Assembly, Bern 1967,* Pub. No. 75, 117–139 (1968)
182. Odd, N. V. M. and Owen, M. W., 'A two-layer model of mud transport in the Thames estuary', *Proc. Supplement (IX) Instn Civil Engrs,* Paper 7517S, 175–205 (1972)
183. Lindley, E. S., 'Regime Channels', *Proc. Punjab Engineering Congress,* **7,** 63 (1919)
184. Lacey, G., 'Flow in alluvial channels with sandy mobile beds', *Proc. Instn Civil Engrs,* **9,** 145–164 (1958)
185. Inglis, C. C., 'Historical note on empirical equations developed by engineers in India for flow of water and sand in alluvial channels', *Proc. Int. Assoc. for Hydraulic Research* (1948)
186. Hsieh Wen Shen, 'The regime concept of sediment-transporting canals and rivers', Ch. 30 of *River Mechanics,* Colorado State University (1973)
187. Ackers, P. and Charlton, F. G., 'The geometry of small meandering streams', *Proc. Supplement (XII) Instn Civil Engrs,* Paper 7428S, 289–317 (1973)
188. Nixon, M., 'A study of the bank-full discharges of rivers in England and Wales', *Proc. Instn Civil Engrs,* **12,** 157–174 (1959)
189. Johnson, R. H. and Paynter, J., 'The development of a cutoff on the River Irk at Chadderton, Lancashire', *Geography,* **52 (1),** 41–49 (1967)

190. Edwards, A. M. C., 'Dissolved load and tentative solute budgets of some Norfolk catchments', *J. Hydrology*, **18**, 201–218 (1973)
191. United States Geological Survey, *Technique of water resources investigations of the USGS*, Book C, Ch. C1 to C3 on Fluvial Sediments, US Govt Printing Office, Washington (1970)
192. Binnie and Partners, *Report on River Tame purification scheme, Appendix D, Sediment load in the river*, Trent River Authority (1973)
193. Fleming, G., 'The application of a continuous monitoring instrument in sediment transport and water pollution studies', *Bull. IASH*, **12**, 34–41 (1967)
194. Muir, T. C., 'Sediment transport in the River Tyne', *unpublished paper presented to the Northern Section*, Instn Civil Engrs (1967)
195. Thomas, A. R., 'Hydrological and hydraulic aspects of the design of Strathclyde Park', *Scottish Hydrological Group Discussion Note*, Instn Civil Engrs (1971)
196. Fleming, G., 'Design curves for suspended load estimation', *Proc. Instn Civil Engrs*, **43**, 1–9 (1965)
197. Gregory, K. J. and Walling, D. E., *Drainage basin form and process*, Edward Arnold (1973)
198. Cummins, W. A. and Potter, H. R., 'Rates of erosion in the catchment area of Cropston Reservoir, Charnwood Forest, Leicestershire', *Mercian Geologist*, **6**, 149–157 (1972)
199. Painter, R. B., Blyth, K., Mosedale, J. C. and Kelly, M., 'The effect of afforestation on the erosion processes and sediment yield', *UNESCO/WMO/IAHS Symp. on effects of man on the interface of the hydrological cycle with the physical environment*, Pub. No. 113, 62–67 (1974)
200. Clayton, C. L., 'The problem of gravel in highland watercourses', *J. Instn Water Engrs*, **5**, 400–406 (1951)
201. Muir, T. C., 'Bed load discharge of the River Tyne, England', *Bull. IASH*, **15**, 35–39 (1970)
202. White, W. R., Milli, H. and Crabbe, A. D., *Sediment transport: an appraisal of available methods, Vol. 1*, Hydraulics Research Station, Wallingford, Report INT 119 (1973)
203. Freshwater Biological Association, *Fortieth Annual Report for the year ended March 1972* (1972)
204. Radford, P. J., Peters, J. and Farmer, H. R., 'Digital simulation of the upstream movement of migratory salmonids', *Symp. on mathematical modelling techniques in water resources systems*, Ottawa (1972)
205. Winter, T. S. R., 'The silting of impounding reservoirs', *J. Instn Civil Engrs*, **35 (2)**, 65–88 (1950)
206. Cummins, W. A. and Potter, H. R., 'Rate of sedimentation in Cropston Reservoir, Charnwood Forest, Leicestershire', *Mercian Geologist*, **2**, 31–39 (1962)
207. Binnie and Partners, 'Siltation of Deep Hayes Reservoir', *Report to Staffordshire Potteries Water Board* (1966)
208. Armstrong, A. F., '*Regulation of the River Wyre*', Section 1, dissertation submitted in part fulfilment of requirements for MSc degree, Birmingham University (1967)
209. Young, A., 'A record of the rate of erosion on Millstone Grit', *Proc. Yorkshire Geological Society*, **31**, Part 2, 149–156 (1958)
210. Potter, H. R., 'Sediment studies in the Trent Basin', *Assoc. of River Authorities Year Book*, 169 (1973)
211. Somerset River Authority, *First Water Resources Survey Report* (1970)

ANNEX: NOTABLE FLOODS

The great Till flood, 16 January 1841

River Till, a tributary of the Wylye, in the county of Wiltshire.
Communities affected. Tilshead, Orcheston, Maddington, Shrewton and Winterbourne Stoke.
Catchment geology. Chalk; the valley is a bourne.
Preceding conditions. Autumn 1840 was wet and the Till at Shrewton was bank-full by early January. Heavy snow 9 Jan. Slight temperature rise on 10–11 Jan brought glazed ice with heavy rain. Frost and occasional snow 12–15 Jan, with Salisbury temperatures between 23° and 33°F.
Storm of 16 Jan. Eastward movement of a warm front with a temperature rise to perhaps 42°F. Heavy rain led to a rapid thaw over frozen sub-soil.
Flood hydrograph. Topographic catchment 72 km²; typical slope of 1 in 150 down-valley.

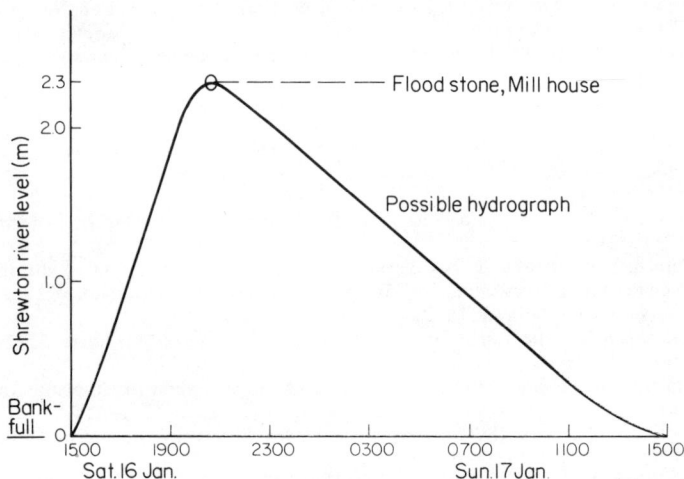

Consequences. 3 drowned; 72 houses destroyed, leaving 200 homeless; £10000 damage in affected communities (at original prices). Sunday morning service at Salisbury cancelled as Avon floods. A move against building cottages of clay and rammed chalk rubble; there was a tax on bricks at that time.
Source material. D. A. E. Cross, 'The great Till flood of 1841', *Weather,* **22,** 430–433 (1967). He drew on local newspapers (Salisbury, Wiltshire Independent) and on *An account of the late flood, 1841,* published by Brodie, Salisbury.

The Louth flood, 29 May 1920

River Lud, flowing east from the Lincolnshire Wolds.
Community affected. Louth.

Catchment geology. Chalk, with some boulder clay cover mainly in valleys.
Preceding conditions. May had been an unexceptional month after a very wet April.
Storm of 29 May. A depression moved from Cornwall to Northumberland. Cyclonic rainfall was associated with a warm south-westerly air current meeting a cool south-easterly air stream, both of which were very moist. Thunder and rain commenced at 1300 GMT and continued until about 1600. 116 mm fell at Elkington Hall but an unofficial measurement at Hallington exceeded 150 mm. The rainfall averaged over the catchment was probably 98 mm.
Flood hydrograph. Topographic catchment 52 km², typical slope 1 in 70.

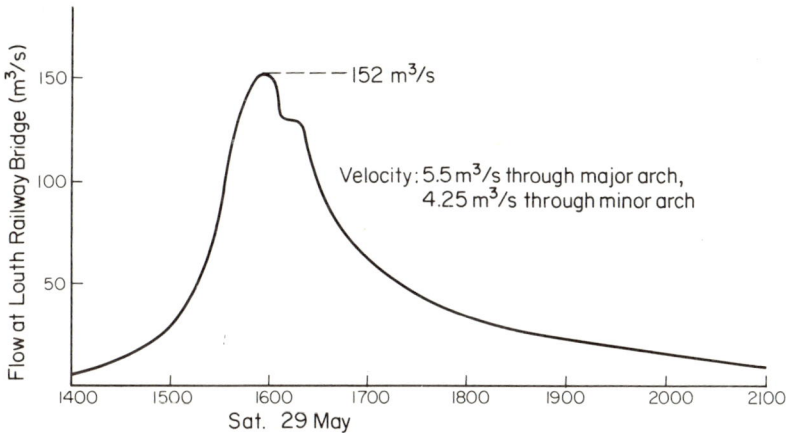

Consequences. 24 drowned when the flood level in Louth reached to the first floor of houses. 46 houses were wrecked and 173 others uninhabitable. Major soil erosion in normally dry valleys. Damage put at £100000. Relief Fund raised £90000.
Source material. (a) P. M. Crosthwaite, 'The Louth flood of 1920', *Trans. Instn Water Engrs*, **26,** 204ff (1921). (b) P. R. Latter, 'The Louth flood', *The Lincolnshire Magazine*, **1** No. 2, 73–76 (1932). (c) *British Rainfall 1920*, 64–68, HMSO.

The Lynmouth flood, 15 August 1952

Rivers East and West Lyn, in the county of Devon.
Communities affected. Lynmouth, Parracombe, Challacombe, Exford, Dulverton and Simonsbath.
Catchment geology. Slates, shales and grits of the Devonian.
Preceding conditions. Just over the average rainfall for August fell in the fourteen days preceding the flood, and the ground was approaching saturation. The early summer was notable for thunderstorm frequency but was not unusually hot. The flow of the Lyn had been low at the end of July but had reached normal summer flow by 14 Aug.

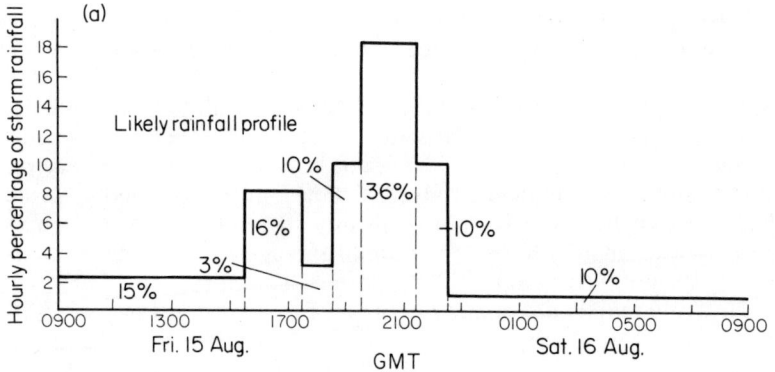

Storm of 15 Aug. Maximum rainday fall 228 mm at Longstone Barrow, 472 m O.D. Unofficial maximum rainfall at Simonsbath 231 mm from 1845 hours onwards; unmeasured previous rainfall up to 1845 must bring the total into the range 250–290 mm. Average storm rainfall on the Lynmouth Catchment 143 mm. Areal distribution: 44 km², more than 200 mm; 110 km², more than 150 mm; 363 km², more than 100 mm.

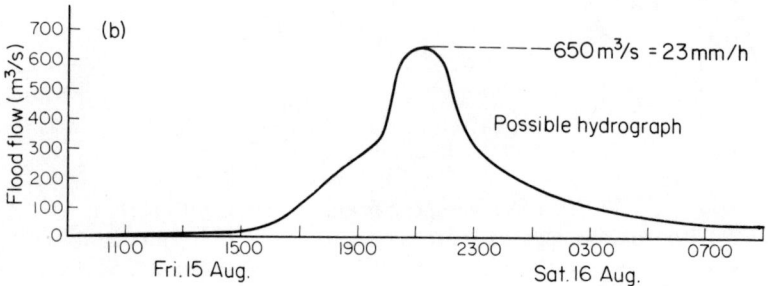

Flood hydrograph. Topographic catchment 101.5 km², typical slope 1 in 20 to 1 in 60 but steeper near coastline.

Consequences. 30 killed plus 4 presumed drowned; 93 buildings destroyed, 28 bridges damaged or swept away, 132 vehicles ruined. Relief Fund raised £1 336 425. 114 000 tons of debris removed from Lynmouth channels, roads and harbour. Several minor dams and a high railway embankment washed out.

Source material. (a) E. R. Delderfield, *The Lynmouth Flood Disaster,* Raleigh Press, Exmouth (reprinted July 1960). (b) A. Bleasdale and C. K. M. Douglas, 'Storm over Exmoor on August 15th, 1952', *Meteorological Magazine,* **81,** 353–367 (1952). (c) C. Kidson and J. Gifford, 'The Exmoor storm of 15th August 1952', *Geography,* **38,** Part I, 1–17 (1953). (d) G. W. Green, 'North Exmoor floods, August 1952', *Bulletin of the Geological Survey of Great Britain* No. 7, 68–84 (1955). (e) C. H. Dobbie and P. O. Wolf, 'The Lynmouth flood of August 1952', Hydraulics Paper No. 1, *Proc. Instn Civil Engrs,* Part III Vol ii, 522–588 (1953)

The Martinstown storm, 18 July 1955

River Wey in the county of Dorset.
Communities affected. Weymouth, Upwey, Broadwey and Coryates.
Catchment geology. Chalk hills but locally impermeable deposits at Weymouth.
Preceding conditions A fine dry summer; temperature on 17 July about 29°C (84°F). Cold air came in aloft from a shallow depression moving from NW. France. Instability thunderstorms initiated.
Storm of 18 July (largest recorded British daily rainfall). The storm moved over the south Dorset hills and became quasi-stationary, possibly enhanced by a weak cold front. Wind speeds about 5 knots. Main rain began 1430 GMT until 1900, being more intense in latter half. Light rain to 2100 then very heavy rain until 0500 on 19 July. Maximum official rainfall 279 mm at Martinstown (unofficial measurement of 355 mm at Hardy Monument).

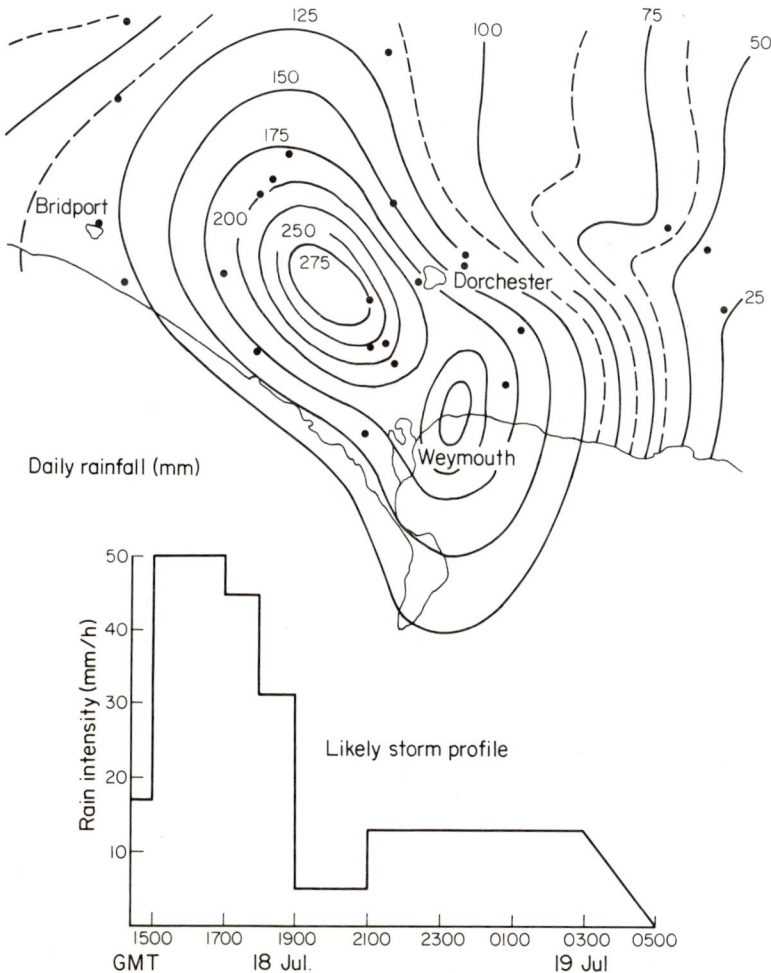

Daily rainfall (mm)

Likely storm profile

Flood hydrograph. Shallow flooding along Winterbourne valley. Radipole Lake, Weymouth, reached 2.7 m O.D. by midnight and high tides restricted outflow worsening flooding. Osmington Mills stream rose 3.6 m below Upton road bridge.

Consequences. Groundwater levels in Chalk Downs rose 12 m at Upwey in one week. Stream beds discharging over coastal cliffs eroded beds by about 1–1.5 m, i.e. greatly exceeding total for previous 20 years. No casualties but property flooded.

Source material. (a) A. Bleasdale, 'The rainfall and flooding in Dorset on July 18th 1955', *British Rainfall 1968,* HMSO (1974). (b) D. J. Paxman, 'The exceptional rainfall of July 18th 1955', *Proc. Dorset Nat. Hist. and Archeol. Soc.,* **77,** 86–89 (1955). (c) W. J. Arkell, 'Geological results of the cloudburst in the Weymouth district', ibid, 90–96 (1955). (d) I. C. Forbes, unpublished communication (1968).

The Forest of Bowland floods, 8 August 1967

River Roeburn, River Dunsop, Claughton Brook and Pendle Water on the Lancashire/Yorkshire border.

Communities affected. Wray, Hornby, Farleton, Claughton, Dunsop Bridge and Barrowford.

Catchment geology. Gritstones and some limestones; thin soil cover.

Preceding conditions. Heavy rain on 13 July was followed by near-average conditions. Soil moisture at commencement of storm close to field capacity.

Storm of 8 Aug. Several intense cells of rainfall occurred along a cold front in the sultry afternoon; the strongest, perhaps 8 km × 5 km, travelled at roughly 8 km/h from SE. to NW. A rainfall of 116 mm in about 1½ hours was recorded at Middle Knoll, Dunsop; an unofficial bucket estimate was between 125 and 150 mm at Whitendale. Rain commenced soon after 1630 BST. The average rainfall over three adjacent catchments of between 10 and 30 km² each was roughly 100 mm.

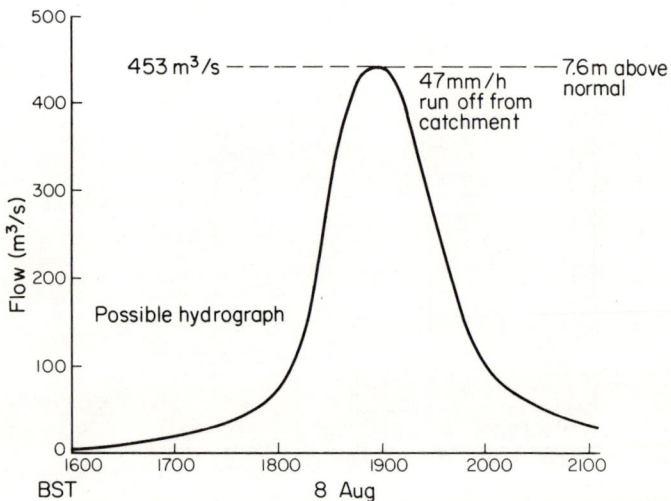

Flood hydrograph. Considering the main flood in Wray village, where damage was most significant: catchment 34.4 km².

Consequences. Wray: 14 houses wrecked and 15 others flooded but no loss of life. Hornby, Farleton and Claughton: 70 properties flooded up to 1.8 m in depth; 0.8 km² of agricultural land suffered deposition of debris. Dunsop Bridge: 14 properties flooded up to 1.5 m in depth; Dunsop water supply intake for Blackburn destroyed. Barrowford: 600 properties flooded, cars swept away. Total damage at least £750 000.

Source material. (a) J. A. Duckworth, 'Bowland Forest and Pendle floods', *Association of River Authorities Yearbook 1969.* (b) 'Bowland Forest and Pendle floods of August 1967', *Hydraulics Research Station Report EX382.* (c) G. Manley, 'Some notes on the Bowland cloudburst of 8 August 1967', unpublished. (d) *British Rainfall 1967,* HMSO. (e) C. Clay, 'Investigations following a catastrophic flood', Hydrological Group discussion, *Proc. Instn Civil Engrs,* **42,** 431–434 (1969).

The Bristol floods, 10 July 1968

River Chew, Cheddar Springs, Malago Stream, and River Yeo in the county of Somerset.

Communities affected. Bath, Bedminster (Bristol), Cheddar, Pensford and Keynsham.

Catchment geology. Carboniferous Limestone headwaters, Marl valleys.

Preceding conditions. Soil moisture deficit initially only 17 mm following a poor summer. 20-minute storm at about 1300 gave roughly 7 mm rain. Afternoon generally dry.

Storm of 10 July. Heavy thunderstorms ahead of a warm front commenced about 1830 and lasted until 0100 next day. Just over 165 mm recorded at Chew Stoke to give a daily total of 172 mm; peak intensity approx. 37 mm/h from 2100 to 2200. 83 km² experienced over 150 mm, 535 km² over 125 mm; the storm had subsidiary centres near Peterborough and in Lincolnshire. Winds north-easterly.

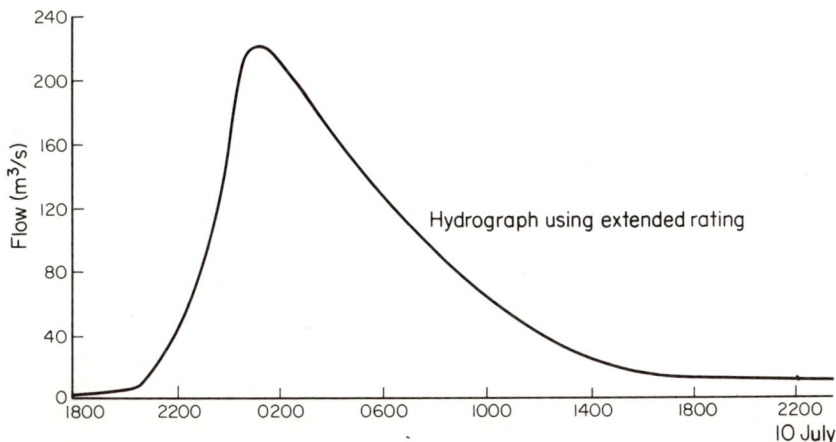

Hydrograph using extended rating

Flood hydrograph. Topographic catchment of River Chew at Compton Dando 72 km² (omitting Chew Valley Lake, which did not spill); typical slope 1 in 14. Chew Valley Lake, catchment 58 km², was 0.9 m down prior to the flood and rose 0.5 m by Thursday morning, an inflow of 2.14 million m³ = 37 mm runoff (some headwater springs diverted to Barrow). Avon peak passed Bath at 2320 on 11 July (but below peak in 1960). Mendip dry valleys became raging torrents.

Consequences. Eight people drowned. Cheddar Gorge devastated *(Figure 6.7).* Over 15 bridges washed out, cost at least £500000. Chew Magna reservoir spillway damaged. Small mudflows, one moving a total of 160 m in 45 minutes. Stimulus given to £3.5 million Malago (Bristol) drainage scheme. Treasury approval of UK Flood Studies project.

Source material. (a) J. D. Hanwell and M. D. Newson, 'The great storms and floods of July 1968 on Mendip', *Wessex Cave Club Occasional Pubn.* 1(2) (1970). (b) Bristol Avon River Authority records. (c) A. B. Hawkins, 'The geology and slopes of the Bristol region', *Q.J. Engng Geol.,* **6,** No. 3 and 4, 199–203 (1973). (d) 'The big flood', *Bristol Waterworks Co. Mag.* (Jul/Aug 1968). (e) P. R. S. Salter, 'A further note on the heavy rainfall of 10th July 1968', *Meteorological Magazine,* **98,** 92–94 (1969).

Chapter 7

Water Quality

Introduction

The chemical composition of natural waters in rivers, lakes and below the surface of the ground depends upon many interrelated factors, including geology and soil, climate, topography, biological processes and time[1]. However, geology is generally, although not always, the dominant factor.

The mechanical and chemical weathering of rocks ultimately produces clay minerals and other secondary minerals, and in these processes substances are partly or completely dissolved by water circulating in the hydrosphere. Although rock weathering is the main source of dissolved constituents in water, precipitation also contributes since rain is a dilute chemical solution. The constituents of surface waters draining igneous and metamorphic rocks are mainly derived from precipitation.

The composition of natural waters is modified by chemical reactions, including ion exchange and chemical precipitation, and by the life processes of plants and organisms, including bacteria; in addition, in the last 200 years the activities of man have had an increasing influence in areas of urban and industrial development and in areas where intensive agricultural practices have incorporated the use of pesticides and artificial fertilisers.

Groundwater quality is commonly modified by ion-exchange reactions. These involve the transfer of ions between the soil or rocks and the water they contain. Most ion-exchange reactions involve cations. The process is a reversible reaction, and reversal of the exchange direction may arise from small changes in the composition of the aqueous solution. Natural ion-exchange media include clay minerals, recent organic material and zeolites, which are all of widespread occurrence in nature. A common reaction is the replacement of calcium and magnesium in groundwater by sodium held in exchangeable media.

The concentration of substances in solution may be modified by changes in the physical situation and in physical properties such as temperature and pressure. Examples are the precipitation of calcium carbonate following the release of carbon dioxide from solution at a spring outlet, the precipitation of iron when ferrous iron in groundwater is oxidised on contact with the atmosphere, and the precipitation of silica at the outlet of a thermal spring.

Biological processes may change the composition of a water; a well-known example is the precipitation of calcium carbonate following the use in photosynthesis of carbon dioxide dissolved in water. Bacteria involved in the sulphur and nitrogen cycles are other modifying agents. Sulphates are reduced to sulphides by anaerobic bacteria using oxygen from sulphate ions to oxidise hydrocarbons, and similarly nitrates are converted to ammonia or nitrogen by nitrate-reducing bacteria. Under oxidising conditions sulphur-oxidising bacteria assist in the formation of sulphates, while the oxidation of nitrogenous organic matter by bacteria is a source of nitrate in natural waters. The direction and end-products of these reactions depend upon the oxidation-reduction state of the local environment.

The principal ions in natural waters are calcium, magnesium, sodium, potassium, bicarbonate, carbonate, sulphate, chloride and nitrate. Iron, aluminium and silicon may also be present in significant amounts and these elements may occur in colloidal form. Certain trace elements are important because of their influence in other fields, for example fluoride in preventing dental caries and boron in agriculture.

Waters are often classified according to the dominant cation and anion present or, if a single ion is not dominant in each or both cases, the two principal cations or anions present—for example, calcium bicarbonate water or sodium bicarbonate-sulphate water *(Figure 7.1)*. In temperate climates, such as that of the UK, calcium bicarbonate waters are most commonly found.

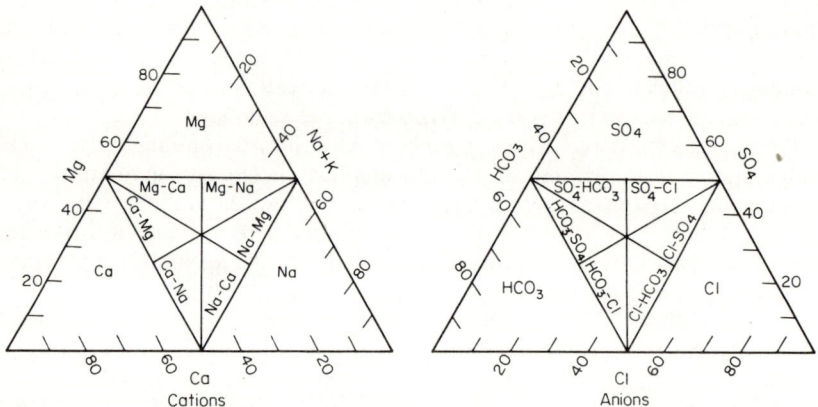

Figure 7.1   A classification of natural waters

Man's activities have modified water quality in the UK in many ways, including the discharge of effluents into surface waters and aquifers, the use of artificial fertilisers, the disposal of acid mine drainage, the increase in surface runoff from urban areas, and the overdevelopment of coastal aquifers causing sea-water intrusion. The recovery of water to its natural state after it has been polluted can be a long process, especially if the replacement of storage takes place slowly as is the case with groundwater.

As a result of nuclear explosions, which commenced in 1945, man has caused widespread radioactive pollution. In the world as a whole this artificial radioactivity is equivalent to only a very small part of natural radioactivity, but the manner in which some products of nuclear explosions, such as strontium-90, are concentrated by living organisms is a matter for concern.

The chemical nature of a natural water is in a dynamic state, and it is changing continuously to maintain physico-chemical equilibrium with its environment. From a knowledge of the environment the type of water can often be forecast.

Rainfall

Surface waters derive part of their chemical composition from the atmosphere through rainfall, snowfall and dry fallout. The sources of the atmospheric constituents are the sea, the soil, atmospheric pollution and naturally derived gases including those from volcanoes. Thermo-nuclear explosions in the atmosphere have produced a marked increase in the radioactive constituents of rainfall, a trend that has been reversed since most tests have been conducted underground.

The total ionic concentration of rainfall is about 10–20 mg/l, although near coastlines values are higher. The principal ions are sodium, magnesium, calcium, potassium, chloride, sulphate and nitrate; very little bicarbonate is present. The sources of chloride and sulphate are the sea and industrial contamination. Sodium and most of the magnesium also come from the sea, while calcium and potassium and the remainder of the magnesium come from the soil. Enrichment of both calcium and potassium may be related to industrial contamination and, in the case of potassium, to agricultural contamination through the use of potash fertilisers. The nitrate is possibly formed by the photochemical oxidation of ammonia[2] derived from the soil, domestic sewage and industrial processes. The gases in the atmosphere that contribute to the composition of rainfall are carbon dioxide, hydrogen sulphide and sulphur dioxide, and it has been suggested that hydrogen chloride is also present[3].

Although the widespread dispersal of small particles and gases occurs in the atmosphere (for example the dispersal of the products of volcanoes), local factors predominantly determine the chemical composition of rainfall.

Table 7.1 ANNUAL MEDIAN VALUES OF VARIOUS CONSTITUENTS (mg/l) IN RAINFALL OVER THE UK FOR 1959–1964 (FROM STEVENSON[4], COURTESY ROYAL METEOROLOGICAL SOCIETY)

	Aberdeen	Eskdalemuir	Leeds	Rothamsted	Newton Abbot	Stornoway
Calcium	1.1	0.3	2.7	1.4	1.5	1.1
Magnesium	0.6	0.2	1.0	0.3	0.6	1.9
Sodium	4.6	1.8	2.3	1.8	5.5	17.0
Potassium	0.4	0.2	0.5	0.3	0.7	0.9
Chloride	5.7	2.8	5.3	3.1	6.3	28.0
Sulphur	2.1	1.2	3.9	2.1	2.9	2.4
Nitrogen in nitrate	0.3	0.1	0.6	0.4	0.7	0.2
Nitrogen in ammonia	0.3	0.2	1.1	0.5	1.4	0.2

In the UK the influences of the sea and industrial pollution are very marked *(Table 7.1)*. The role of the sea is clearly indicated by the decrease inland of chloride values *(Figure 7.2)*. Values between 15 and 30 mg/l recorded along the exposed north-west coast decline to about 3 mg/l in rural areas in the centre of Britain. Winter values are about two to three times greater than summer values due to an increase in the spray content of the air and the salt content at cloud level caused by winter storms[4]. Industrial pollution is also a source of chloride but a high proportion is believed to fall out near its point of origin[3].

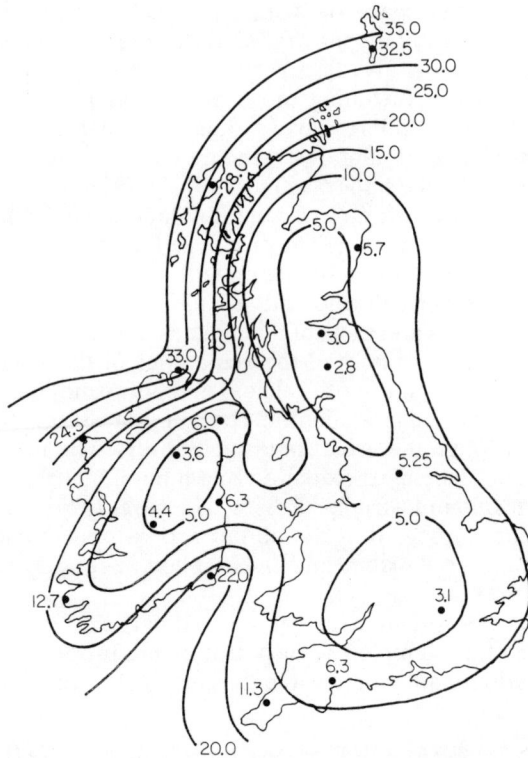

Figure 7.2   Annual chloride medians of rainfall over the British Isles (mg/l) (from Stevenson[4], courtesy Royal Meteorological Society)

The distribution of sulphur in rainfall *(Figure 7.3)* illustrates the influence of both the sea and industrial contamination from the burning of coal. Values do not decline over the inland areas of England as they do over Ireland and Scotland. Gorham[3] indicated that about 25 per cent (1.0 mg/l) of the total sulphate content of rain in the Lake District may come from a source other than sea spray and the combustion of fossil fuels. Conway[5] had previously suggested that the source of the 'additional' sulphate was hydrogen sulphide from decaying organic matter in muds on the continental shelf, which was oxidised to sulphuric acid in the atmosphere. However, it

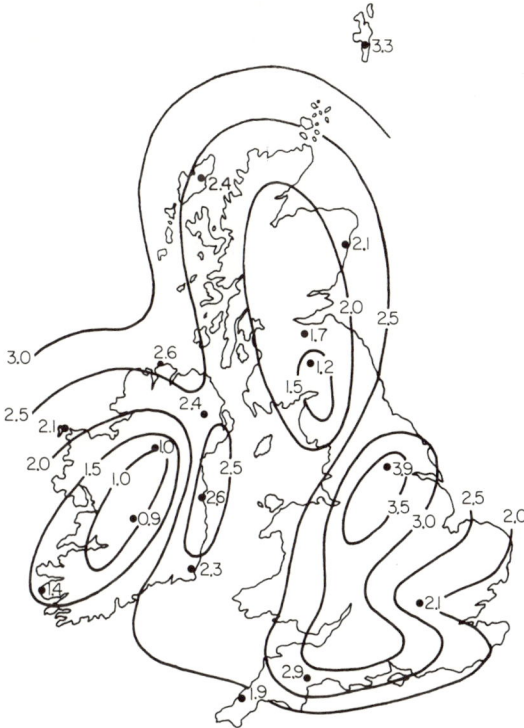

Figure 7.3   Annual sulphur medians of rainfall over the British Isles (mg/l) (from Stevenson[4], courtesy Royal Meteorological Society)

has been estimated more recently[6] that on a global scale hydrogen sulphide from decay in the seas only provides 20 per cent of the natural supply of sulphur to the atmosphere, while some 50 per cent is due to hydrogen sulphide from biological decay on land and 30 per cent from sulphate in sea spray. To complete the picture, the sulphur emitted to the atmosphere by man in 1971 has been estimated to be about 40 per cent of the natural emission[7]. The relative importance that each source makes to the composition of rain and fallout depends upon geographical factors. In the UK industrial sources are dominant in some areas and maritime sources in others. Although artificial emission of sulphur in the UK increased from 9.1 to 11.4 mg/km^2 between 1950 and 1970, the sulphate content of rain has not increased, possibly because there is an effective limit to the solubility of sulphur dioxide in rain[7].

The distribution of sodium and potassium in rain over the UK is closely related to maritime sources but industrial pollution clearly contributes to the calcium content[4].

The factors controlling the composition of rain have been clearly indicated in a study of rainfall over the Lake District[3]. The composition is closely related to wind direction. When the wind is from the sea the proportion of chloride in the rain is higher, but sulphate and nitrate show maximum accumulations when the wind is from the industrial centres to the south-east. Wind speed influences the distribution of constituents derived from both the sea and pollution, but of course the entry of pollution into the atmosphere is independent of this factor whereas, as already indicated, it is related to the formation of sea spray.

The rain over the Lake District that is contaminated by air pollution is acidic, with pH values below 4.5 due to the presence of sulphuric acid[8]. Similarly in west Yorkshire the rainfall is more acidic when the wind is from industrial areas to the south-east and east[9].

Information about dry fallout in the UK is limited; the proportion recorded for sodium, calcium and chloride in the Lake District, an area of relatively heavy rainfall, is about 5 per cent, but it varies between 1 and 40 per cent for other elements[10].

Groundwater

Principal chemical changes in British aquifers

The chemical quality of groundwater is determined by the composition of the soil and soil air through which the water has infiltrated, and the composition of the rocks with which it is or has been in contact. The ionic content of a groundwater tends to increase with the length of the flow path through the ground.

As indicated earlier, rain is not a pure water but contains small amounts of dissolved mineral matter. The ionic content of rainfall is increased by evaporation and, when allowance is made for this, rainfall can account for significant proportions of all the main ions, except calcium and bicarbonate, in actively circulating groundwaters *(Table 7.2)*.

Table 7.2   COMPARISON OF CHEMICAL COMPOSITION OF RAIN AND GROUNDWATER IN THE CHALK (mg/l) (AFTER WATER RESOURCES BOARD[19])

	Rainfall at Rothamsted*	*Composition of the infiltrate†*	*Average composition of Chalk water at outcrop*
Calcium	1.4	3.9	100
Magnesium	0.3	0.8	2
Sodium	1.8	5.0	10
Potassium	0.3	0.8	1
Bicarbonate	—	—	280
Sulphate	6.3	17.6	15
Chloride	3.1	8.7	15
Nitrate as NO_3	1.8 ⎫	11.2	20
Ammonia as N	0.5 ⎭	—	

*After Stevenson[4]
†Following concentration by evaporation and transpiration

As water infiltrates through the soil layer and comes into contact with soil air considerable changes occur. One of the most important reactions is that involving calcium and magnesium carbonates and carbonic acid. Carbonates are relatively insoluble in pure water but their solubility increases in the presence of hydrogen ions. The main source of these is carbon dioxide in the soil, and the reactions may be expressed as:

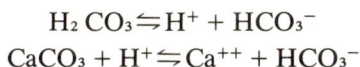

$$H_2CO_3 \rightleftharpoons H^+ + HCO_3^-$$
$$CaCO_3 + H^+ \rightleftharpoons Ca^{++} + HCO_3^-$$

Soil air contains as much as thirty to forty times the proportion of carbon dioxide in the atmosphere. The partial pressure of the gas in air is only about 3×10^{-4} atmospheres, which is sufficient to dissolve no more than 20–30 mg/l of calcium[11]; as a result of respiration of roots and microbiological decay of organic matter, however, soil air contains 1–5 per cent carbon dioxide. This can dissolve about 100 mg/l of calcium, a more typical value for actively circulating groundwater in contact with calcium carbonate.

The solution of calcium carbonate by carbonic acid and also by organic acids from the soil is a dominant reaction in limestones and in sandstones where the grains are cemented by calcium carbonate. As a consequence groundwaters in such aquifers at outcrop in temperature latitudes are generally of the calcium bicarbonate type. Any magnesium carbonate present is dissolved in a similar manner, although this mineral is more soluble than calcium carbonate. The double carbonate, $CaMg(CO_3)_2$, is less soluble than either calcium or magnesium carbonate.

A considerable proportion of carbonate solution occurs in the upper few metres of the unsaturated zone, particularly if the flow is intergranular or through a fine fissure system. In fact the composition of water in the upper part of the unsaturated zone often closely resembles groundwater in the aquifer at depth[12]. Below the water table water is generally considered to be saturated with respect to calcium carbonate. However, solution of carbonates does occur in the saturated zone because of changes in temperature and pressure, and as a consequence of chemical reactions. In this context, the mixing of saturated solutions of calcium bicarbonate (containing different concentrations of bicarbonate) releases carbon dioxide, which is then available to dissolve more carbonate. This is because less carbon dioxide is required to maintain the equilibrium of the mixture than the total required by the separate solutions[13]. Solutions following different paths are constantly mixing underground, and this phenomenon is important in the solution of limestones[9].

Small amounts of sulphides of heavy metals occur in rocks, particularly clays and mudstones and soils derived from them. These are oxidised by weathering processes, assisted by sulphur-oxidising bacteria. Sulphuric acid is a product of the reactions and, in the presence of calcium and magnesium carbonate, calcium and magnesium sulphates are formed and are dissolved by circulating waters. The reactions are:

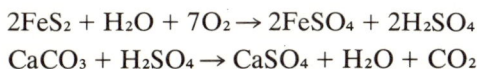

$$2FeS_2 + H_2O + 7O_2 \rightarrow 2FeSO_4 + 2H_2SO_4$$
$$CaCO_3 + H_2SO_4 \rightarrow CaSO_4 + H_2O + CO_2$$

A further consequence is that the carbon dioxide produced dissolves more carbonate. The first reaction is discussed in more detail later, in the section on mine-drainage waters.

Chloride in groundwater is derived from rainfall and the solution of small amounts of chlorides in rocks, but it is also derived from residual modified

Table 7.3 REPRESENTATIVE ANALYSES OF GROUNDWATER (mg/l)

	Chalk at outcrop	Chalk below boulder clay	Chalk below Eocene	Chalk contaminated by saline intrusion	Triassic ssts at outcrop	Jurassic lst at outcrop	L. Greensand at outcrop	Connate water Millstone Grit
Calcium (Ca)	104	150	9	225	40	108	103	6 980
Magnesium (Mg)	2	15	2	245	12	12	10	978
Sodium (Na)	3	} 40	181	2 000	} 8	11	11	36 230
Potassium (K)	2		—	82		—	3	142
Bicarbonate (HCO_3)	282	392	270	306	112	270	236	126
Sulphate (SO_4)	4	85	62	525	25	70	80	559
Chloride (Cl)	14	46	96	3 700	19	21	23	70 800
Nitrate (NO_3)	19	11	—	50	29	16	—	—
Total dissolved solids	310	555	505	—	213	373	380	115 752
Carbonate hardness (as $CaCO_3$)	235	325	32	—	93	225	197	—
Non-carbonate hardness (as $CaCO_3$)	30	100	Nil	—	97	91	102	—
Alkalinity (as $CaCO_3$)	235	325	225	—	93	225	197	—

sea water occluded in marine sediments. The source of nitrates is primarily the bacterial oxidation of nitrogenous organic matter and the production of nitrates, from nitrogen in the air, by leguminous plants.

In summary, rainfall and the chemical changes outlined above account for the composition of groundwater in limestones and sandstones at outcrop *(Table 7.3)*.

These outcrop waters are modified as they flow through aquifers below impermeable or semi-permeable confining beds. The basic change, from a hard calcium bicarbonate water to a soft alkaline sodium-rich type, due to ion exchange, has been known for many years[14], but more recently zones of different water quality have been delineated in some aquifers in England[15]. Under ideal conditions five zones can be differentiated.

The solution reactions that predominate in the outcrop area continue for some distance downgradient below confining beds. These reactions are encouraged if acidic runoff from the confining strata recharges the aquifer. This runoff often contains sulphate ions balanced with calcium or magnesium ions, and recharge of this type of water increases the non-carbonate hardness. For example the non-carbonate hardness of Chalk groundwater, downgradient of the Eocene base in parts of the London Basin, increases to as much as 100 mg/l[12]. Thus the initial change as the water flows below confining beds is often an increase in bicarbonate and sulphate ions associated with an increase in alkaline earth ions.

With increasing distance from the outcrop the calcium and magnesium ions are replaced by sodium as ion exchange becomes a dominant reaction.

Maximum bicarbonate values are attained in the outcrop area or within some 5 km from the outcrop. Further downgradient, bicarbonate tends to decline, but still further downgradient it increases again. This increase may be due to carbon dioxide produced by sulphate reduction or by the oxidation of organic matter, although further solution of calcium carbonate also occurs as calcium ions are removed from the water by ion exchange. The end product of these various processes is the formation of a soft sodium-bicarbonate water. However, with increasing distance from the outcrop such waters pass into a saline zone as discussed later.

The principal characteristics of the various zones are summarised below[15]:

Zone 1 This zone coincides with the outcrop of the aquifer; solution reactions predominate.

Zone 2 The first zone within the confined area. Again solution reactions predominate. Carbonate hardness and non-carbonate hardness commonly rise to maximum values. The downgradient limit is marked by maximum values for the bicarbonate ion.

Zone 3 Carbonate hardness declines and the bicarbonate ion decreases to minimum values.

Zone 4 A zone of very soft water, the hardness being almost entirely due to carbonate hardness. Ion exchange is a dominant reaction and sodium the principal cation. The bicarbonate ion, and hence alkalinity, increases.

Zone 5 Saline zone with water in excess of 250 mg/l of chloride.

These changes have been described for the Chalk[15], the Lincolnshire Limestone[16], and are apparent in the Lower Greensand and other aquifers in England. *Figure 7.4* illustrates the change of Chalk water in north Kent from

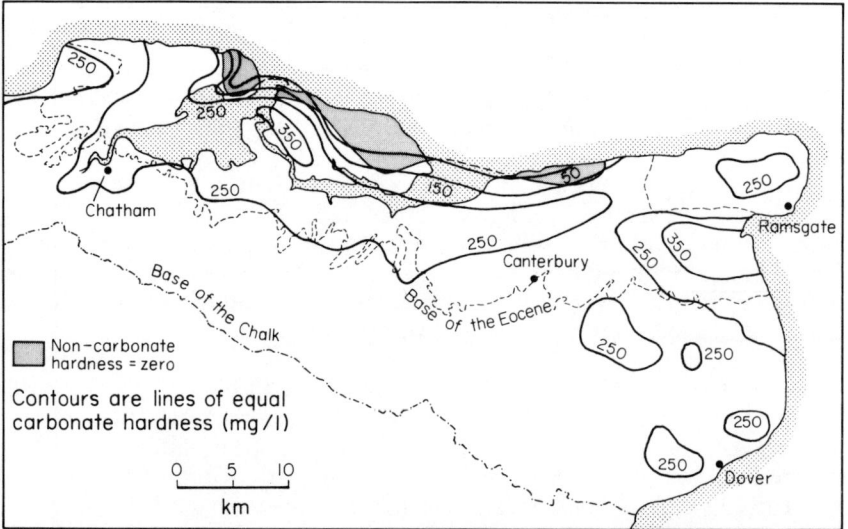

Figure 7.4   Hardness of groundwater in the Chalk in Kent (after 'Hydrogeological map of the Chalk and Lower Greensand of Kent', published by the Institute of Geological Sciences)

a hard calcium bicarbonate type at and near the outcrop to a soft alkaline water some 10 km from the base of the Eocene.

The oxidising capacity of groundwaters decreases downgradient, and in the Lincolnshire Limestone a sharp oxidation-reduction 'barrier' occurs some 12 km from the onset of confined conditions[17]. Downgradient of this sulphide is present in the water mainly as HS^-, hydrogen sulphide being readily detected at well-heads. Some 4 km east of the 'barrier' ion exchange and sulphate reduction become significant.

The onset of the various chemical changes in a confined aquifer is related to the permeability of the aquifer and hence the rate of groundwater flow and whether it is through fissures or intergranular. This may be illustrated using the Jurassic limestones as examples. In north Lincolnshire, the Lincolnshire Limestone has a lower transmissivity than in south Lincolnshire, and the isopleth for zero non-carbonate hardness (which indicates approximately where ion exchange becomes significant) occurs 5 km and 14 km, respectively, from the outcrop in these two areas; the same point is recognised from a comparison of the chloride content of the groundwater[16]. Similarly saline waters are found at variable distances from the outcrop of the Great and Inferior oolites in the Thames Basin[18].

Calcium bicarbonate waters extend much further downgradient in the London Basin along the lines of the main tributary valleys where permeability is higher *(Figure 7.5)*. In the confined part of the London Basin the chemistry of Chalk groundwater is modified by the leakage of sulphate and chloride-rich waters from the London Clay, due to the reduction of artesian pressure in the Chalk[19].

The change from a calcium-bicarbonate water to a sodium-rich alkaline type is typical of many aquifers throughout the world, but the sequence represents only the initial stages of an extended series of chemical changes.

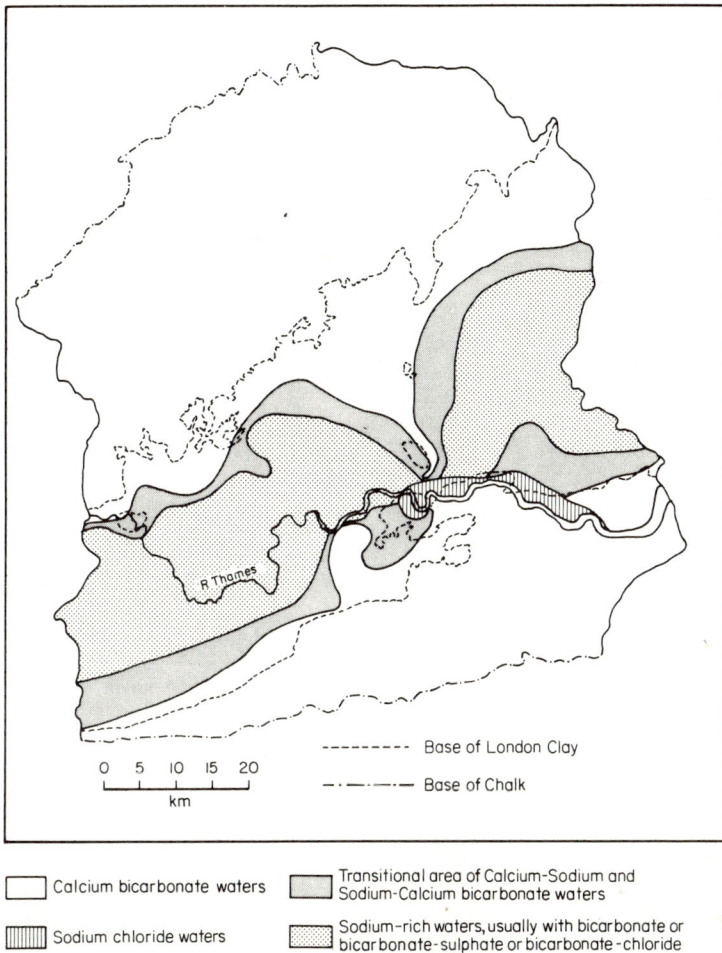

Figure 7.5 Chemical types of groundwater in the Chalk in the central part of the London Basin (after WRB publication[19])

Chebotarev[20] pointed out that a general sequence involves the passage from bicarbonate waters at outcrop through sulphate waters at intermediate depths to chloride waters at greater depths, the sequence being due to the precipitation of the least soluble salts first.

This sequence occurs in the Carboniferous Limestone, where it is overlain by Upper Carboniferous and younger rocks east of the Peak District[21], and in the sandstones of the Coal Measures in the east Midlands[22]. In the latter case, the types of water with increasing distance from the outcrop are:

1. Calcium bicarbonate type at outcrop.
2. Sulphate waters, excessively hard (500–1600 mg/l) mainly due to non-carbonate hardness.
3. Ion-exchange waters, due to modification of types 1 and 2.
4. Sodium chloride waters, which develop 5 km from the outcrop.

Ion exchange does not always attain the same importance in hard compact rocks such as the Carboniferous Limestone, where active movement of groundwater is virtually entirely through fissures[21].

It is important to realise that the ideal sequence may be modified by variations in a number of factors in the environment. For example the development of a particular type of water often depends upon the distribution of minor constituents in the rock.

The Chalk is overlain by glacial deposits in East Anglia, and these influence considerably the quality of water in the aquifer. Much of the central part of the region is covered by thick boulder clay. In Suffolk this has increased the carbonate and non-carbonate hardnesses and the chloride content as a result of relatively high carbon dioxide contents in the clay and clay-derived soils, and the presence of sulphate minerals (such as selenite) and small amounts of chloride in the clay[23]. The escape of hydrogen sulphide from wells drilled to the Chalk through boulder clay implies that sulphate reduction may be a further factor increasing the carbonate hardness.

Chalk groundwater below boulder-clay covered areas of Norfolk does not have the high non-carbonate hardness and chloride content characteristic of Suffolk. The line separating the two types of water coincides with the boundary separating the boulder clay into a Chalky drift over Norfolk and a Chalky-Jurassic drift over Suffolk[24]. The former was deposited by an ice sheet that had crossed the Chalk outcrop of the Lincolnshire Wolds, while the Suffolk clay was derived from the clay belt of eastern England.

The partial pressure of carbon dioxide in the soil air of soils with small organic contents is low, and as a consequence the carbonate hardness of groundwaters in aquifers underlying such soils is also low. The sandy soils of the Breckland, near Thetford, are reflected in carbonate hardnesses of Chalk water of 100–150 mg/l [23]. Similar soils exist in north-west Norfolk, where the carbonate hardness is less than 200 mg/l. In north-east Norfolk the boundary between boulder clay and sandy glacial drift is marked by the 200 mg/l carbonate hardness value for groundwater in the underlying Chalk. A further example is in the Bunter Sandstone of Nottinghamshire, which contains water with a carbonate hardness of 50–100 mg/l[25].

In unconsolidated sands overlain by an organic soil the water may be soft because of the absence of carbonate cement, but the water in such rocks (the Upper Greensand and glacial sands are typical examples) is corrosive because of the relatively high free carbon dioxide content, the pH being about 6. Where sandstones are overlain by boulder clays (for example the Triassic sandstones in the Midlands and northern England, and the Lower Greensand in Cambridgeshire) the clay increases the availability of carbon dioxide, and if carbonates are present in the clay or the aquifer they will control the carbonate hardness. The presence of overlying clays, of course, may also increase the non-carbonate hardness. Thus the nature of a sandstone depends primarily upon whether the grains are cemented by a carbonate, upon the nature of any cover, and upon the extent to which an organic soil layer has been formed by decay of vegetation.

The Triassic rocks of Britain were formed in a desert environment in which the natural waters would have been predominantly chloride- and sulphate-rich types. Salts containing these ions precipitated in the rocks on both a large and small scale. The climatic conditions at the time these rocks formed is reflected in the chemistry of the groundwater they now contain although, in the zone of active circulation in the sandstones, the sulphates and chlorides have been essentially flushed out by meteoric waters. Three main types of water may be recognised in the Triassic sandstones. A calcium-bicarbonate water occurs in the outcrop area and persists for some distance downgradient below the Keuper Marl. The concentration of water increases both vertically with increasing depth in the aquifer and horizontally in a downgradient direction below the overlying Marl, as the lengths of the flow lines increase. The sulphate concentration tends to rise, and sulphate may become the dominant anion to give a calcium-sulphate water. At greater depths and distances from the outcrop a non-potable sodium-chloride type is found[25-27]. The onset of the latter type depends upon the nature of the aquifer, particularly the permeability; it occurs about 17–27 km from the outcrop in Nottinghamshire, but the distance is only a few kilometres south-east of the Warwickshire coalfield.

Ion exchange occurs in the sandstones at depth and below the Keuper Marl, but the softening effects are often masked by the high total hardness. South-west of the Warwickshire coalfield the total concentration of groundwater in the Keuper Sandstone decreases towards the deeper parts of the Worcester Basin, and near Stratford-upon-Avon a soft sodium-sulphate water occurs[28]. The presence of such a water so remote from the outcrop is unusual for the Triassic sandstones.

Iron compounds are of widespread occurrence in rocks. Sands, sandstones and gravels commonly contain iron as oxides, often associated with manganic oxides. If organic matter is present and oxygen is absent, the organic matter is oxidised by the oxygen in the oxides and the metal ions go into solution. These reactions occur below the water table. Iron and manganese are found in groundwaters in the ferrous and manganous forms. They are precipitated on contact with air, manganese less readily than iron.

Groundwaters containing iron and manganese in solution are found in river gravels, the Lower Greensand, the Crag and other sandstones. When present in concentrations in excess of 0.1 mg/l treatment is necessary if the water is to be used for public supply, but even lower maximum limits are necessary for some industrial requirements.

During the summer in areas where evaporation exceeds rainfall, soluble matter accumulates in the soil. This is dissolved and flushed out of the soil in the autumn, when infiltration begins to move down towards the water table as rainfall once again exceeds evaporation and begins to eliminate soil moisture deficits. Because of this the composition of groundwater may show a seasonal variation. The chloride and total hardness of water in the Chalk near Brighton increased in the autumn of 1960 from 30 to 45 mg/l and from 220 to 290 mg/l respectively[29]. Similarly cyclic patterns have been recognised in the Jurassic limestones[12,16]. These responses to infiltration fluxes emphasise the rapid movement of water in these aquifers through fissures in the unsaturated zone.

Recent preliminary work[30] on the interstitial water extracted from the Chalk by a centrifuge has shown that it can be more mineralised than the water in the fissures by as much as a factor of ten and that the ionic composition also differs from the water in the fissures.

Saline connate waters

Groundwaters in aquifers at depth are saline where circulation is restricted by low permeabilities of either the aquifer itself or the confining beds. Information about the chemical nature of such waters is primarily obtained in the course of mining operations and oil-field exploration and development.

Groundwater from the concealed Kent coalfield has a chloride content of 20 000 mg/l[31], while in the east Midlands coalfield values exceed 60 000 mg/l about 5 km from the outcrop, and in the Durham coalfield waters with two or three times the concentration of sea water are known[32]. During the exploration for oil in the Midlands saline waters were encountered in Carboniferous rocks with maximum values in excess of 200 000 mg/l *(Figure 7.6)*, the concentration increasing with increasing depth of the aquifer[33].

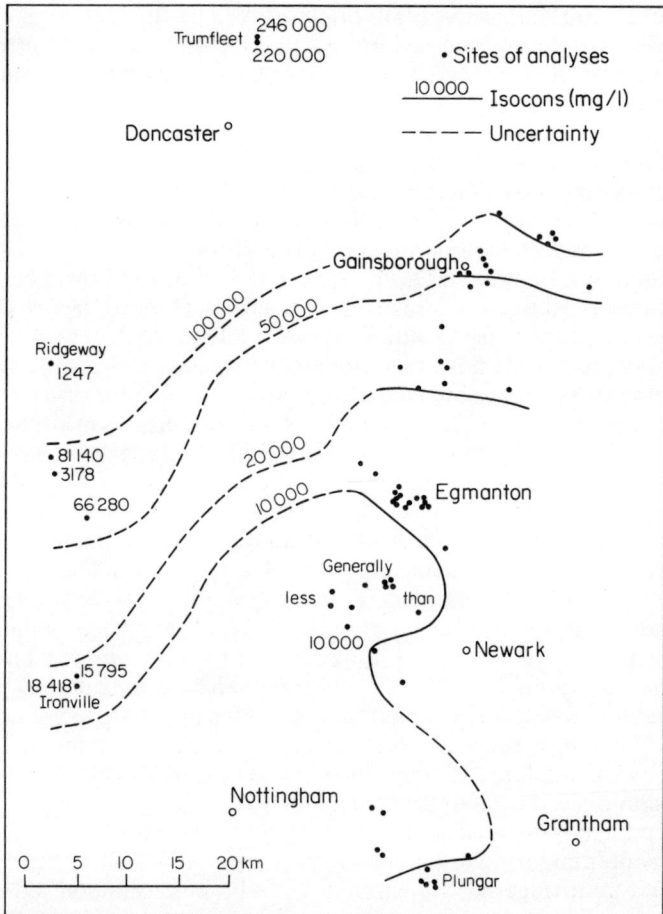

Figure 7.6   Variation in the concentration of groundwaters in Millstone Grit sandstones in the east Midlands (from Downing and Howitt[33], courtesy Geological Society of London)

It is generally accepted that waters of this nature have developed from sea water trapped in marine sediments at the time of their formation[34], although it should not be overlooked that high salinities can be attained by meteoric water slowly dissolving soluble constituents as it moves through the ground, for example if the groundwater comes in contact with evaporite deposits as in the Keuper Marl or Upper Permian. During the course of time the original interstitial water of sediments is redistributed by fluid potential gradients, including those induced by the consolidation of the sediments and later deposits.

Connate brines have different ionic ratios from sea water. The Na/Cl, K/Cl, Mg/Cl, SO₄/Cl and HCO₃/Cl ratios are lower but the Ca/Cl ratio is higher. These changes are due to reactions between saline pore water in the sediments and the sediments themselves, commencing soon after deposition and continuing for an extended period thereafter as the rocks are steadily consolidated by burial beneath younger deposits[35]. The increase in concentration to levels above that of sea water is due to rocks acting as semi-permeable membranes[36]. The concentrating mechanism is considered to be that, as fine-grained sediments are consolidated, negatively charged clay mineral particles prevent the passage of anions in the water but not the uncharged water molecules, which move in the direction of decreasing pressure. The cations cannot move through the membrane as a chemical and electrical balance must be maintained. The result is that the concentration of water on the input side increases[35].

As the total concentration increases in highly concentrated connate brines, there is an increase in Ca/Cl and Mg/Cl ratios and a decrease in the Na/Cl and K/Cl ratios. This appears to be because fine-grained shale membranes selectively concentrate calcium and magnesium ions. It is possible that the ease with which ions move through a shale membrane is inversely proportional to valency and ionic size, the rock acting in a similar manner to cellulose acetate membranes used to desalinate sea water[33,35].

From the previous discussion in this chapter, it will be evident that there are two principal sequences controlling the chemistry of groundwaters. One involves the modification of sea water, once it becomes the interstitial pore water of sediments, and its concentration by ultra-filtration as the sedimentary succession is consolidated. The other sequence is the modification of rain as it moves through the ground. This begins with a new cycle of erosion, which results in the gradual flushing of connate waters from the rocks.

Thermal and mineral waters

Mineral waters in the UK have been defined as those having a total ionic concentration in excess of 1000 mg/l [37]. This includes connate waters and natural brines associated with evaporite deposits. However, most mineral waters exploited for therapeutic purposes now and in the past are (or were) derived from springs or relatively shallow wells. Deep connate waters were only used if they naturally discharged at the surface. In a recent review[37] more than eighty mineral-water sources were listed. Many were associated with argillaceous rocks, such as the London, Kimmeridge and Oxford clays, and the Lower Lias, all of which invariably contain water of a mineral character with a high sulphate content. However, they have only been developed as mineral waters at a limited number of localities mainly on the Lower Lias, for example at Cheltenham. The Permian and Triassic rocks were also important sources of such waters (for example Leamington and Droitwich) and further examples are from Palaeozoic rocks (for example Llandrindod Wells). Most waters are either sulphate waters, associated with calcium and magnesium, or sodium-chloride waters.

At Harrogate 88 springs issue from the Millstone Grit and are located for the most part along an anticlinal axis. The waters are of the sodium-chloride type and are probably deep-seated connate waters diluted by meteoric waters, the degree and seasonal variation of the dilution accounting for changes in the chemical composition of some springs[37,38].

Thermal springs occur in Derbyshire (Buxton, Bakewell, Matlock and Stoney Middleton), north Somerset (Bristol, Bath and Batheastern) and at Taff's Well in south Wales. All have surface temperatures in excess of 15°C, the highest being 28 and 49°C at Buxton and Bath respectively. They do not have high ionic contents and are considered to be meteoric waters that have circulated to depths of about 2000 m or more in the Carboniferous Limestone, although the Bath, Batheastern and Taff's Well springs do not issue at the surface from this formation.

Natural brines with concentrations of more than 100 000 mg/l occur where salt deposits in the Keuper Marl are in contact with the zone of active groundwater circulation in the marl.

In the Cheshire saltfield groundwater circulates in the marl to depths of 75–150 m below the base of the superficial deposits that cover the area. Where brine is produced by natural groundwater circulation the salt is said to have a 'wet rock-head' and the gradual solution of the salt causes surface subsidence.

Commercial exploitation of brine takes place in Cheshire, Staffordshire and Worcestershire. It may be obtained by pumping brine from the wet rock-head, which encourages the inflow of fresh water to dissolve more salt, but nowadays most brine is produced by introducing fresh water at a controlled rate into the salt layers at depth[37].

Mine-drainage waters

Coal seams and mudstones in the Coal Measures contain pyrite and marcasite (ferrous sulphide); in the course of mining the coal the water table is artificially lowered, air gains access to these minerals and they are oxidised. The various processes and reactions involved are complex, but the initial stage[39,40] is believed to be a chemical oxidation represented by:

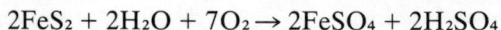

$$2FeS_2 + 2H_2O + 7O_2 \rightarrow 2FeSO_4 + 2H_2SO_4$$

The ferrous sulphate is then oxidised to ferric sulphate and, as the chemical oxidation of iron in an acid medium is slow, the reaction is primarily due to bacteria:

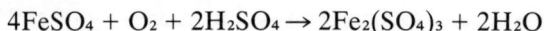

$$4FeSO_4 + O_2 + 2H_2SO_4 \rightarrow 2Fe_2(SO_4)_3 + 2H_2O$$

Part of the ferric sulphate reacts with more pyrite, converting it to ferrous sulphate and producing sulphuric acid, while part is hydrolysed to ferric hydroxide, again with the production of sulphuric acid. Bacteria are also involved in these reactions:

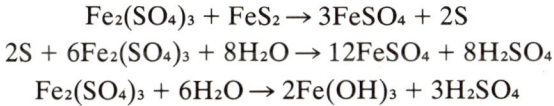

$$Fe_2(SO_4)_3 + FeS_2 \rightarrow 3FeSO_4 + 2S$$
$$2S + 6Fe_2(SO_4)_3 + 8H_2O \rightarrow 12FeSO_4 + 8H_2SO_4$$
$$Fe_2(SO_4)_3 + 6H_2O \rightarrow 2Fe(OH)_3 + 3H_2SO_4$$

Barnes and Clarke[41] have suggested that the reduction of water may be a factor in the chemical oxidation of iron sulphides in view of the virtual absence of dissolved oxygen in many mine waters:

$$FeS_2 + 8H_2O \rightarrow Fe^{++} + 2SO_4^{--} + 2H^+ + 7H_2$$

However, bacteria are generally considered to play a dominant role. Investigations in Scottish coal mines indicated that for every tonne of sulphuric acid produced by chemical means approximately four tonnes are produced through bacteria[42]. There is usually a delay between the beginning of the mining process and the production of excessive volumes of acid waters.

Similar reactions to those just discussed occur in metalliferous mines and in mine wastes, when other metallic sulphides (for example copper, lead, zinc and antimony sulphides) may also be involved.

As a consequence of these reactions, mine-drainage waters from argillaceous rocks are acidic, with pH values of 2 or 3, due to the presence of sulphuric acid. They have high sulphate and iron concentrations, they contain appreciable amounts of calcium, magnesium, aluminium and manganese (from the effect of the acid environment on clays and silicates in the rocks), and they may also contain toxic metals. The iron and sulphate contents may be several thousand milligrams per litre. If such waters are discharged to rivers they are a serious form of pollution. The iron precipitates as ferric hydroxide when the pH rises above 6, following reactions with sediments or mixing with alkaline waters in rivers, or when chemical or bacterial oxidation of ferrous compounds occurs. The oxidation of ferrous compounds in a river reduces the oxygen content, leading eventually to oxygen depletion. The iron precipitate discolours the water and leaves unsightly deposits.

This pollution problem, encountered in many mining areas including Scotland, Lancashire, Yorkshire and south Wales, is difficult to solve satisfactorily. The acidic waters may be neutralised by the addition of lime or by passing them over limestone, but the removal of iron from such waters is complicated, requiring the addition of alkalis to raise the pH before aeration.

Streams draining coal-mining areas in Britain generally have a low bicarbonate content, because of the absence of carbonate rocks, and therefore have a limited neutralising action. This would be increased by placing limestone upstream of the discharge point of the mine water[41]. Various attempts have been made to limit the pollution problem by sealing abandoned mines to prevent large volumes of polluted waters issuing at localised outlets, and also to prevent the access of oxygen, although this is unlikely to be successful if the reduction of water is a factor in the oxidation of pyrite. Antibacterial substances have also been used[40]. Because of the large volumes of rock affected by mining and the overall increase in mass permeability due to both the mining process and subsequent subsidence, it is not feasible to prevent access of water to coal mines and the problem remains intractable.

Sea-water intrusion

Where sea-water intrusion has taken place, the saline groundwater may not have the composition of a theoretical mixture of the natural groundwater and sea water because of ion exchange, sulphate reduction, solution and precipitation reactions[43,44].

In the Wirral Peninsula, where the Triassic sandstones are contaminated by intrusion from the Mersey, the increase in chloride is not matched by an equivalent rise in sodium; instead calcium as well as sodium rises. This is due to ion exchange involving the replacement of calcium ions in calcium-saturated clay minerals, which are probably in the aquifer, by magnesium and sodium ions in the saline water moving into the aquifer. Where the sediments contain magnesium as well as calcium in exchange positions, both are exchanged for sodium in the water[45]. Similar modifications have been observed in the Chalk where saline intrusion has occurred near the mouth of the Essex Stour[46] *(Figure 7.7)*.

The gains and losses of cations due to ion exchange do not balance in the saline groundwater in the Wirral Peninsula, there being a loss of calcium. This is caused by reduction of sulphates as the saline estuary water infiltrates through the sediments on the river bed, and, under the physico-chemical conditions existing, calcium carbonate precipitates. High bicarbonate values in groundwater in the Chalk adjacent to the Thames estuary, where intrusion has occurred, have also been related to sulphate reduction[19].

During the early stages of saline intrusion it may be difficult to distinguish whether the increase in chloride is caused by the ingress of saline water or by pollution. To overcome this problem changes in various ionic ratios are used to detect intrusion, for example Cl/HCO_3, Br/Cl and the sum of the cations/Cl.

Surface water

Chemistry of inland waters

The chemical nature of rivers and streams is subject to seasonal and shorter-term variations. These are related to flow rates, which in turn are dependent on the varying contributions of the different sources making up the total flow at any particular time. There is an inverse relationship between volume of river flow and ionic concentration of the water. During periods of dry weather the ionic concentration of a river is higher because groundwater makes a significant contribution to the total flow at such times, whereas during periods of high effective rainfall river flow is derived mainly from

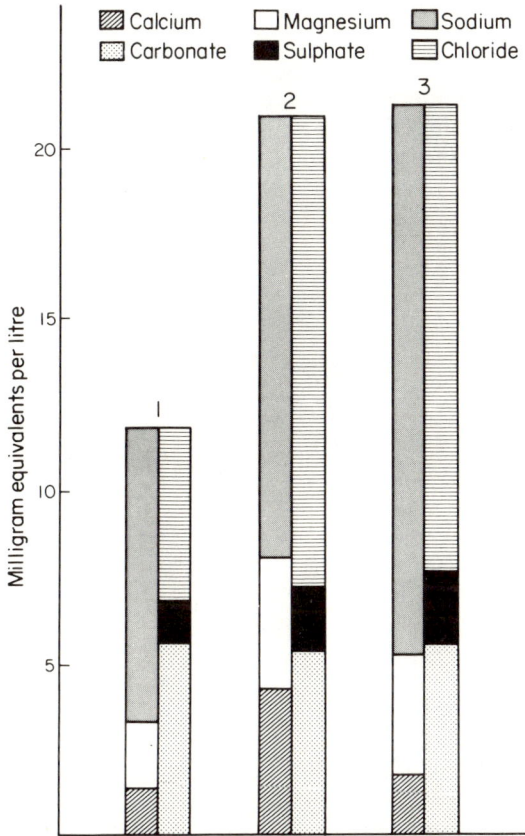

Figure 7.7 *To illustrate ion-exchange reactions as a result of saline intrusion at Lawford, Essex: (1) natural groundwater, (2) groundwater following saline intrusion, (3) calculated analysis assuming mix of natural groundwater and seawater to give observed chloride in (2) (from Hoather[46], courtesy Institution of Water Engineers)*

direct runoff, which generally has a lower ionic concentration. This generalisation may not apply if high flows are dominated by direct runoff from a young incompletely leached clay formation.

The actual amount of dissolved material carried by a river is directly related to the volume of flow, up to a certain level of runoff, as defined by:

$$T = KD^f$$

where T = the amount of dissolved solids, D = discharge, K is a constant and f is less than unity[47]. Therefore, during floods the total dissolved matter is greater although the concentration is lower *(Figure 7.8)*.

Studies of the relationship between river water quality and discharge have shown that the fall in the concentration of naturally occurring constituents in a river lags behind the increase in flow[48]. This is because the velocity of the flood wave is greater than the mean water velocity and there is, therefore, a lag in the arrival of the dilute surface water. The effect is only likely to be observed in catchments larger than $100 \, km^2$ in area[49].

In small catchments the relationship between the ionic concentration and discharge may not be identical for the rising and falling stages of a runoff event[50]. At the beginning of storm flow there tends to be an increase in ionic concentration due to the displacement of water with a higher concentration from the soil zone.

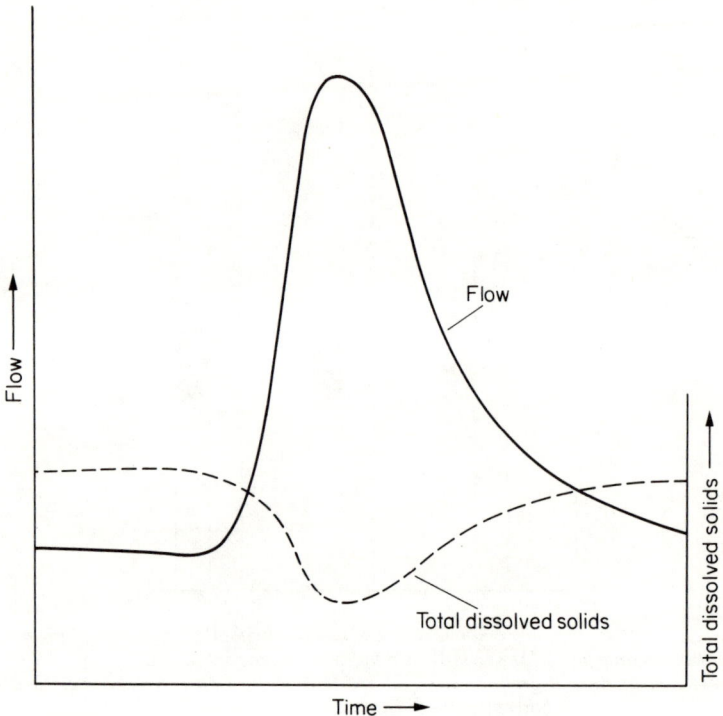

Figure 7.8   The relationship between flow and solute content of a river

The factors controlling the quality of surface waters are similar to those given earlier for natural waters as a whole. The relationship between rainfall and geology has been demonstrated in the Lake District[51], an area of Lower Palaeozoic strata surrounded by the more recent Carboniferous and Triassic rocks *(Figure 7.9)*.

The Palaeozoic rocks in the north and centre of the Lake District are the Skiddaw Slates and Borrowdale Volcanic Series, but in the south Silurian grits, limestones and shales outcrop. The composition of the surface waters of the Lake District, in the lakes themselves, the tarns and the rivers, is related to these main rock groups, the waters associated with each group being characterised by differences in total ionic concentration and the proportions of the major ions they contain. Waters from the Skiddaw Slates and Borrowdale Volcanic Series are very dilute with concentrations between 0.15 and 0.3 milliequivalents per litre (meq/l). These waters are

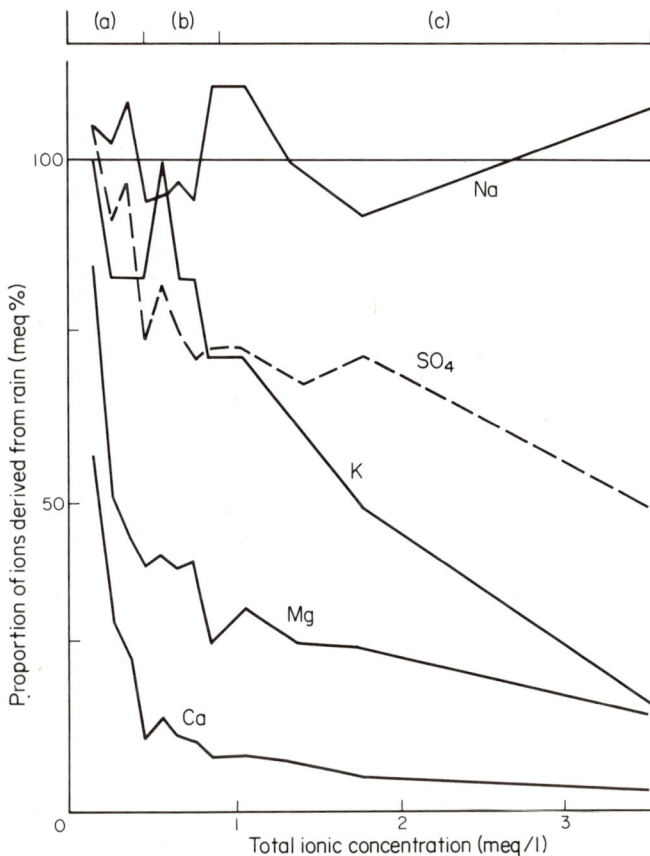

Figure 7.9   The proportion of ions derived from rain in tarns and lake waters of different ionic concentration on various geological strata in the Lake District: (a) upland tarns on hard volcanic rocks, (b) upland and lowland tarns and lakes on Silurian slates and flags, (c) lowland tarns on softer sedimentary rocks of the Carboniferous and Trias (from Gorham[3], courtesy Royal Society of London)

dominated by sodium, magnesium, hydrogen, chloride and sulphate ions. The more soluble Silurian rocks in the south give rise to waters with ionic concentrations between 0.5 and 2.0 meq/l. In these waters hydrogen is no longer a major ion and calcium and bicarbonate have increased to become as important as sodium, chloride and sulphate, and calcium has become the dominant cation. In the third group, associated with Carboniferous and Triassic rocks, the ionic concentrations vary between 2.0 and 5.0 meq/l, the waters are of the calcium bicarbonate type, with sodium, chloride and sulphate assuming less importance.

The most dilute waters found in the high level tarns are similar in composition to rainfall *(Figure 7.9)*. They are essentially dilute sea water acidified with sulphuric acid from industrial pollution, but this basic water is modified by the solution of calcium carbonate. Thus the differences between the various waters in the Lake District are primarily due to geological factors, but the influence of rainfall is clearly evident particularly in the more western areas where sodium and chloride remain the dominant ions at higher ionic concentrations than further east.

In Scotland, the northern part of the catchment that includes Loch Lomond is formed of metamorphic rocks of the Dalradian series, while the southern or lower part is in Old Red Sandstone and Carboniferous strata. Slight differences in the concentration of the water in the loch reflect these differences, the northern part of the loch having a concentration of 22 mg/l compared with 28 mg/l in the lower part. The rivers draining into the loch over rocks of the Dalradian series have ionic concentrations between 2 and 7 mg/l whereas the River Endrick, which crosses the Old Red Sandstone and Carboniferous, carries water of about 16 mg/l into the loch[52].

Similar environmental controls to those found to apply in the Lake District have been described[53] for waters in the Cairn Gorm–Strath Spey area of Scotland. Waters derived from granite have concentrations of 0.25 to 0.4 meq/l and atmospheric precipitation is the principal factor determining their composition.

Further examples of low ionic contents of surface waters may be quoted for rivers draining the high areas of other parts of Scotland and Wales. In the upper parts of the Teifi, which drains Silurian rocks, the total hardness averages about 30 mg/l, the chlorides 10 mg/l and the electrical conductivity 60–70 micromho/cm [54]. The River Towy is similar.

The ionic concentration of rivers increases from source to mouth due to the greater thickness of soil and weathered rocks in the lower parts of river basins. However, where the rocks are hard, compact and relatively impermeable, the increase may be quite small. In the Teifi, electrical conductivity, hardness and chloride increase from 60 to 100 micromho/cm, 30 to 45 mg/l and 8 to 13 mg/l respectively between the upper and lower reaches. Similarly in the Usk, which drains extensive areas of Old Red Sandstone, the chloride content is about 10 mg/l in the upper reaches and 15 mg/l in the lower reaches, although the total hardness increases from 50 to 100 mg/l and conductivity from 150 to 300 micromho/cm. The chemical nature of rivers with such low concentrations is influenced by the change in composition of rainfall as the coast is approached.

The predominantly sandstone-shale sequences of the Carboniferous rocks of the south Pennines give rise to soft waters, the carbonate hardness generally being about 20 mg/l. Further north in the Pennines, limestones are more important and comprise some 30 per cent of the Carboniferous sequence. This is reflected in a slightly harder water but the average carbonate hardness of the upper part of the River Tees, as recorded by Butcher *et al.*[55] between November 1929 and September 1930, was only about 40 mg/l and the chloride value only 8 mg/l.

Extensive areas of the Midlands and north of England are formed by Triassic rocks, which are overlain in many areas by glacial deposits. The River Blithe, a tributary of the Trent, drains a catchment of Triassic rocks, particularly Keuper Marl, overlain by glacial deposits. The average carbonate and non-carbonate hardnesses of the river are about 150 and 100 mg/l, and the chloride and electrical conductivity are 25 mg/l and 500 micromho/cm respectively; the range of the latter parameter between 1968 and 1970 was 340–640 micromho/cm [56]. These figures are higher than those found where pre-Triassic rocks occur, although the influence of any glacial deposits overlying pre-Triassic rocks has to be considered when applying this general conclusion.

Near Northwich, in Cheshire, a series of meres occurs in an area where the Keuper Marl is overlain by various types of glacial and recent drift. The average ionic concentration of seven of these lakes is 4.53 meq/l [57]. Again they are more concentrated than waters draining pre-Triassic rocks. A relatively high proportion of sodium chloride and magnesium sulphate indicates their relationship to the Keuper Marl.

The Chalk crops out over a large part of south-east England, and because of the high infiltration capacity of this rock there is little direct runoff. The rivers are largely fed by groundwater discharge and as such are similar in character to a typical groundwater from the Chalk, although the carbonate hardness may be lower due to the precipitation of calcium carbonate following the loss of carbon dioxide through photosynthesis. The seasonal variation of most constituents is small[58], and the total ionic concentration is between 300 and 400 mg/l.

Rivers draining clay deposits are generally characterised by a high non-carbonate hardness due to the presence of calcium and magnesium sulphates in clays. The chloride content also tends to be relatively high. Extensive clay outcrops occur in central and eastern England, and in the London and Hampshire basins. Much of East Anglia is covered by glacial deposits with boulder clay particularly evident in the higher central areas. The runoff from a typical clay catchment may have a hardness in excess of 450 mg/l, of which 150 mg/l may be of the non-carbonate form. Marked seasonal variations in non-carbonate hardness, with maximum values in the winter and minimum values in the summer, have been detected in East Anglia and a rising trend is superimposed on this seasonal cycle *(Figure 7.10)*. As the increase in hardness is due to calcium sulphate, changes in the sulphate ion are an index for change in hardness. In the Chelmer and Blackwater rivers between 1931–32 and 1962–63 the sulphate content (expressed as SO_4) doubled from about 70 mg/l. The increase is believed to be due to changes in agricultural practice since World War II, involving better drainage of the clay soils, a change to arable farming and the widespread use of artificial fertilisers[59].

Chloride values from clay catchments often tend to rise with the onset of winter runoff but a distinct seasonal cycle associated with a higher proportion of sewage effluent in dry weather flows can also be identified *(Figure 7.10)*. In the Blackwater values of about 50 mg/l are recorded in the winter, when runoff dilutes the sewage effluents discharged to the river, but in the summer maximum values are between 100 and 150 mg/l [59]. However values have increased from the early 1930s when the mean was about 32 mg/l with maxima and minima of about 27 and 38 mg/l. Similar cyclic patterns have been recorded for hardness and chloride for the Thames and Lee[60], the Essex Stour[61] and the Great Ouse[62].

Edwards[63,64] studied the variation of the principal dissolved constituents in four rivers in central Norfolk where extensive boulder clays overlie Pleistocene sands and the Chalk, the sands and the Chalk being exposed only in the valleys. In these rivers the nitrate and sulphate increased with discharge, but magnesium, bicarbonate and phosphate were usually diluted

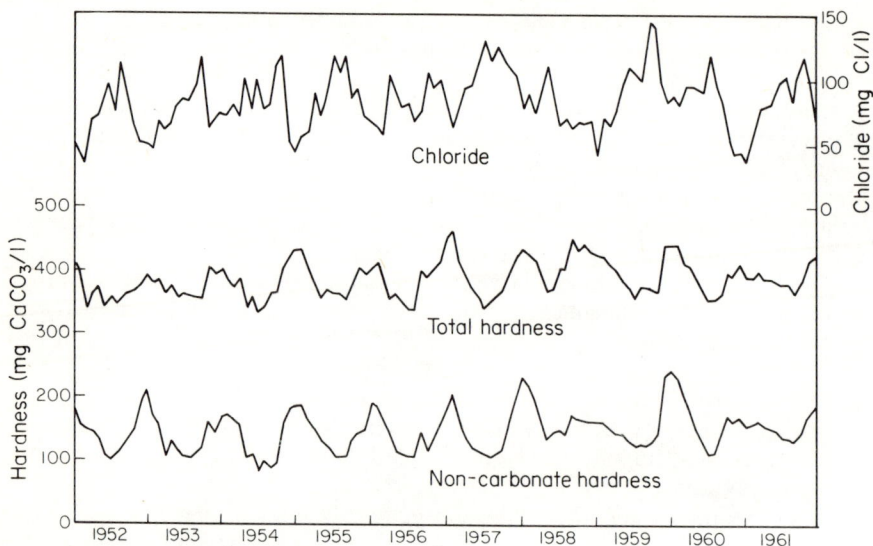

Figure 7.10　Variation in the chemistry of the River Blackwater, Essex (from Davis and Slack[59], courtesy Society for Water Treatment and Examination)

and sodium and potassium showed a variable relationship, although potassium increased at the beginning of storm flows. There was an apparent lack of correlation between calcium and discharge, probably because calcium is derived from calcium sulphate in the clay and also from bicarbonate in the groundwater, the former increasing with discharge and the latter decreasing. Probably the absence of a well-defined dilution relationship between flow and ionic concentration is because seasonal extremes of discharge are relatively small, extending over only two orders of magnitude, and because the surface deposits contain soluble minerals. These relationships are likely to be typical of East Anglia as a whole.

The above discussion has summarised differences in the chemical quality of rivers draining the UK, but there is an underlying general pattern of change in chemical composition as rivers follow their courses to the sea. Groundwaters entering rivers tend to lose carbon dioxide and become aerated. The loss of carbon dioxide, together with photosynthesis, results in the precipitation of calcium carbonate until a new equilibrium is established. The ionic concentration of river water increases by solution of soluble minerals in the soil and rocks and on the stream bed; this is more accentuated where rivers are draining the softer, less compact, post-Carboniferous rocks. The waters become enriched in nitrogen and phosphate from sewage effluents and fertilisers in land drainage.

Except where indicated, the examples of river quality quoted so far in this section have been for relatively unpolluted rivers. Typical analyses are given in *Table 7.4*. The increase in the chloride content above natural levels is a measure of the increase in ionic content due to pollution. In a typical river draining a predominantly sandstone or limestone catchment the natural chloride content is unlikely to exceed 15–20 mg/l, but a higher value of

Table 7.4 REPRESENTATIVE ANALYSES OF INLAND WATERS (mg/l)

Locality:	Thirlmere	Loch Einich	Meres, Cheshire	River Lambourn	River Great Ouse
Geology:	Lower Palaeozoic	Granite	Trias	Chalk	Clays and Chalk
Calcium (Ca)	3.3	1.3	49.3	105	161
Magnesium (Mg)	0.7	0.2	14.7	2	2
Sodium (Na)	3.1	2.0	17.2	8	26
Potassium (K)	0.3	0.2	4.3	1.6	7
Bicarbonate (HCO_3)	4.1	1.0	132.4	295	201
Sulphate (SO_4)	6.0	3.1	71.2	13	188
Chloride (Cl)	5.4	2.7	29.4	15	50
Nitrate (NO_3)	0.6	0.3	2.4	22	53

25–30 mg/l would be more typical of a catchment dominated by post-Carboniferous clay formations, and about 10 mg/l for rivers from the older harder rocks forming the upland areas of central and north Wales, the Lake District and the Highlands of Scotland. By comparison, sewage effluents may contain 50–100 mg/l of chloride ion.

As a general indication the natural total ionic content of an upland river is likely to be up to about 100 mg/l, whereas 250–350 mg/l would be more typical of a lowland river. It is interesting to compare these figures with the estimated average ionic content of 100 mg/l for the world's rivers[5,65]. The lowland drainage areas in the UK yield surface runoff with values in excess of the world average, reflecting the distribution of the younger post-Palaeozoic rocks and the fact that these rocks contain significant aquifers.

The natural concentration of some constituents has been increased by pollution, particularly in lowland rivers flowing through urban and industrial areas. This is indicated by typical values for chloride in the lower non-tidal reaches of the following rivers:

Aire at Swillington (near Leeds)	80 mg/l
Great Ouse at Offord	55 mg/l
Irwell at Salford	85 mg/l
Lee at Chingford	50 mg/l
Mersey at Howley Weir	160 mg/l
Severn at Upton	60 mg/l
Thames at Walton	40 mg/l
Trent at Nottingham	100 mg/l

Many factors control the detailed variations of the chemistry of an inland water, for example local variations in composition are influenced by the growth and decay of aquatic plants, including algae. The biochemical reactions involve gains and losses of oxygen and reactions within the nitrogen, phosphorus and sulphur cycles; these in turn influence the concentration of other ions.

Eutrophication

The enrichment of water by nutrients is referred to as *eutrophication*. The process occurs naturally during the slow aging of lakes, but it has been accelerated by runoff from fertilised agricultural land and the discharge of domestic sewage and industrial effluents into lakes and also reservoirs or into the rivers feeding them. This commonly leads to excessive growths of algae followed in some cases by a serious depletion of dissolved oxygen as the algae decay after death. Algal blooms on reservoirs, particularly pumped storage reservoirs fed from lowland rivers, are an increasing problem. They can give the water objectionable tastes and odours, reduce its amenity value, and block filters of water treatment works.

The factors controlling the growth of algae are complex and include physical factors (for example area, shape and depth of a lake) as well as chemical factors[66]. Various trace elements are necessary for their growth including manganese, iron, boron, molybdenum, vanadium, cobalt and copper[67], but the concentrations of nitrogen and phosphorus are believed to be particularly important in limiting growth. Diatoms have a specific requirement for silicon.

Algal blooms can occur when inorganic nitrogen and phosphorus concentrations are as low as 0.3 and 0.01 mg/l respectively. Efforts have been made to control both nitrogen and phosphorus inputs to inland waters (particularly phosphorus because certain blue-green algae can use atmospheric nitrogen dissolved in water, and in many waters the supply of phosphorus is generally more limited than nitrogen).

Studies have been made of several rivers to determine the source of nutrients that adversely affect river quality[68]. Two main sources can be recognised: diffuse sources, including rainfall, land drainage and urban runoff, and point sources such as sewage effluents, industrial wastes and in some cases farm wastes.

A study of the nutrient balance of the Great Ouse above Bedford revealed that sewage effluents contribute most of the phosphorus in the river but only a small proportion of the nitrogen, potassium, silicon, chloride and sulphate; land drainage was the main source of these[69], and this is probably typical of other rivers draining agricultural areas[64,68,70].

Lough Neagh in Northern Ireland, the biggest expanse of inland water in the UK, shows many characteristics of a highly eutrophic lake, being extremely rich both chemically and biologically[71]. Phosphorus is considered to be the key factor influencing the growth of algae in the lough. Domestic and industrial sewage contributes 80 per cent of this element and agriculture the remaining 20 per cent. The total dissolved solids in the lough are increasing at a rate of 2 mg/l per year[72]. In this case sewage effluents rather than land drainage are the cause of accelerated eutrophication.

In rivers draining urbanised and industrial catchments, effluents are the main source of nitrogen, chloride and phosphorus[73]. There has been a marked increase in the phosphorus content of some rivers since the early 1950s, an increase related to the use of detergents[74] *(Figure 7.11)*; the increase in nitrates, in at least some situations, is due to more efficient nitrification of ammonia in sewage effluents.

In agricultural areas river waters derive nitrogen from fertilisers, animal wastes, fixation of nitrogen, mineralisation of organic matter in the soil, and rainfall. The average use of nitrogen fertilisers in England and Wales, on land under crops and grass, has increased more than four times since the mid-1950s; it has therefore been suggested that fertilisers are responsible

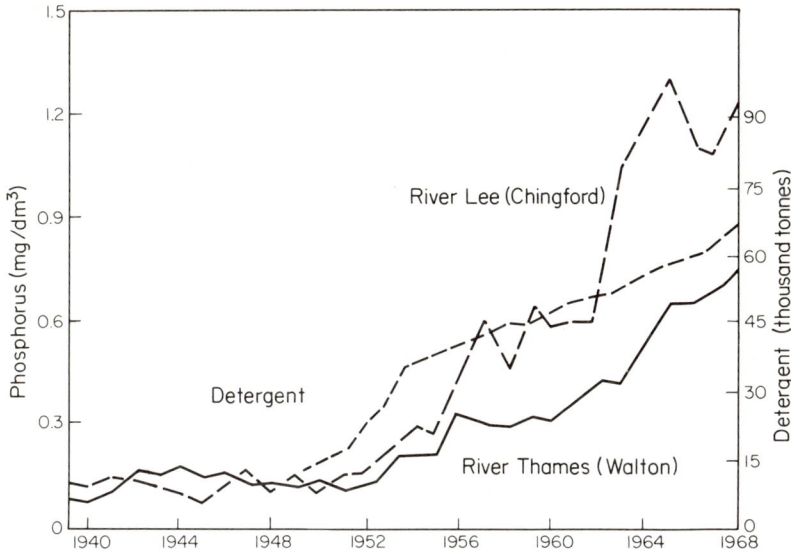

Figure 7.11 Comparison of phosphate content of rivers Thames and Lee with annual consumption of detergents (from Royal Commission on Environmental Pollution[74], reproduced by permission of the Controller, Her Majesty's Stationery Office)

for the increase in nitrate concentrations in river water, but in fact there is not a direct correlation between their use and nitrate concentrations in rivers. Some rivers do not show an increase despite increased use of nitrogen fertilisers in their catchments[70,73].

Tomlinson[75] examined the change in nitrate content of eighteen rivers between 1953 and 1967 and only in six did the concentration increase significantly with time. He concluded that there was no evidence for a general increase in nitrate in rivers during the 15-year period. This may be related to the large storage capacity for nitrogen in the soil.

The mean nitrogen load derived from the land in a number of rivers in England and Wales has been estimated at 11.3 kg of inorganic nitrogen per hectare per year, a value believed to be relatively independent of the use of fertiliser[68]. It has been shown that in Norfolk the loss of nitrogen fertilisers to some rivers by direct leaching cannot be more than 10 per cent of the total applied[64].

The concentration of nitrate in drainage water is related to intensity of rainfall. High concentrations occur in the autumn, as infiltration removes nitrate released by decay of roots and plant residues, and during wet periods in the spring before crops have developed sufficiently to use any fertiliser applied[76]. Drainage from well-managed arable farmland in England contains an average of 45 mg NO_3/l, but concentrations from heavy soils in spring may average 90 mg NO_3/l and on occasions considerably more. Drainage from land not treated with fertilisers contains 20 mg NO_3/l, derived from the extensive natural nitrogen reserves of the soil.

The variation in the concentration of nitrate in drainage water depends upon many factors including the husbandry and the nature of the soil and sub-soil. Nitrate releases are less from grassland than from arable land, and are likely to be greater from permeable soils and sub-soils than from less permeable deposits[76] through which water movement is slower giving greater opportunity for denitrification. Land use may be a more important factor controlling variation in nitrate releases than fertiliser application. In many situations only a small proportion of the inorganic nitrogen applied to the land as fertiliser appears to be lost to rivers through land drainage. Nevertheless, in recent years there have been many cases of significant increases in nitrate contents in inland waters, and it would appear that this is related to more intensive agricultural practices including the increased use of fertilisers.

In both the Thames and the Lee the nitrate contents are steadily increasing and between 1971 and 1973 there were marked increases. Part of the reason for the marked rise is believed to be the below-average rainfall in 1972 and 1973. A major factor in both rivers is the more complete nitrification of sewage effluents at treatment works. The high concentrations during winter, however, are derived from land drainage.

Rising nitrate concentrations in rivers used for water supply cause concern because, in addition to the relationship between nitrate content of inland waters and eutrophication (leading to problems with water treatment and river management), a high nitrate content is recognised as a health hazard. It can cause methaemoglobinaemia in babies, and because of this the World Health Organisation recommends that concentrations should not exceed 45 mg NO_3/l in drinking water supplies (the European Standards for drinking water recommend 50 mg NO_3/l). More recently it has been shown that in some human beings bacteria can produce nitrosamines, which are carcinogens, from nitrates. Evidence has been presented that implies that, where the water supply contains an abnormally high nitrate content (of the order of 90–100 mg NO_3/l), the death rate from gastric cancer is also abnormally high[77].

Thermal and chemical stratification of lakes and reservoirs

Lakes and reservoirs in temperate climates, such as that of the UK, become thermally stratified in summer if more than 10 m deep. A warm upper layer develops (the epilimnion) overlying a colder layer (the hypolimnion). Between the two there is a layer referred to as the thermocline in which the temperature changes markedly. The epilimnion is generally 10–15 m deep and the thermocline 3–6 m deep. The waters of both the epilimnion and the hypolimnion are well mixed, but the two layers remain distinct until the temperature of the upper layer begins to decline with the onset of cooler conditions in the autumn and eventually strong winds mix the two layers, this being referred to as the 'turnover'.

The tendency for lakes to stratify depends upon the relationship of depth to surface area and the exposure to wind. Shallow lakes with a large surface area do not stratify. Lough Neagh, Bassenthwaite and Derwentwater are possibly examples in which only slight stratification occurs under ideal conditions[78], but stratified lakes as shallow as 6–8 m are known in the Lake District.

A body of water that is thermally stratified also becomes chemically stratified *(Figure 7.12)*. There is a reduction in the dissolved oxygen content with depth, and anaerobic conditions may develop in the lower part of the water column. Other chemical changes that may follow as a consequence of deoxygenation were summarised by Hammerton[79] using Chew Valley Lake in the Mendip Hills as an example. They include the liberation of ammonia by bacterial decomposition of protein in organic matter (if oxygen were present this would be converted to nitrate), the release of phosphate and silica into solution, the production of hydrogen sulphide, methane and carbon dioxide, and the release of iron and manganese. If calcium carbonate is

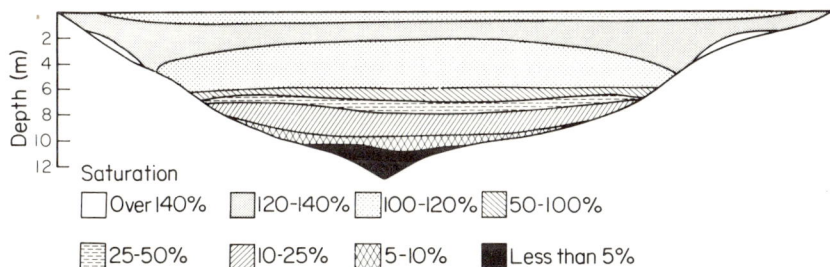

Figure 7.12 Diagrammatic cross-section of Chew Valley Lake showing dissolved oxygen content in August 1957 (from Hammerton[79], courtesy Society of Water Treatment and Examination)

present (for example as a result of precipitation due to photosynthesis), the presence of carbon dioxide will raise the alkalinity. Nutrients available in the hypolimnion are produced from organic matter enclosed within it when stratification began and from material falling from the epilimnion[80]. The complexity of the chemistry may be illustrated by considering the changes affecting iron. When oxygen is present in lower layers of the hypolimnion, iron at the mud surface is in the ferric state and forms a colloidal complex. If deoxygenation occurs, the ferric iron is reduced to the ferrous state and goes into solution. The colloidal complex breaks down releasing silica, phosphate and ammonia, as well as iron and manganese, into the water. The ferric complex reforms when the lake 'turns over' in the late autumn, but some of the ions released rise nearer the surface of the lake as mixing occurs[81]. In oligotrophic lakes such as Loch Lomond, ferric iron is not reduced to the ferrous state. This limits the concentration of nutrients, which is further reduced by the removal of phosphate, either as ferric phosphate (in the presence of adequate oxygen) or by adsorption of the phosphate ion by ferric hydroxide. The phosphate content (as PO_4) in Loch Lomond was about 0.001 mg/l in 1957[52].

The hypolimnion may represent a large proportion of the total storage in a reservoir, and therefore means of modifying the natural thermal conditions by artificially inducing circulation have been developed. The general policy of the Thames Water Authority at major reservoirs in the Thames valley is to prevent thermal stratification either by reducing retention periods or by using artificial turbulence systems[82]. This prevents or reduces anaerobiosis of the bottom layers but a consequence can be that nutrients are transferred to the surface layers, as would occur naturally in the autumn, tending to advance the date of algal blooms[83]. An alternative method is to aerate the lower portion of the reservoir in order to eliminate anaerobic conditions[84]. Artificial turbulence may increase the turbidity of the water, thereby reducing the development of algal blooms as penetration of light into the water is restricted.

Pollution

Inland waters

Pollution of inland waters is associated with the social and industrial developments since the industrial revolution. These led to the significant modification of the chemistry of rivers due to the discharge into them of increasing amounts of sewage and industrial effluents. Within the last 25 years the extension of main drainage systems into rural areas has had marked effects on the quality of some inland waters, for example in the Lake District[85]. The extent of pollution in rivers and canals in England and Wales has been assessed[86] by dividing their waters into four classes based on one or more of the following criteria:

1. Unpolluted—have a biological oxygen demand (BOD) generally less than 3 mg/l, are well oxygenated and do not receive significant discharges of toxic materials or suspended matter.
2. Mildly polluted—of doubtful quality needing improvement, have reduced oxygen content at times of normal dry weather flows, receive significant toxic or turbid discharges but with no significant effect on the ecology.
3. Badly polluted or of poor quality—have dissolved oxygen saturation below 50 per cent for considerable periods, contain substances believed to be actively toxic at times and have been changed in character by discharge of solids in suspension.
4. Grossly polluted—have a BOD of 12 mg/l or more under average conditions, are incapable of supporting fish, are completely deoxygenated at any time other than during exceptional drought, have offensive appearance and offensive smell.

On this basis about 75 per cent of non-tidal river courses were unpolluted in 1972, but only 50 per cent of tidal rivers and 45 per cent of canals were in this state. Eight per cent, 27 per cent and 15 per cent respectively were heavily polluted *(Table 7.5)*. Very serious widespread pollution occurs in the Yorkshire, Trent, Glamorgan, Mersey and Weaver and Lancashire river basins *(Figure 7.13)*, and in the Thames downstream of London.

It is clear from *Figure 7.13* that, although only about 8 and 27 per cent of the lengths of non-tidal and tidal rivers are grossly or badly polluted, the actual percentage volumes of water that are in these states is much greater because pollution tends to occur in the lower reaches. Rivers in classes 3 and 4 are not usually suitable for public water supply, and in some exceptional situations rivers in classes 1 and 2 may also be unsuitable.

Table 7.5 EXTENT OF POLLUTION IN RIVERS AND CANALS IN
ENGLAND AND WALES IN 1972 (DATA FROM REF. 86)

	Non-tidal rivers		Tidal rivers		Canals	
	km	%	km	%	km	%
Unpolluted	27 807	77.4	1 416	49.4	1 136	45.7
Mildly polluted	5 258	14.7	666	23.2	988	39.7
Badly polluted	1 511	4.2	407	14.2	237	9.5
Grossly polluted	1 339	3.7	380	13.2	126	5.1

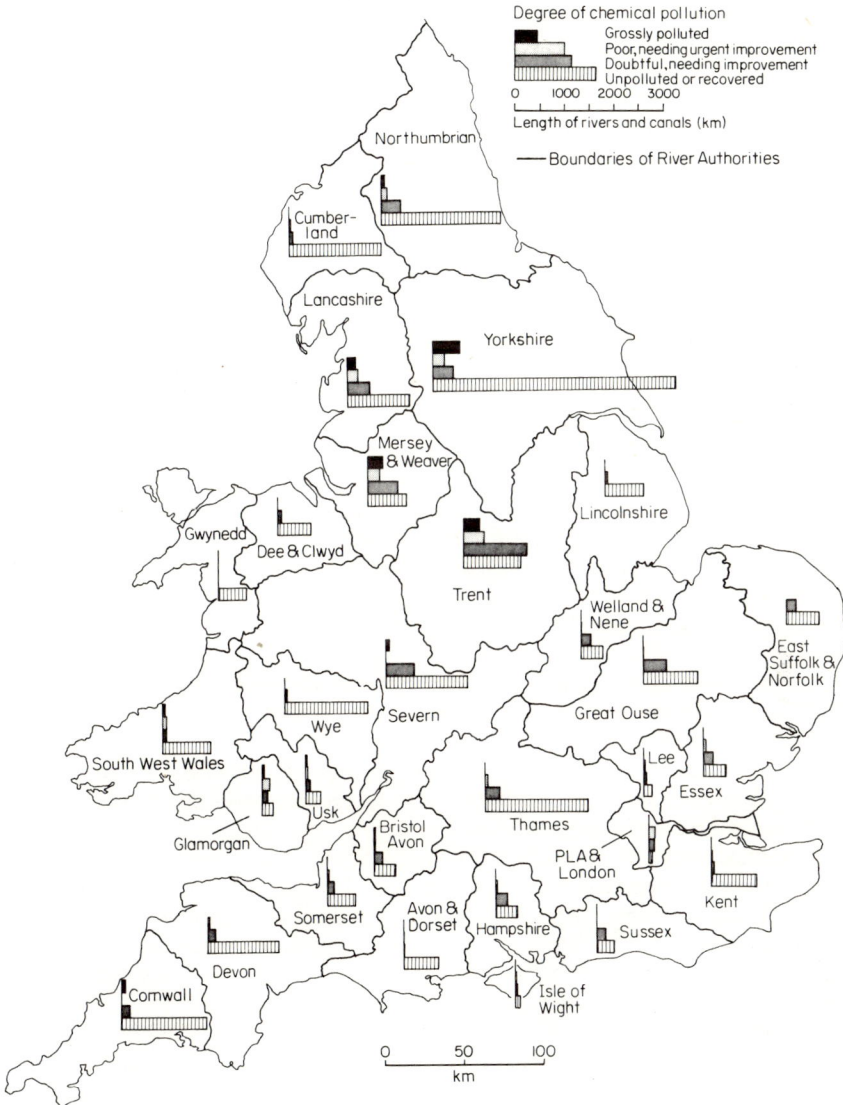

Figure 7.13   Pollution of waterways in England and Wales in 1972; all canals, and all rivers and streams (tidal and non-tidal) having a summer flow of more than 1 million gal/d, were included (reproduced by permission of the Controller, Her Majesty's Stationery Office)

Industry and population in Scotland are concentrated in the lowland central parts of the country. Public water supplies are largely derived from upland sources and only about 8 per cent is taken from the lower reaches of rivers, although this will increase when schemes currently under development are completed. The prevention of pollution of rivers in the central and southern parts of the country is exercised by nine river purification boards; in the remaining less populated northern areas, the county and town councils are responsible. (As from May 1975, seven river purification boards are responsible for the Scottish mainland; see *Figure 1.4* and Appendix 2.)

The Scottish river purification boards control about 5100 km of rivers in central and southern Scotland; a classification, on the same basis as that just described for England and Wales, indicated that in 1968 about 3 and 5 per cent were grossly and badly polluted, respectively, and 78 and 14 per cent unpolluted or only mildly polluted. The pollution problems are more important in the central region covered (after 1975) by the Clyde and Forth river purification boards, where over 20 per cent are either badly or grossly polluted[87]. In the remaining parts of the country, drained by some 2200 km of principal rivers, there are only a few stretches requiring improvement, and in Scotland as a whole only about 6 per cent of rivers are at present badly or grossly polluted.

The pollution of rivers has been examined in detail by Klein[88]. The commonest form results from the discharge of inadequately treated sewage. The Royal Commission on Sewage Disposal recommended that sewage effluents should have a BOD of not more than 20 mg/l and a suspended solids content of less than 30 mg/l, and should be diluted by at least eight volumes of clean river with a BOD of not more than 2 mg/l, the dilution factor to apply under conditions of dry-weather flow. Maintenance of these standards is not always possible, particularly because the dilution requirement is not attainable. In some situations where rivers are used for water supply this has led to an increase in the stringency of standards, requiring the introduction of tertiary treatment of effluents.

In industrial areas the discharge of industrial wastes from many diverse processes is a source of pollution equal in importance to sewage *(Table 7.6)*, while in some areas discharge of mine-drainage water has significant effects on river water quality. Pollution by farm wastes is sometimes a local problem, accentuated by the modern practice of keeping animals permanently confined in relatively restricted areas on farms that do not have arable land on which to use their manure.

Effluents have been divided into six groups[90]:

1. *Inert suspensions,* for example as found in some rivers of Cornwall, due to the working of china clay, and in many coal-mining areas.
2. *Poisons* such as chromium salts from electro-plating, phenols from the chemical industry, copper, lead and zinc in industrial wastes and from mine tips, insecticides etc.
3. *Inorganic reducing agents* including sulphides and sulphites.
4. *Oil* and oil waste products.
5. *Organic residues* from agricultural wastes, breweries, tanneries, paper mills etc.
6. *Warm water* discharged to rivers after use for cooling, in particular from power stations.

Table 7.6 MAIN SOURCES OF MAJOR POLLUTING SPECIES (FROM DRAKE[89], COURTESY THUNDERBIRD ENTERPRISES LTD)

Pollutants	Metal finishing	Iron and steel	Chemicals	Food	Agriculture	Mechanical engineering	Domestic	Other
Solids				*	*		*	Crown Establishments
High BOD			*	*	*		*	Crown Establishments, textiles
DO depletion			*	*	*		*	Paper
Organic solvents			*					Paint, dry cleaning
Ammonia			*		*			
Cyanide	*	*						Fruit canning
Nitrate			*		*			
Phosphate			*				*	
Chromium	*							Tanneries
Cadmium	*	*						
Nickel	*							
Manganese	*					*		
Copper	*	*	*		*			
Alkalinity or acidity	*					*		
Chloride			*	*				Mining
Phenols		*	*			*		
Boron			*					
Borates								
Oil and grease						*	*	Garages
Pesticides			*		*			Textiles
Herbicides			*		*			
Mercury			*					
Sulphide		*	*					Tanneries, paper
Detergents							*	Hospitals, laundries
Excess temperature								Electricity generation

BOD = Biochemical oxygen demand; DO = Dissolved oxygen; * = Origin of pollutant

To these may be added *radioactive wastes*.

The effects of pollution on rivers has been summarised by Klein as follows[88]:

1. Organic matter decomposes and deoxygenates the water, leading ultimately in severe conditions to the production of hydrogen sulphide, mercaptans, organic amines etc.
2. Suspended matter is deposited on the river bed. This may affect the ecology of the river. If the suspended matter is organic, putrefaction may result.
3. Corrosive or toxic substances may produce a sterile river incapable of natural self-purification.
4. Pathogenic micro-organisms may be discharged with sewage.
5. Undesirable physical effects may result, for example increase in turbidity, radioactivity, temperature. Taste and odour may become noticeable.
6. Undesirable biological effects may occur, for example excessive growth of sewage fungus.
7. Certain mineral constituents, for example calcium and magnesium (which give excessive hardness) and chloride, may reach undesirable concentrations.

When considering the problem of pollution, the capacity of a river for self-purification has to be borne in mind. Some of the changes in the chemistry of a water following the addition of sewage effluent are illustrated in *Figure 7.14*. The antipathetic relationships between dissolved oxygen and BOD and between ammonia and nitrate are clearly seen, and also the improvement in quality with distance from the outfall. Where possible, advantage should be taken of natural purification processes by increasing the distance between sewage outfalls and water supply intakes, and by further increasing the time between successive reuses of water by incorporating storage.

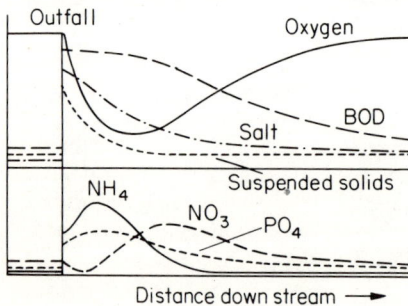

Figure 7.14 *Diagrammatic representation of the physical and chemical changes in a river downstream of an organic effluent outfall (from Hynes[90], courtesy Liverpool University Press)*

In recent years the use of salt for de-icing roads in winter has increased the chloride content of runoff. For example in Lough Neagh the chloride content has increased from 12 mg/l in 1958 to a maximum of 23 mg/l in 1969[72,91]. The boron concentration has also increased in rivers because of the use of perborates in detergents, but the concentration is below the limit for water used for irrigation in the UK. Reference has already been made to the relationship between increasing phosphate concentrations and the use of detergents.

The increase of river temperatures caused by the discharge of warm effluents is referred to as *thermal pollution;* its main source is condenser cooling water from electricity generating stations. The extent of such pollution can be judged by comparing the resulting temperatures with those normally encountered in rivers. Rivers that contain a large groundwater component have a small seasonal temperature range because groundwater has a constant temperature of about 11°C. For this reason the temperature of Chalk streams and the spring-fed headwaters of many rivers appear abnormally warm in winter when compared with air temperatures and, conversely, are cool in summer. The temperature of rivers that include a large direct runoff component tends to follow air temperatures. Relatively slow-moving lowland rivers on occasions attain temperatures of 25°C but 20–22°C is the more general maximum. Diurnal variations of large rivers tend to be small (i.e. about 2°C in the Severn) but up to 10°C may occur in small streams.

Power stations require large volumes of water for cooling purposes. For example a large station of 2000 megawatt capacity uses some 270 Ml/hour, which cannot be provided for direct cooling by any British river (although smaller stations are cooled directly). Cooling towers are therefore necessary for large power stations, and only 3 per cent of the water required need be abstracted from the river, two-thirds of this being returned to the river and one-third evaporated[92].

Water discharged from power stations has been heated some 6–9°C but it usually has a temperature of less than 30°C. Below the Ironbridge Power Station, on the Severn, measurements indicated that the temperature of the river increased 0.5°C during floods and 8°C at times of low flow[93]. The temperature of the River Trent at Shardlow and Trent Bridge (about 5 and 20 km, respectively, below the Castle Donnington Power Station) is compared with that of a tributary of the river in *Figure 7.15*[94]. When only part of a river is used by a generating station, and also when the effluent is discharged intermittently, the warm water tends to occur as a surface layer, the lower part of the river being unaffected or only slightly warmed[95].

Although a rise in the temperature of a river will increase the biological activity and rate of chemical reactions, very little deoxygenation is found below power-station outfalls, in fact dissolved oxygen content tends to be higher than above the outfall because of aeration in cooling towers[96]. If the water for cooling is taken from a polluted river the water will be returned with a higher nitrate but a lower ammonia and organic nitrogen content[97,98], and this is of course beneficial. Because part of the water extracted for cooling evaporates, the effluent has a higher concentration of dissolved salts. The concentration factor is usually between 1.15 and 1.3[97].

The small size of rivers in the UK means that future power stations will incorporate fully-closed cooling systems requiring cooling towers, and over the next 20–25 years the gradual replacement of direct-cooled stations will reduce considerably the heat discharge to rivers[92]. Essentially thermal pollution is only a problem in the immediate vicinity of power stations. The fauna and flora of the river are modified near outfalls[99] but fish are killed only occasionally, for example if temperatures increase suddenly. The British

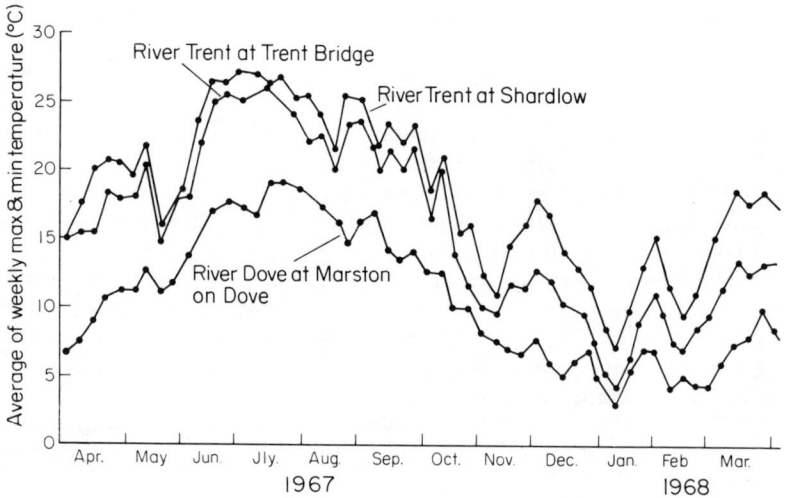

Figure 7.15   River temperatures in the Trent river basin (from Alabaster[94], courtesy Vanderbilt University Press)

river particularly affected by thermal pollution is the Trent, which has over twenty power stations along its length or on its tributaries.

One of the most important problems now affecting water quality is the greater complexity of industrial processes and the increasing use of synthetic organic chemicals that are non-biodegradable, for example organo-chlorine compounds (as in some pesticides). This has led to the presence of small amounts of possibly hazardous substances in surface waters. Some of these are partially absorbed and concentrated by aquatic plants and organisms, possibly in some cases as much as 10 000 times. They then move through the food chains into higher organisms. In some circumstances human health could be endangered.

This brief discussion of the pollution of inland waters has emphasised the chemical aspects, but pollution is essentially a biological problem[90]. A number of attempts have been made to assess the extent of pollution by relating the distribution of plant and animal species to the state of the water. A recent example was discussed in the river pollution survey of England and Wales[86] previously referred to. In studying the extent of pollution of an inland water both chemical and biological assessments have a role to play, for the two approaches often complement each other.

Biological aspects of pollution have been discussed by Hynes[90] and reference should be made to his book for further details. He pointed out that biological studies can reveal the occurrence of intermittent pollution and possibly the source of such pollution, which would have been detected otherwise only by continuous chemical sampling. The ecology of a river also provides a record of the typical state of the water, for plant and animal life are not necessarily affected by transient changes.

Chemists and engineers involved in the management of river water quality are making increasing use of mathematical models that relate river flow, abstractions and effluent discharges to quality parameters, so that predictions of the quality state of a river can be made. With such models the quality standards required in a particular river reach, and the amounts and quality of effluent discharges, can be determined. The success of management policies will be assessed by automatic continuous monitoring of water quality. An important factor determining river quality is likely to be the extent of the use of the river for water supply, which will, in turn, set effluent standards.

Groundwater

Permeable rocks are very vulnerable to pollution, and adequate precautions have to be taken to protect groundwater in such deposits from the consequences of waste disposal.

Groundwater may become contaminated from:
1. sewage effluents, for example from septic tanks, cess-pits, fractured sewers and overflows from sewers;
2. effluents from refuse-disposal tips or mine-waste tips;
3. agricultural and industrial wastes;
4. infiltration from agricultural land that has been treated with fertilisers.

Factors requiring consideration when assessing the potential risk of polluting an aquifer are:
1. the nature of the waste material, the solubility of its constituents and their mobility in the ground;
2. the permeability of the aquifer and whether it is primarily due to fissure or intergranular permeability;
3. the hydraulic gradient and the direction of groundwater flow, and whether these are likely to be modified by groundwater abstraction;
4. the composition of the aquifer in so far as this affects the adsorption of pollutants, for example ammonium ions or various other cations;
5. the depth of the water table and the extent of its seasonal fluctuation, in other words the thickness of the unsaturated zone as this influences the extent to which pollutants are oxidised;
6. the influence of any less permeable deposits overlying the aquifer;
7. the extent to which the pollutant will be diluted by groundwater.

Flow through limestone aquifers in the UK is predominantly through fissures. In these circumstances groundwater can be readily polluted and the contaminants transmitted considerable distances very rapidly (Chapter 5). Correlations can commonly be established between heavy rainfall and bacterial contamination of water from wells in limestones. Contamination may occur in wells where the water table is many metres below ground level and the wells are lined to depths of 25–30 m. A fact of some importance in pollution control is that bacterial pollution is often evident before chemical pollution[100].

Pollution of deep Chalk wells has been reported in north Kent three to four days after sewers overflowed, following heavy rain, and allowed polluted water to infiltrate the aquifer[100,101]. In one case the well had to be pumped for twelve days before pollution was no longer detectable, but in another case some two months elapsed before this was so.

Bacteria do not persist to any extent after polluted water has passed through sand or unfissured sandstone. Experiments have indicated that the filtering of effluents from domestic refuse through about 8 m of sand reduces bacteria by four orders of magnitude, and water treated in this manner is unlikely to contain significant bacterial contamination[102]. But the effective distance depends obviously upon the nature of the sand, and while some sands would remove bacterial contamination in less than 8 m others, including sands mixed with gravel, would require greater thicknesses. To ensure adequate bacterial purification, water in the saturated zone should flow through at least 30 m of unfissured rock although no more than 3 m may be necessary in the unsaturated zone[103,104].

These precautions are probably adequate in unconsolidated sands but many sandstones are intersected by fissures and natural purification by filtration does not occur. A well in the Bunter Sandstone in Shropshire was polluted by bacteria from river water entering the aquifer almost 500 m from the well[105].

Chemical pollution can travel considerably greater distances than bacteria in the saturated zone, even in unfissured sands, before dilution by groundwater reduces it to undetectable amounts.

Because of ion exchange the movement of pollutants in an aquifer may be retarded, different ions being retarded by different amounts because the capacity of minerals in the ground to adsorb ions is selective. As a consequence the water front moves in advance of the pollutant fronts. The rock depletes the pollutant in the water until the exchange capacity of the rock is consumed. For example, the soil derived from the Bunter Sandstone will adsorb ammonia from water recharged on its surface. However, if this continues the exchange capacity becomes saturated and ammonia begins to penetrate to lower layers. On the other hand, if the recharge is intermittent, oxygen enters the pores of the soil and the ammonia is oxidised to nitrate and the exchange capacity is regenerated[106].

As a pollutant moves through an aquifer it mixes with the native groundwater as a result of mechanical dispersion and, to a more limited extent, diffusion. The pollutant front is not sharp but is marked by a transition zone that increases in thickness with the distance travelled.

Examples of chemical pollution in limestones are well known but sandstones are also affected; for example a well in the Bunter Sandstone was polluted by seepage from a colliery waste tip that was more than 750 m from the well. Over a period of fourteen years an increase in chloride from 12 to more than 300 mg/l led to the well being abandoned as a source of public water supply[105]. A similar example was the increase of chloride ion from 15 to 30 mg/l in a borehole in the Folkestone Beds about 500 m from a source of pollution with a chloride content of 300 mg/l[107].

A degree of control exists over the siting of domestic refuse tips, and advice is often sought from hydrogeologists regarding the location of suitable sites. Domestic refuse gives rise to extremely polluted effluents[108]. Ammoniacal nitrogen, organic matter and many inorganic constituents, including sulphate and chloride, are high. Hardness is excessive and iron may be present in considerable amounts[102,109]. For some months after disposal of refuse, effluents from tips are likely to contain large numbers of harmful bacteria, although analyses of effluents from refuse tipped under controlled dry conditions indicated that the quality improved rapidly with time. Experiments carried out under such conditions showed that, during the first winter after tipping, the BOD and ammoniacal nitrogen were 6000 and 700 mg/l respectively and that these declined to less than 50 and 40 mg/l after two years. Similarly chloride was initially in excess of 1700 mg/l but after three years the refuse would have been virtually free from chloride[102]. However, this may not always be the case as the nature of the refuse is a controlling factor; for example the effluent from a tip in Essex, which had not been used for disposal for some thirty years, still contained 3500 mg Cl/l and had a hardness of 1350 mg/l[110].

It is desirable to tip under dry conditions for several reasons[102]:

1. There is an opportunity for purification within the tip itself and also while the effluents are moving through the unsaturated zone, although the overall extent of purification in this zone may not be very great.
2. The effluent issues from the tip only after considerable delay as no drainage occurs until the tip reaches field capacity.
3. The quantity of polluting matter extracted is less than if tipping occurs in wet pits.

Because of the nature of the effluents from refuse tips, it is necessary to ensure that groundwater is not abstracted for water supply in the vicinity of the tip before purification and dilution have eliminated or reduced the concentration of any undesirable constituents. Precautions taken to prevent pollution of groundwater from such tips include[101,102,109]:

1. placing an impermeable layer at the base of the tip and draining the effluent to a foul-water sewer;
2. tipping refuse at a rapid rate, so that the tip does not reach field capacity, and then covering with an impervious layer;
3. placing at least 1 m of sand at the base of the tip;
4. ensuring that the base of the tip is always above the groundwater level;
5. excluding contaminants such as oil products, poisonous or toxic materials, radioactive wastes, acids or materials likely to produce acid conditions.

Groundwaters in aquifers at outcrop commonly contain up to 10–20 mg NO_3/l, but shallow wells in superficial deposits may have much higher concentrations. This is the case in East Anglia, where the cause is believed to be intensive use of animal and, more recently, artificial nitrogenous fertilisers. Twenty-one per cent of 600 wells examined in Essex[111] and 41 per cent of more than 1300 wells examined in East Suffolk[112] had nitrate contents greater than 88 mg/l (i.e. 20 mg/l nitrate nitrogen).

In recent years cases have been quoted of increasing nitrate contents of groundwaters in the Chalk. Near Eastbourne values increased to between 45 and 65 mg NO_3/l in the late 1960s and in some wells there was a close relationship between nitrate content, bacterial pollution and heavy rain[113]. During 1970–72 a marked increase from between 10 and 15 mg NO_3/l to between 35 and 50 mg NO_3/l occurred in the Chalk of East Yorkshire[114], and similar increases have been reported in north Lincolnshire[115] and other areas.

The principal sources of nitrates in groundwater are the natural nitrogen content of the soil (supplemented by the use of fertilisers), farm wastes and sewage effluents. Studies of the tritium concentration in the unsaturated zone have indicated that water movement in this zone in many rock media is by displacement and that the average rate of flow is slow (see Chapter 4). It has been suggested that if this is the case a nitrate 'front', due to the increased use of fertilisers since the 1950s, may exist in the unsaturated zone where this is more than about 15 m thick. However, whether such a 'front' exists is at present unknown although some evidence has been presented for high nitrates at shallow depths below fertilised land[114].

The concentration of nitrate in the unsaturated zone is likely to depend upon the physical and chemical conditions in the overlying soil, the geology of the zone and the chemical and biological processes, including denitrification, that occur in the zone. It seems probable that (as with nitrate concentrations in surface waters) land use, together with rainfall intensity and the timing of its occurrence in relation to the farming seasons, is a factor exerting considerable influence on the nitrate content of groundwaters.

The use of sulphate fertilisers has been suggested as a possible cause of the increase in total ionic concentration (due almost entirely to calcium and sulphate ions) of some groundwaters from the Lincolnshire Limestone[16].

The increasing use of petroleum products presents a potential risk, particularly to groundwater, from accidental spillages. When oil is spilled on the ground the lighter fractions tend to evaporate and soluble constituents are removed by infiltrating water, leading to differential movement of the soluble and immiscible fractions. Oil products tend to be adsorbed on fine-grained or argillaceous material[116,117]; they also tend to accumulate at the top of the saturated zone. Many examples of groundwater pollution in British aquifers have been summarised by Ineson[116]. In most cases the effects were noticed only within a few tens of metres or at most a few hundred metres of the source of the pollution. This is a measure of the diluting effect of groundwater, but in some cases it is due to the slow rate of movement of the pollutant and its gradual biodegradation.

The pollution of groundwater is often a slow gradual process not readily apparent in its early stages. For example, water from a pumping station in the Triassic sandstones at Aston, Birmingham, steadily declined in quality between 1873 and 1935 and during that time the total dissolved solids increased from 190 to 560 mg/l and chloride from 20 to 40 mg/l due to pollution from the River Tame[26]. Groundwater beneath towns tends to be of lower quality, presumably due to slow, gradual pollution. This is the case for example in Nottingham and Watford. Once groundwater pollution is recognised correction may take many years, and this emphasises the need to assess potential pollution hazards in advance of waste disposal or accidental spillages. To emphasise the point with one further example, it has taken up to twenty years to reduce the pollution of water sources by gas liquor to a level that would permit the water to be taken into supply again.

Instrumentation

Samplers and sampling

The simplest method of obtaining a sample of a river or effluent is to take a 'grab' sample using an appropriate container. If this method is used the results can be of greater value if samples are collected in a consistent manner. The container should be heavy enough to sink below the surface but should not reach the bottom[118].

It is often necessary to collect samples of water from a river, reservoir, or well at a particular depth. The various depth samplers that have been designed are of two basic types. Either a watertight container is lowered to the required depth and then opened to allow water to enter, or an open cylinder is lowered to the required depth and closed by a signal or 'messenger', thereby enclosing a volume of water.

Samplers for use in wells have to be of small diameter to overcome limited access because of the pump and rising main. A suitable device has been designed by the Institute of Geological Sciences. It is based on the 'flow-through principle', which is generally more satisfactory. The sampler, which is suspended on an electrical cable, includes a solenoid or motor operated latch. When this is released it allows a spring to close the plungers[119]. It has a capacity of 1.25 litres.

Individual samples are often an inadequate means of assessing variations in the quality of a river or effluent when changes may occur with time and flow rate. Automatic samplers, either time-dependent or flow-dependent, have been designed to overcome such variations[118,120]. Time-dependent models incorporate a clock that actuates a mechanism (for example opens a valve or operates a suction pump) to abstract a sample and discharge it to a bottle. Such samplers may include 24 or 48 sampling bottles, which are filled in turn at the appropriate time interval.

Ideally the volume of the sample should be proportional to the flow at the time, and instruments exist that take flow variations into account, the sampler being operated in conjunction with a flow recorder[118,120].

Obtaining representative samples of a river water for analysis is difficult, but it is most likely to be realised if the sampling point is where the river is well mixed. Stagnant pools or areas of excessive weed growth should be avoided. Glover and Johnson studied river-water mixing processes[49] and concluded that for naturally occurring dissolved substances the concentration across a river cross-section is virtually uniform at any one instant, and a single sample is representative of the mean concentration across the cross-section. Considerable care is required when collecting water samples; the procedure and precautions necessary and the frequency of sampling for various purposes are discussed in standard texts[121,122].

Water-quality monitoring

The analysis in a laboratory of samples of a river indicates the past state of the water. Obviously for efficient management and adequate control of its quality, information is required about the 'true-time' state of the river. Because this can change rapidly through both natural and artificial factors, there is a need in some circumstances for accurate instruments to monitor continuously the chemical state and indicate by an alarm system any undesirable changes. This is particularly necessary where a lowland river is used for public water supply.

Continuous monitoring is also required when analysis of individual samples does not indicate the range of variation of a particular constituent (for example dissolved oxygen or suspended solids), where discharges of trade wastes may occur for limited periods only, where there is a possibility of the accidental discharge of toxic wastes, and where it is necessary to monitor variations in river quality caused by discharge of both sewage effluents and industrial or agricultural wastes[123].

The Water Resources Board provided financial and technical assistance to river authorities to encourage the establishment of water quality monitoring systems. The Board's Instrument and Data Group considered the requirements for such systems, advised on technical problems involved in their development and evaluated commercial products[124,125]. (Since April 1974 the Water Research Centre and the Water Data Unit have continued this work.) The Group suggested that ideally the following parameters should be measured on a continuous or semi-continuous basis[123,125]: temperature, dissolved oxygen, ammoniacal nitrogen, organic matter, suspended matter, conductivity, chloride.

In some circumstances information about the following may be required: nitrate, hardness, pH, dissolved carbon dioxide, sunlight intensity, cyanide, phenol, copper, zinc, nickel, cadmium, chromium, pesticide and herbicide residues, colour, chemical oxygen demand, toxicity, and substances responsible for taste and odour.

Where water is used for irrigation, details of the concentrations of sodium, boron, heavy metals and suspended solids may be necessary, while bacterial content should be monitored if the water is used for public supply and many industrial processes. It is invariably necessary to be able to relate quality data to accurate measurements of river flow.

At the present time land-based units are available for monitoring temperature, dissolved oxygen, suspended matter, conductivity and pH, and temperature. Many of the instruments have been developed as a result of work at the Water Pollution Research Laboratory (since April 1974 part of the Water Research Centre), and recently the laboratory has developed a sensor that records the concentration of organic matter and suspended solids in river water and effluents.

The two measurements are based on the absorption of ultra-violet light and the scattering of light in the visible spectrum by a sample of water.

In the future, selective-ion electrodes will probably become available for reliable continuous monitoring of individual ions. About twenty have been developed but present disadvantages include short working life, susceptibility in some cases to interference from other ions and the need for frequent calibration[89]. In fact careful maintenance and regular recalibration are necessary for all automatic monitoring equipment if reliable results are to be obtained.

In most cases automatic monitoring stations need only collect data at hourly intervals, although in some circumstances 15-minute intervals may be more appropriate. The information is recorded on magnetic tape. Situations where monitoring stations could be installed with advantages are on both streams above a confluence, below outfalls, and above intakes[125].

Continuous monitoring of water quality will ultimately be carried out by telemetry. A system installed on the River Lee by the Water Research Centre consists of a computer-controlled central station connected to three outstations (although up to twelve are possible). The outstations are linked to a water-quality monitor or group of sensors that record various parameters. The central control station can call for data from the outstations, control equipment for the verification, storage and presentation of data, and raise alarms.

A harmonised monitoring network for river waters has been operated in England and Wales since 1974 and in Scotland since 1975; see page 336.

Stations that automatically measure river temperatures exist on most of the main rivers and streams in England and Wales but to a lesser extent in Scotland except in the central valley[126]. The density of the network in Great Britain as a whole is one station per 2300 km².

REFERENCES

1. Gorham, E., 'Factors influencing supply of major ions to inland waters, with special reference to the atmosphere', *Bull. Geol. Soc. Am.,* **72,** 795–840 (1961)
2. Dhar, N. R. and Ram, A., 'Variations in the amounts of ammoniacal and nitric nitrogen in rainwater of different countries and the origin of nitric nitrogen in the atmosphere', *J. Indian Chem. Soc.,* **10,** 125–133 (1933)
3. Gorham, E., 'The influence and importance of daily weather conditions in the supply of chloride, sulphate and other ions to fresh waters from atmospheric precipitations', *Phil. Trans. R. Soc. Lond., B,* **241,** 147–178 (1958)
4. Stevenson, C. M., 'An analysis of the chemical composition of rain water and air over the British Isles for the years 1959–1964', *Q. J. R. Met. Soc.,* **94,** 56–70 (1968)
5. Conway, E. J., 'Mean geochemical data in relation to oceanic evolution', *Proc. R. Ir. Acad., B,* **48,** 119–159 (1942)
6. Robinson, E. and Robbins, R. C., *Sources, abundance and fate of gaseous atmospheric pollutants,* Stanford Research Institute (1968)
7. Prince, R. and Ross, F. F., 'Sulphur in air and soil', *Water, Air and Soil Pollution,* **1,** 286–302 (1972)
8. Gorham, E., 'On the acidity and salinity of rain', *Geochim. Cosmochim. Acta,* **7,** 231–239 (1955)
9. Sweeting, M. M., *Karst Landforms,* Macmillan, London, 362 (1972)
10. Peirson, D. H., Cawse, P. A., Salmon, L. and Cambray, R. S., 'Trace elements in the atmospheric environment', *Nature,* **241,** 252–256 (1973)
11. Hem, J. D., 'Study and interpretation of the chemical characteristics of natural water', *Water Supply Paper 1473,* US Geol. Surv. (1959)
12. Buchan, S., 'Variations in mineral content of some groundwaters', *Proc. Soc. Water Treat. Exam.,* **7,** 11–21 (1958)
13. Bogli, A., 'Mischungkorrosion: ein Beitrag zum Verkarstungsproblem', *Erdkunde,* **18,** 83–92 (1964)
14. Whitaker, W. and Thresh, J. C., 'Water supply of Essex', *Mem. Geol. Surv.* 509 (1916)
15. Ineson, J. and Downing, R. A., 'Changes in the chemistry of groundwaters of the Chalk passing beneath argillaceous strata', *Bull. Geol. Surv. Gt Brit.,* No 20, 176–192 (1963)
16. Downing, R. A. and Williams, B. P. J., 'The groundwater hydrology of the Lincolnshire Limestone', *Pub. No. 9,* Water Resources Board, Reading, 160 (1969)
17. Edmunds, W. M., 'Trace element variations across an oxidation-reduction barrier in a limestone aquifer', *Proc. Tokyo Symp. on hydrogeochemistry and biogeochemistry,* **1,** 500–526, The Clarke Company, Washington, DC (1973)
18. *Section 14, report of survey,* Thames Conservancy, Reading, 121 (1969)
19. *The hydrogeology of the London Basin,* Water Resources Board, Reading, 139 (1972)
20. Chebotarev, I. I., 'Metamorphism of natural water in the crust of weathering', *Geochim. Cosmochim. Acta,* **8,** 22–48, 137–170 and 198–212 (1955)
21. Downing, R. A., 'The geochemistry of groundwaters in the Carboniferous Limestone in Derbyshire and the east Midlands', *Bull. Geol. Surv. Gt Brit.,* No 27, 289–307 (1967)
22. Ineson, J., in Downing, R. A., Land, D. H., Allender, R., Lovelock, P. E. R. and Bridge, L. R., 'The hydrogeology of the Trent River Basin', *Hydrogeological Rep. No. 5,* Inst. Geol. Sci., 104 (1970)
23. Woodland, A. W., 'Water supply from underground sources of Cambridge-Ipswich district', *Wartime Pamphlet No. 20* Pt 10, Geol. Surv. Gt Brit., 88 (1946)

24. Baden-Powell, D. F. W., 'The Chalky Boulder Clays of Norfolk and Suffolk', *Geol. Mag.*, **85,** 279–296 (1948)
25. Land, D. H., 'Hydrogeology of the Bunter Sandstone in Nottinghamshire', *Hydrogeological Rep. No. 1,* Inst. Geol. Sci., 38 (1966)
26. Land, D. H., 'Hydrogeology of the Triassic sandstones in the Birmingham-Lichfield district', *Hydrogeological Rep. No. 2,* Inst. Geol. Sci., 30 (1966)
27. Downing, R. A., Land, D. H., Allender, R., Lovelock, P. E. R. and Bridge, L. R, 'The hydrogeology of the Trent River Basin', *Hydrogeological Rep. No. 5,* Inst. Geol. Sci., 104 (1970)
28. Lyon, A. L., 'The hydrogeology of the Coventry district', *J. Instn Water Engrs.*, **3,** 209–250 (1949)
29. Warren, S. C., 'Chemical aspects of controlled pumping and automation', *Proc. Soc. Water Treat. Exam.*, **13,** 7–11 (1964)
30. Edmunds, W. M., Lovelock, P. E. R. and Gray, D. A., 'Interstitial water chemistry and aquifer properties in the Upper and Middle Chalk of Berkshire, England', *J. Hydrology*, **19,** 21–31 (1973)
31. Plumptre, J. H., 'Underground waters of the Kent coalfield', *Trans. Instn Min. Engrs*, **119,** 155–169 (1959)
32. Anderson, W., 'On the chloride waters of Great Britain', *Geol. Mag.*, **82,** 267–273 (1945)
33. Downing, R. A. and Howitt, F., 'Saline groundwaters in the Carboniferous rocks of the English east Midlands in relation to the geology', *Q. J. Engng Geol.*, **I,** 241–269 (1969)
34. White, D. E., 'Magmatic, connate and metamorphic waters', *Bull. Geol. Soc. Am.*, **68,** 1659–1682 (1957)
35. White, D. E., 'Saline waters of sedimentary rocks', from *Fluids in subsurface environments,* Mem. No. 4, Am. Assoc. Petrol. Geol. (1965)
36. Sitter, L. U. de, 'Diagenesis of oil-field brines', *Bull. Am. Assoc. Petrol. Geol.*, **31,** 2030–2040 (1947)
37. Edmunds, W. M., Taylor, B. J. and Downing, R. A., 'Mineral and thermal waters of the United Kingdom', *23rd Int. Geological Congress,* **18,** 139–158 (1969)
38. Hudson, R. G. S., 'The geology of the country around Harrogate: VI The Harrogate mineral waters', *Proc. Geol. Assoc.*, **49,** 349–352 (1938)
39. Temple, K. L. and Delchamps, E. W., 'Autotrophic bacteria and the formation of acid in bituminous coal mines', *Applied Microbiology*, **1,** 255–258 (1953)
40. Kuznetsov, S. I., Ivanov, M. V. and Lyalikova, N. N., *Introduction to geological microbiology,* McGraw-Hill, London, 252 (1963)
41. Barnes, I. and Clarke, F. E., 'Geochemistry of groundwater in mine drainage problems', *Prof. Paper 473–A,* US Geol. Surv. (1964)
42. Ashmead, D., 'The influence of bacteria in the formation of acid mine waters', *Colliery Guardian*, **190,** 694–698 (1955)
43. Revelle, R., 'Criteria for recognition of sea water in groundwaters', *Trans. Am. Geophys. Union*, **22,** 593–597 (1941)
44. Love, S. K., 'Cation exchange in groundwater contaminated with sea water near Miami, Florida', *Trans. Am. Geophys. Union*, **25,** 951–955 (1944)
45. Hibbert, E. S., 'The hydrogeology of the Wirral Peninsula', *J. Instn Water Engrs*, **10,** 441–469 (1956)
46. Hoather, R. C., 'Increase of hardness by cation-exchange associated with infiltration of sea water into Chalk under Woolwich and Reading beds', *J. Instn Water Engrs*, **12,** 185–197 (1958)
47. Leopold, L. B., Wolman, M. G. and Miller, J. P., *Fluvial processes in geomorphology,* Freeman, San Francisco, 522 (1964)
48. Davies, A. W., *Changes in river quality associated with storm hydrographs,* MSc thesis, University of Newcastle-on-Tyne, 224 (1971)
49. Glover, B. J. and Johnson, P., 'Variations in the natural chemical concentration of river water during flood flows and the lag effect', *J. Hydrology*, **22,** 303–316 (1974)
50. Gregory, K. J. and Walling, D. E., *Fluvial processes in instrumented watersheds,* Institute of British Geographers, London, 194 (1974)
51. Mackereth, F. J. H., 'Chemical analysis in ecology illustrated from Lake District tarns and lakes: 1 Chemical analysis', *Proc. Linn. Soc. Lond.*, **167,** 159–164 (1957)
52. Slack, H. D., *Studies on Loch Lomond,* **I,** Blackie, Glasgow, 133 (1957)

53. Gorham, E., 'The chemical composition of some natural waters in the Cairn Gorm–Strath Spey district of Scotland', *Limnol. and Oceanogr.*, **2**, 143–154 (1957)
54. *Sixth statutory annual report,* South West Wales River Authority, Llanelli, 93 (1970–71)
55. Butcher, R. W., Longwell, J. and Pentelow, F. T. K., 'Survey of the River Tees, Pt III The non-tidal reaches—chemical and biological', *Water Pollution Research Tech. Paper No. 6,* HMSO, London (1937)
56. *River water quality triennial statistics 1968–70,* Trent River Authority, Nottingham, 88 (1972)
57. Gorham, E., 'The ionic composition of some lowland lake waters from Cheshire, England', *Limnol. and Oceanogr.*, **2**, 22–27 (1957)
58. Casey, H., 'The chemical composition of some southern English chalk streams and its relation to discharge', *River Boards Assoc. Yb.*, 100–113 (1969)
59. Davis, A. L. and Slack, J. G., 'The rivers Blackwater and Chelmer—hardness, sulphate, chloride and nitrate content', *Proc. Soc. Water Treat. Exam.*, **13**, 12–19 (1964)
60. Carter, G., 'The rivers Thames and Lee—chloride content and hardness', *Proc. Soc. Water Treat. Exam.*, **12**, 226–229 (1963)
61. Houghton, G. U., 'The River Stour (Essex and Suffolk)—hardness, chloride, and nitrate content', *Proc. Soc. Water Treat. Exam.*, **13**, 145–152 (1964)
62. Hoather, R. C., 'Chemical characteristics of river waters', *Proc. Soc. Water Treat. Exam.*, **15**, 34–42 (1966)
63. Edwards, A. M. C., 'The variation of dissolved constituents with discharge in some Norfolk rivers', *J. Hydrology*, **18**, 219–242 (1973)
64. Edwards, A. M. C., 'Dissolved load and tentative solute budgets of some Norfolk catchments', *J. Hydrology,* **18**, 201–217 (1973)
65. Clarke, F. W., 'The data of geochemistry', 5th edn, *Bull. 770,* US Geol. Surv. (1924)
66. Sawyer, C. N., 'Factors involved in disposal of sewage effluents to lakes', *Sewage and Industrial Wastes,* **26**, 317–328 (1954)
67. Lund, J. W. G., 'Chemical analysis in ecology illustrated from Lake District tarns and lakes: 2 Algal differences', *Proc. Linn. Soc. Lond.*, **167**, 165–171 (1957)
68. Owens, M., Garland, J. H. N., Hart, I. C. and Wood, G., 'Nutrient budgets in rivers', *Symp. Zool. Soc. Lond., No. 29,* 21–40 (1972)
69. Owens, M. and Wood, G., 'Some aspects of the eutrophication of water', *Water Research,* **2**, 151–159 (1968)
70. Owens, M., 'Nutrient balances in rivers', *Proc. Soc. Water Treat. Exam.*, **19**, 239–252 (1970)
71. Wood, R. B. and Gibson, C. E., 'Eutrophication and Lough Neagh', *J. Instn Water Engrs*, **27**, 409–421 (1973)
72. Gibson, C. E. and Stewart, D. A., 'Changes in the water chemistry of Lough Neagh over a ten-year period', *Limnol. Oceanogr.*, **17**, 633–635 (1972)
73. Owens, M., 'Resources under pressure—water', *CICRA North-western European Region, Symp. on intensive agriculture and the environment,* 33–39, Newcastle on Tyne (1973)
74. Royal Commission on Environmental Pollution, *First Report,* HMSO, 52 (1971)
75. Tomlinson, T. E., 'Trends in nitrate concentrations in English rivers in relation to fertiliser use', *Proc. Soc. Water Treat. Exam.*, **19**, 277–293 (1970)
76. Cooke, G. W. and Williams, R. J. B., 'Losses of nitrogen and phosphorus from agricultural land', *Proc. Soc. Water Treat. Exam,* **19**, 253–276 (1970)
77. Hill, M. J., Hawksworth, G. and Tattersall, G., 'Bacteria, nitrosamines and cancer of the stomach', *Br. J. Cancer,* **28**, 562–567 (1973)
78. Macan, T. T. and Worthington, E. B., *Life in lakes and rivers,* Collins, London, 320 (1972)
79. Hammerton, D., 'A biological and chemical study of Chew Valley Lake', *Proc. Soc. Water Treat. Exam.*, **8**, 87–125 (1959)
80. Lund, J. W. G., 'Primary production', *Proc. Soc. Water Treat. Exam.*, **19**, 332–349 (1970)
81. Mortimer, C. H., 'The exchange of dissolved substances between mud and water in lakes', I and II, *J. Ecology*, **29**, 280–329; III and IV, *J. Ecology*, **30**, 147–201 (1941, 1942)
82. Ridley, J. E., 'The biology and management of eutrophic reservoirs', *Proc. Soc. Water Treat. Exam.*, **19**, 374–393 (1970)
83. Ridley, J. E., 'Thermal stratification and thermocline control in storage reservoirs', *Proc. Soc. Water Treat. Exam.*, **13**, 275–296 (1964)

84. Eunpu, F. F., 'Control of reservoir eutrophication', *J. Am. Waterworks Assoc.,* **65,** 268–274 (1973)
85. Macan, T. T., *Biological studies of the English Lakes,* Longman, London, 259 (1970)
86. *A river pollution survey of England and Wales, Vol I,* HMSO, 39 (1971)
87. Scottish Development Department, *Towards cleaner water, Rivers Pollution Survey of Scotland,* HMSO, 37 (1972)
88. Klein, L., *River Pollution: 1 Chemical analysis,* 206 (1959); *2 Causes and effects,* 456 (1962); *3 Control,* 484 (1966), Butterworths, London
89. Drake, J. F., 'The present and future role of instrumentation in water pollution control', in *Water pollution manual 1972,* Thunderbird Enterprises, Harrow, 220 (1972)
90. Hynes, H. B. N., *The biology of polluted waters,* Liverpool University Press, 202 (1966)
91. *Lough Neagh Working Group Advisory Report,* Govt of Northern Ireland, 96 plus appendices (1971)
92. Langford, T. E. 'A comparative assessment of thermal effects in some British and North American rivers', in *River Ecology and Man,* ed. R. T. Oglesby, C. A. Carlson and J. A. McCann, Academic Press, London, 463 (1972)
93. Langford, T. E., 'The temperature of a British river upstream and downstream of a heated discharge from a power station', *Hydrobiology,* **35,** 353–375 (1970)
94. Alabaster, J. S., 'Effects of heated discharges on freshwater fish in Britain', in *Biological aspects of thermal pollution,* ed. P. A. Krenkel and F. L. Parker, Vanderbilt University Press, 407 (1969)
95. Alabaster, J. S., 'The effect of heated effluents on fish', *Int. Conf. on Water Pollution Research, London 1962,* Pergamon Press, 261–283 (1964)
96. Ross, F. F. 'Warm-water discharges in rivers and the sea', *Water Pollution Control,* **70,** 269–273 (1971)
97. Davies, I., 'Chemical changes in cooling water towers', *Int. J. Air Water Pollution,* **10,** 853–863 (1966)
98. Alabaster, J. S., Garland, J. H. N. and Hart, I. C., 'Fisheries, cooling-water discharges and sewage and industrial wastes', *Symp. on freshwater biology and electrical power generation,* Pt I, 3–7, Central Electricity Research Laboratories, Leatherhead (1971)
99. Hawkes, H. A., 'Ecological changes of applied significance induced by the discharge of heated waters', in *Engineering aspects of thermal pollution,* ed. F. L. Parker and P. A. Krenkel, Vanderbilt University Press, 351 (1969)
100. Shepherd, J. M., 'Pollution of groundwater supplies by sewage', *Proc. Soc. Water Treat. Exam.,* **11,** 12–14 (1962)
101. Maclean, R. D., 'The effect of tipped domestic refuse on groundwater quality: a survey in north Kent', *Proc. Soc. Water Treat. Exam.,* **18,** 18–34 (1969)
102. *Pollution of water by tipped refuse,* Ministry of Housing and Local Govt, HMSO, 141 (1961)
103. Romero, J. C., 'The movement of bacteria and viruses through porous media', *Ground Water,* **8,** 37–48 (1970)
104. Hutchinson, M., 'Microbiological aspects of groundwater pollution', *Groundwater Pollution Conference,* **1,** 69–104, Water Research Association, Medmenham (1972)
105. Nicholls, G. D., 'Pollution affecting wells in the Bunter Sandstone', *Groundwater Pollution Conference,* **1,** 187–195, Water Research Association, Medmenham (1972)
106. Wheatland, A. B. and Borne, B. J., 'Some changes in polluted water during percolation through soil', *Water Waste Treat. J.,* **8,** 330–336 (1961)
107. Swales, G. M. and Davison, A. S., in discussion on Furness, 1962 (Ref. 108)
108. Furness, J. F., 'The effect on surrounding underground water of tipping household refuse into wet gravel pits at Egham', *Proc. Soc. Water Treat. Exam.,* **11,** 23–29 (1962)
109. Davison, A. S., 'The effect of tipped domestic refuse on groundwater quality', *Proc. Soc. Water Treat. Exam.,* **18,** 35–41 (1969)
110. Hoather, R. C., in discussion during symposium on the effects of tipped domestic refuse on groundwater quality, *Proc. Soc. Water Treat. Exam.,* **18,** 56 (1969)
111. Hoather, R. C., 'Methaemoglobinaemia', *Lancet,* **I,** 1324–1325 (1951)
112. Wood, E. C., 'Some chemical and bacteriological aspects of East Anglian waters', *Proc. Soc. Water Treat. Exam.,* **10,** 76–86 (1961)
113. Greens, L. A. and Walker, P., 'Nitrate pollution of Chalk waters', *Proc. Soc. Water Treat. Exam.,* **19,** 169–178 (1970)
114. Foster, S. S. D. and Crease, R. I., 'Nitrate pollution of Chalk groundwater in east Yorkshire: a hydrogeological appraisal', *J. Instn Water Engrs,* **28,** 178–194 (1974)

115. Davey, K. W., *An investigation into the nitrate pollution of Chalk borehole water supplies,* North Lindsey Water Board, Scunthorpe, 167 (1970)
116. Ineson, J., 'Pollution of water and soil by miscellaneous petroleum products', *J. Brit. Waterworks Assoc.,* **46,** 307–326 (1964)
117. Ineson, J., 'The significance of oil pollution in the water resources field', *Seminar on water pollution by oil, Aviemore, Scotland,* Inst. Water Pollution Control, 143–151 (1970)
118. Little, A. H., 'Sampling and samplers', *Water Pollution Control,* **72,** 606–615 (1973)
119. Tate, T. K., 'Variations in the design of depth samplers for use in groundwater studies', *Water and Water Engng,* **77,** 223 (1973)
120. Wood, L. B. and Stanbridge, H. H., 'Automatic samplers', *Water Pollution Control,* **67,** 495–520 (1968)
121. Holden, W. S. (ed.), *Water treatment and examination,* J. & A. Churchill, London, 513 (1970)
122. Skeat, W. O. and Dangerfield, B. J. (editors), *Manual of British water engineering practice, Vol III Water quality and treatment,* Instn Water Engrs, London, 407 (1969)
123. Permanent water quality monitoring and recording stations, *Paper Auto 7,* Water Resources Board, Reading (1968)
124. *Conference on data retrieval and processing,* Water Resources Board, Reading (1969)
125. Briggs, R., 'Water quality monitoring networks—practice in Great Britain', *Casebook on hydrological network design practice,* WMO, Geneva (1972)
126. Herschy, R. W., 'The measurement of river water temperature', *Tech. Note 5* (revised), Water Resources Board, Reading, 18 (1971)

Chapter 8

The Role of Hydrology

Introduction

Almost every country in the world is concerned about its water resources and the UK is no exception[1]. How the demand for water relates to the natural supply is the chief concern, in terms of the current relationship between these factors and how it will develop in the future, both on the countrywide and regional scales. The supply is represented in broad terms by the difference between precipitation and actual evaporation—what is left to run off the land surface in streams and rivers. The distribution of this 'residual rainfall' for the average year is shown in *Figure 8.1*. Totals vary from less than 125 mm a year around the Thames Estuary and the Wash, to well over 2500 mm in North Wales, the Lake District and the Highlands of Scotland. Totals also vary from year to year, while the distribution within years is subject to radical change. Many water-resource problems hinge on these spatial and temporal variations; others relate to the fact that the pattern of population distribution is largely the reverse of that for residual rainfall. For the UK as a whole the supply (average annual residual rainfall) is about 780 mm, or approximately 515 million m³/d. These figures need to be set against the demand for water. The present UK demand for water is approximately 16 million m³/d and is expected to rise to about 33 million m³/d by the end of the century. The figures for England and Wales only are 14 million m³/d in 1972 and 28 million m³/d by about AD2000[2].

Recent forecasts, influenced by the decline in the rate of population growth, suggest these estimates of demand at the end of the century may be too high. But forecasts of future demand are notoriously difficult to make; they must take into account not only the growth of population and changes in its areal distribution, but also other factors including the increase in the demand per head of population (which in the long term is certain to rise as modern standards of water services and sanitation are extended) and the rising demands from agriculture and industry, including water used for cooling purposes in generating stations. There is also need to provide adequate flows in rivers to assure sufficient dilution of effluents discharged from sewage works and by industry. Amenity and recreational interests also need to be safeguarded by maintaining adequate river flows. An estimate of the pattern and amounts of the extra demands for water that are likely to arise by the start of the 21st century is shown in *Figure 8.2*. As might be

Figure 8.1 'Residual' rainfall (mm) in the UK (courtesy Institute of Hydrology)

anticipated, the main demand areas coincide with the major industrial regions, the wide belt extending north and west of London being in a slightly different category. The latter area is one where demand can be expected to rise rapidly, but it is also the area where the resources are already under the greatest stress.

Traditionally water demands in the UK have been satisfied by augmenting storage capacity. Increasingly this necessitates development at sites considerable distances from demand centres because locally available surface storage sites have been used. Generally the objective of these schemes is to meet peak requirements during a drought of a given severity, usually with a return period of 1 in 50 or 1 in 100 years. The introduction of a more realistic pricing policy for water together with consideration, where appropriate, of alternative methods of meeting demands could modify this attitude and thereby influence the rate of development of new sources.

Figure 8.2　Estimated pattern of the extra demand for water by AD 2001 (after Rydz)

Of course, water resources should not be thought of solely in terms of water supply, but in the rather wider context of more or less any large-scale activity or project involving water and its use in some way. These activities range from transport to flood control and from hydroelectric production to recreational uses of water such as boating and fishing *(Table 8.1)*. In terms of the capital invested in the various water resources projects in the UK, some of these items involve only small sums, but others represent a considerable outlay. For example, during the financial year 1971–72, capital expenditure on water supply, sewerage and sewage disposal in England and Wales was expected to be about £280 million[3]. Some 3–4 per cent of the annual expenditure on road construction and maintenance (which amounted to £630 million in 1969[4]) is usually allocated to hydraulic structures, such as bridges and culverts. When it comes to motorway construction, 7–10 per cent of the cost, about £1 million per mile, goes on structures of this type. On the other

Table 8.1 WATER RESOURCES PROJECTS (AFTER DIXON[119])

No.	Purpose	Works involved
1	Public water supply	Dams, reservoirs, wells, conduits, pumping stations, treatment works, desalting plants, distribution systems
2	Irrigation	Dams, reservoirs, wells, canals, pumping stations, desilting works, distribution systems
3	Flood control	Dams, regulating reservoirs, levees, floodwalls, channel improvements, by-pass channels, pumping stations, zoning, flood forecasting
4	Drainage	Ditches, tile drains, levees, pumping stations, soil treatment, culverts
5	Hydroelectricity	Dams, reservoirs, penstocks, power plants, tunnels and shafts
6	Navigation	Dams, reservoirs, canals, locks
7	Pollution and safety control	Regulating reservoirs, treatment facilities, barriers, groundwater recharge facilities
8	River and basin conservation	Soil-conservation practices, headwater control structures, fish ladders and hatcheries, land management practices
9	Recreation	Reservoirs and canals with access for fishing, boating and scenic areas

hand, until recent years few projects have been undertaken in this country solely for the recreational facilities they provide. For example, the expenditure on canal maintenance by British Waterways[5] amounts to over £3 million a year.

The hydrological network

Investment in water-resources projects demands hydrological data as a basis for the decisions and designs that have to be formulated. Hence the information collected through the countrywide network is of prime importance for new projects and also for the operation and management of existing ones. The network consists of not only the gauging stations on rivers and streams where discharge is measured continuously, but also the rainfall stations, the climatological stations and the wells and boreholes where groundwater levels are measured. In addition, the quality of water in rivers, lakes and reservoirs is monitored in terms of the principal dissolved constituents, and in some cases suspended sediment load, biological characteristics and radioactive constituents.

Of course it might be argued that, if a network is accepted as being any system for the acquisition of hydrological data, the term network should not be limited to these station-type time series observations. Surveys should be included, such as the River Pollution Survey[6], as well as questionnaires and even maps (for example, soil maps and geological maps showing features of hydrological importance). In this wider sense, the UK hydrological network is really much more comprehensive than might be thought at first; in fact, it involves nearly all the elements suggested for a classification of networks in *Figure 8.3*. Probably the purpose of the network is one of the more

Purpose of network	Process observed	Type and frequency of observation	Type of field record	Length of record
Resources survey	Precipitation	Continuous	Time Series — Autographic	Long-term primary
Project planning	Storage changes	Periodic	Telemetered	Short-term secondary
Project operation	Evaporation	Intermittent	Manual	Benchmark
Warning and monitoring	Runoff	Repeated field surveys	Samples (e.g. water or soil)	Survey
Statutory accounting	Erosion and sediment transport		Questionnaire and census	
Research	Water quality and pollution			Standards of precision
	Water use			First-order stations or investigations

Maintenance and quality control

None ←	Frequent and regular checks of instruments and observers, strict office codes for checks of data consistency and homogeneity, standard data storage and retrieval practices	Nature of spatial design	Second-order stations or investigations
		None ← Scientific plan	Third-order (reconnaissance)

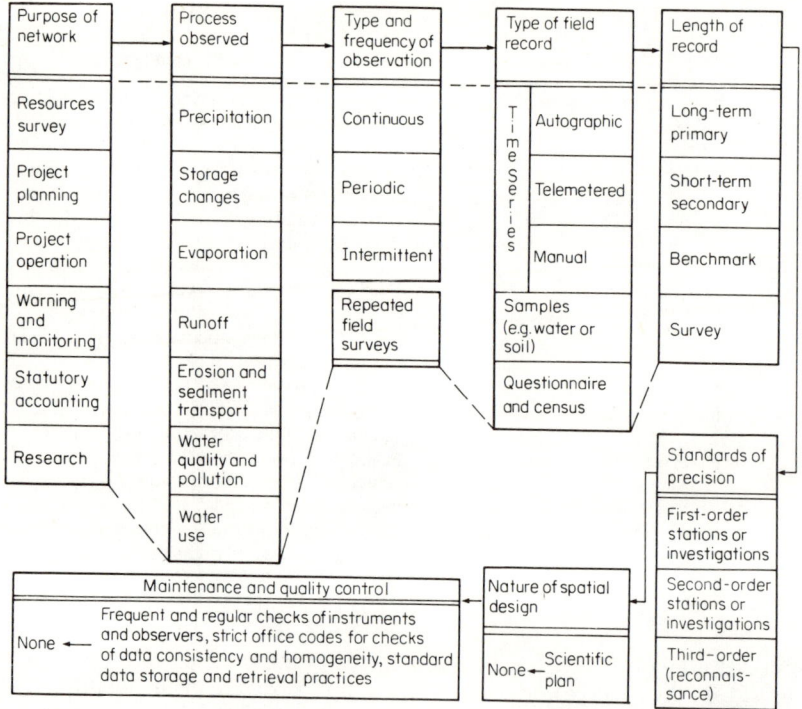

Figure 8.3 Suggested network classification

important of these elements, but it is unusual to find that even part of the network within a single river basin can be identified with just one purpose; on that scale, the network is dual or multipurpose and its growth over the last 100 years has been haphazard rather than to a scientific design.

In terms of number of instruments and length of records the UK raingauge network is one of the world's best. By comparison, it is considered that the network for measuring snow is extremely poor, but as so many more problems arise in assessing snow than rain, perhaps this is not surprising. In the past neither the number of river gauging stations nor the lengths of their records were adequate for a country as developed as the UK. Records from some 450 stations are contained in the *Surface Water Year Book* for 1965, but fewer than half a dozen of these have continuous records for fifty years or more *(Table 8.2)*. During recent years a considerable number of new gauging stations have been installed and the density of the network is improving. Between 1965 and 1971[7], under the aegis of the hydrometric schemes described in the next section, some 737 river gauging stations were approved for construction. The large number of sites where river water samples are collected, automatically in some cases, is another component of the network. These samples are analysed for a dozen or more water-quality characteristics. Then there are sites where continuous records of suspended sediment, electrical conductivity and several other variables are obtained from automatic water quality stations. It is difficult to assess the total

Table 8.2 INCREASE IN THE RIVER GAUGING NETWORK
(FROM DATA PUBLISHED IN THE SURFACE WATER YEAR BOOK)

Year	No. of river gauges	Year	No. of river gauges
1935	28	1958	188
1936–1945	52	1959	205
1946–1952	81	1960	238
1953	102	1961	271
1954	116	1962	311
1955	128	1963	364
1956	147	1964	434
1957	174	1965	450

number of sites where water quality is measured, but there must be between 1000 and 5000 places in the UK where samples are taken with frequencies ranging from once a day to once every six months. The other components of the network are the climatological, agrometeorological and synoptic stations that provide records for evaporation determination, the rain recorder stations and the 1200 or more observation wells.

A rough estimate has been made of the cost of providing and maintaining the UK hydrological network, excluding the costs of the water-quality network *(Table 8.3)*. Much of the spending is on the capital cost of river gauging stations, but even the total sum involved seems small by comparison with the annual investment in water resources projects.

Table 8.3 ESTIMATE OF COST OF THE HYDROLOGICAL NETWORK
(1971 VALUES)

	£	£
A. Capital *(assuming installation in 1971)*		
1. Rainfall: raingauges and records	500 000	
2. Evaporation: climatological stations	600 000	
	1 100 000	1 100 000
3. River flows: river gauging stations	10 000 000	10 000 000
4. Groundwater: wells and boreholes	800 000	800 000
		11 900 000
B. Recurrent	£	£
1. Rainfall: raingauge observations	200 000	
2. Evaporation: climate observations	200 000	
Rainfall and evaporation processing and publication	400 000	
	800 000	800 000
3. River flows: maintenance, collection, analysis and publication	500 000	500 000
4. Groundwater: maintenance, collection, analysis and publication	100 000	100 000
		1 400 000

Hydrometric schemes

In order to improve the hydrological network in England and Wales, the Water Resources Board was empowered under Section 89 of the Water Resources Act to make grants to river authorities for hydrometric schemes. In these schemes, river authorities were required to submit proposals for obtaining and recording measurements of rainfall, evaporation, flow and level of water, and other matters likely to affect water resources. The Board recommended that schemes should aim to gauge every principal river and every large tributary, to obtain water-quality and water-temperature measurements and to measure monthly rainfall, intense rainfall and evaporation. Preference was given to developing integrated networks and to establishing gauging stations on the lower reaches of main rivers, while stress was placed on the need for accuracy. It has already been mentioned that 737 new river gauging stations were completed or were approved for construction, while comparable figures for rainfall, evaporation and water quality were 1747, 189 and 208 stations respectively. The cost of these hydrometric schemes was £6.3 million, over £5 million being spent on gauging flow.

The river authorities, again supported by grants from the Water Resources Board, also established groundwater networks of observation boreholes to monitor changes in groundwater storage and quality (see also Chapter 5). The cost of schemes approved by the Board amounted to £620000.

In order to supplement the River Pollution Survey, a harmonised monitoring system has come into operation for sampling river water at nearly 300 sites in the UK. Samples are taken weekly or monthly and analysed regularly for some 12–15 constituents and at predetermined but less frequent intervals for about 25 others.

In the future the hydrological network should continue to be expanded. More telemetering instruments are likely to be installed for operational purposes, including instruments for quality as well as quantity measurements. Probably soil moisture measurements will be better organised, and radars will provide coverage of the whole country in order to detect intense storms and provide rainfall measurements integrated over complete basins.

Representative and experimental basins

In an attempt to improve knowledge basic to the science of hydrology and understanding of the processes involved, a number of studies of small river basiins were commenced during the 1960s. The usual approach was to make precise measurements of precipitation, evaporation, discharge, changes in soil moisture and, if a permeable catchment was being studied, changes in groundwater level. These measurements were used to establish a water balance for the basin on a monthly, weekly or sometimes daily basis. In some basins, the object was simply to identify the factors controlling the flow of the river and to use the measurements of these factors to predict its discharge. In others, the aim was to determine and quantify the hydrological consequences of land use change, usually by instrumenting two basins side by side, changing the land use on one and employing the other as a control. If the paired basin approach could not be adopted (because of the lack of two similar adjacent catchments, for instance) a single catchment was studied, before and after treatment. These studies attempted to answer such questions as: What are the hydrological effects of building a town or planting a forest? Is the incidence of floods increased, and their severity heightened? Is the sediment load of a river altered? Are evaporation rates changing?

These and other likely consequences of land use change have been widely discussed in the literature and there are several far-ranging reviews of the subject[8-10], but these deal very largely with work undertaken abroad, particularly in the United States, South Africa and East Africa.

The UK is primarily an urban country, so the hydrological effects of urbanisation probably need special attention. On the other hand, the nature of the hydrological role of forests has aroused considerable controversy. Of the several studies that have been conducted, and for which results have been published, the conclusions appear somewhat contradictory. One investigation showed that no changes in stream flow took place that could be ascribed to afforestation[11]. Then the work at Stocks Reservoir[12,13] demonstrated that an increased water use and a reduced volume of runoff were due to the presence of trees. By way of contrast, Green[14] showed that flows in the Brenig were increased by over 10 per cent after the forested area had been extended. This was, of course, the initial reaction to planting and to the ploughing that preceded it, a reaction that other workers[15] have found to include about a hundredfold increase in the suspended sediment load. A similar increase in sediment load and changes in other water-quality characteristics have been found to accompany the urbanisation of a catchment that was previously grass[16,17].

Because of the possibility that the losses at Stocks Reservoir were enhanced by the small size of the plantation where the study was conducted, Law examined the rainfall–runoff relationships for the whole Stocks Reservoir catchment and for some of the sub-catchments. Quoting data assembled by Walsh, Law[18] showed that the average annual loss for the period 1937–1954 was 405 mm compared with 440 mm for the period 1954–1969. A double mass curve technique had been used to determine when there was a discontinuity, and this proved to be around 1954. In fact, afforestation of the Stocks catchment commenced early in the 1950s. Some 25 per cent of the catchment area was planted, the increased loss being ascribed to the presence of these trees.

A comparison of the adjacent headwater catchments of the Wye and Severn was commenced in 1967. The catchments are similar in size and in every respect other than land use, the Wye being sheep pasture and the Severn largely forest. Structures were installed to gauge the flow of the two main streams and specially designed flumes were built in the six sub-catchments *(Figure 8.4)*. The remainder of the instrument network includes about fifty rain gauges, some at canopy level and the remainder at ground level, about the same number of access tubes for soil moisture measurement by neutron probe, automatic weather stations and other instruments. Among the initial results[19], it is interesting to note that from the thirty storms greater than 25 mm the Wye generated larger peak flows than the Severn, while in addition the time to peak was less for this grass catchment.

While these studies seem to indicate that the presence of a forest alters

Figure 8.4   Plynlimon catchments (courtesy Institute of Hydrology)

both the total amount of runoff and its magnitude at times of high flows, the results provide no explanation of why these differences should exist. In order to obtain such an explanation, two different approaches are currently being followed: the first relies on small-scale studies of processes using even more complex instrumentation than the catchment studies themselves; the second uses mathematical models to simulate a catchment's response to rainfall models that employ functions based upon assumptions about the processes involved, such as the relationship between runoff and soil moisture when soil moisture is limited.

While some attention has been given to the way in which rainfall is converted to flow through the various surface and sub-surface routes (for example, flow through natural pipe systems[20]), a greater effort has been devoted to studying the evaporative characteristics of vegetation. It was argued initially[21] that evaporation and transpiration were largely independent of the type of vegetation and that the meteorological conditions were the most important controls of these factors. This is, of course, the concept of potential transpiration (Penman) or potential evapotranspiration (Thornthwaite): the concept that the rate of transpiration from a complete crop cover with a plentiful water supply is governed only by the available energy. This led to the assumption that, because adjacent basins of forest and grass would experience the same weather and, apart from albedo differences, would have the same energy available to them, their water losses would be similar. A further assumption was that this similarity would not extend to periods of soil moisture deficit: at such times the trees, with a much greater root range and available water, would continue to transpire at a high rate long after transpiration from the grass had diminished. It was suggested that, in any case, water losses from vegetation could not exceed the open-water evaporation rate. One study[22], in agreement with this idea, showed that the evaporation from a pine forest was 0.7 that from open water, but there has been some disagreement with this view[23].

Interception studies

How much rainfall is intercepted by vegetation, especially forests, has been the subject of a number of investigations. Law's work in this area has already been referred to, but there are several other examples of studies of this type. The basic approach is to measure the rainfall incident to the top of the forest canopy, the water that falls off the foliage to reach the forest floor, and the water that drains down the trunks of the trees. The difference between the incident or gross rainfall and the sum of the throughfall and stem flow (sometimes known as the net rainfall) gives the amount of interception. Other studies have relied on large numbers of regular measurements of the moisture content of the forest soil together with measurements or estimates of the rainfall on the canopy to determine interception.

The rainfall on the canopy has been measured by aerodynamically shaped gauges mounted on poles level with the canopy. Alternatively, the incident rainfall has been estimated from measurements made in nearby clearings. Raingauges and troughs on the forest floor have been used for measurements of throughfall, but because of the variability in space and time special attention has had to be given to the distribution of these devices[24]. One procedure is to site troughs at random beneath the canopy. If the troughs are maintained in the same positions, a certain level of accuracy will be attained, a level that can be improved by moving the devices to newly selected random positions after a specified amount of throughfall has been caught. A combination of permanent and moving troughs is another possibility, but if it can be shown that the throughfall is systematically distributed (for example, that throughfall increases with distance from the trunk), a stratified sampling procedure can be adopted. Stem flow has been measured by a metal collar fixed around the tree near its base, but aluminium coach guttering wound spirally around the trunk[25] is a preferred method. Again there are problems of sampling because of variations in stem flow, but the same design procedures can be adopted as those employed for assessing throughfall.

The amount of water that can be retained on a canopy is known as the interception storage capacity. By plotting the incident rainfall against the net rainfall for a number of storms, the interception storage capacity can be determined as the intercept on the incident rainfall axis. Interception storage capacity varies with type of forest, age, stand density, season and other factors; typical values are 0.1–7.6 mm for conifers and 0.2–2.0 mm for deciduous trees. Leyton *et al.*[26] found that Norway Spruce, hornbeam, heather and bracken had a similar storage capacity during the summer months (1–2 mm) but that winter values for the hornbeam were much reduced. Transpiration does not occur while the foliage is wet and the air around it is saturated. A little of the intercepted water may be absorbed by the foliage, but most of it evaporates[27]. Several studies[28-30] have shown that for forest this evaporation can take place at a rate up to five times as great as the transpiration rate, a rate exceeding that from grass and open water by a considerable amount. Rutter demonstrated that this difference between evaporation and transpiration rates was due to the fact that the aerodynamic resistance to transport of water away from the surface of the canopy was very small by comparison with the internal resistance to vapour diffusion in the transpiration pathway, particularly the resistance due to the stomatal barrier. Thus for forests in areas of frequent rainfall, the amount of interception will be large and evaporation rates high[31]. In fact one study[32] suggests the existence of a linear relation between duration of rain and loss due to evaporation of intercepted water. It is likely that this linear relation extends to some of the sites with a longer duration of rain, but at the really wet sites where the evaporation opportunity is limited a curvilinear relation may be more appropriate. At dry sites, it is probable that transpiration can be reduced considerably, because of the greater degree of control that trees exercise over their stomatal resistance compared with short vegetation. Spraying trees with chemical transpiration inhibitors[33] that cause stomatal closure could reduce the loss of water even further, should this be considered necessary.

These results are in marked contrast to those obtained where evaporation and transpiration from short vegetation have been studied. For grass, when the soil moisture deficit is small, transpiration and evaporation of intercepted water take place at approximately the same rate, as opposed to a ratio of about 1:4 for pine forest. It is not so clear how the transpiration from trees and grass differs when soil moisture is limited, but probably amounts are similar.

The use of models

Hydrology is still some way from being considered an exact science. As experimental data accrue, so analytical techniques must develop and expand in ways that are helpful to the scientist, engineer and planner. A prime method involves the use of models, whether these be mathematical equations, scale replicas or physical analogues. As modelling techniques are developed to cope with more complex hydrological systems, so a more complete understanding of the science can be expected.

A model is a simplified representation of a system or process in which only the basic aspects are considered. It is used to identify relationships and interrelationships in terms of cause and effect. In essence a model[34] describes the conversion of one phenomenon into another, an 'input' into an 'output', in a situation where continuity in time and in quantity is important. A prime example is a rainfall–runoff model, in whatever form it is constructed. Existing models have already investigated problems on a greatly varying scale:

1. Global—a complete atmospheric model with a simplified hydrological representation of the earth's surface[35].
2. National—fortnightly updating of a map of soil moisture deficiencies in the UK, for subscribers[36].
3. Regional—conjunctive use of water resource systems[37].
4. Basin—an economic model of the Trent Basin with particular reference to water quality[38].
5. Experimental catchment—a detailed 3-hourly water balance for a clay catchment on a tributary of the River Thames[39].
6. Representative plot—water used by a coniferous forest at Stocks Reservoir, Yorkshire[13].
7. Groundwater systems—digital[40] and analogue[41] solutions of groundwater flow.
8. Laboratory—physical simulation of rainfall conversion to runoff[42].
9. Micro-scale—transpiration by a single tree[43].

Types of model

Collinge[44] has identified two types of mathematical model, one 'descriptive' and the other 'optimising'. With the former a given set of system rules is used in a sequence that describes the consequential process of input becoming output. With the latter, it is the rules themselves that are produced in optimum form to meet a stated objective, normally that of least cost.

Clarke[45] is more concerned to group model types by their technique rather than by their aim. He lists all the combinations that result from the following alternatives:

Stochastic—the model variables have time-dependent probability distributions, or

deterministic—the process follows a law of certainty, i.e. the critical factors are determinate or fixed quantities.

Conceptual—following a known physical process, or

empirical—following observed relationship rather than postulated theory, knowledge of the processes causing the relationship may be unknown.

Linear— the output following from input $(x_1 + x_2)$ is the sum of the outputs from x_1 and x_2 separately, or

non-linear—the principle of superposition (as above) does not hold.

Both Collinge and Clarke give examples demonstrating that modelling is a powerful analytical technique. To indicate the scope several examples are briefly referred to in the following paragraphs.

In many situations rainfall records are longer than river-flow records. The latter can be extended if the rainfall can be related to the flow over the period for which both are available. A simple model, developed by Dawdy and O'Donnell[46], may be used to illustrate the mechanism. A river basin was simplified to four interconnected 'reservoirs' representing surface storage, channel storage, soil moisture storage and groundwater storage. Movement between the various storages is controlled by a set of concepts that act as operating rules. For example, surface storage is augmented by rainfall and depleted by evaporation (which has first call on the available water), infiltration and also discharge to channel storage (if rainfall exceeds a threshold value). Similar balances and budgets are calculated for each 'reservoir' and the ultimate output is total discharge, representing channel storage and baseflow.

The use of models to solve the groundwater flow equation generally requires that the groundwater potential is determined throughout an aquifer at a particular time, given the initial water potential, the areal variation of the aquifer properties, the boundaries of the aquifer, and the inflow and outflow rates.

The groundwater flow equation is usually expressed in the finite difference form, and solutions are obtained using analogue and digital techniques. Of the two techniques, the analogue has the advantage that time remains a continuous function and only the space dimensions are divided into discrete intervals. The electrical analogue resistance-capacitance network models rely on the fact that the flow of water through a porous medium is analogous to the flow of electricity through a circuit[47], the various related variables and parameters being:

water potential	—	voltage potential
flow	—	current
permeability	—	resistance
storage coefficient	—	capacitance

Values of resistors and capacitors (representing the aquifer properties) are repeatedly adjusted until the computed water levels satisfactorily match the historical water-level record. The model can then be used to design and manage groundwater development.

A range of numerical methods is available to solve the finite-difference equations using digital computers[48–50]. A map of the aquifer is divided by a grid into convenient areas or nodes, and the equations are solved for groundwater level for all the nodes successively throughout the grid. The estimates of water level at individual nodes are steadily improved and the process is repeated until changes in water level are acceptably small. Although the finite-difference method is usually applied, finite-element methods have found wide application in recent years[51].

Nowadays digital models are often preferred as a means of analysing groundwater flow problems, but if an extensive area is being covered in some detail they are expensive to run, even with modern computers. An electrical analogue on the other hand may be time consuming to build and prove (although this is becoming less so by, using a digital model to derive the aquifer parameters) but in subsequent use it is relatively cheap. The advantages of both digital and analogue methods may be successfully combined by using hybrid computers.

A model has been made of the River Trent[52] that provides forecasts of water quality. The river was divided into sections of approximately similar quality. Annual mean chemical properties were used in the model and sixteen water quality parameters were taken into account. The calculation of the values of quality parameters was based on a mass balance of all the estimated inputs. Mathematical functions were incorporated to represent the rate of chemical reactions, such as changes of biological oxygen demand, ammoniacal nitrogen and variations in oxygen content.

The model accounts for the changes in chemical quality due to the inflow of tributaries, discharges of effluents, abstractions of water, and natural purification reactions as the water moves down the river from one section to another. Relationships were established between the quality of the river and the nature of the fisheries, and between the quality and the type of water treatment required. In this way, the model could be used to forecast annual mean quality of the river resulting from a predicted change in use, and to estimate the effect on fisheries and the type of treatment plant likely to be required. This model of the River Trent was part of a more comprehensive economic model, which was required to evaluate alternative methods of using water resources in the Trent Basin and to assist in identifying the preferred alternative.

The quality model of the River Trent is a steady-state model; that is, it assumes equilibrium conditions exist. In the future river quality is likely to be studied using dynamic models, which will reproduce transient conditions. This is a more realistic approach as it takes account of the fact that river quality is changing continuously.

An appreciation of the factors controlling the chemical composition of groundwater can be obtained by simulating the interrelationships between the physical and chemical nature of the aquifer, the velocity of groundwater flow, and the rate of chemical reactions between the water and the aquifer.

These examples illustrate the use of models to interpret the component processes of the hydrological cycle. Increasingly complex modifications are being made to the hydrological cycle for a variety of purposes. A wide range of questions has to be answered about rivers as potential water resource systems. How much storage is required? How should it be controlled? How should it be allocated? Such problems are investigated by systems engineering[53,54]. It allows a water resource system to be examined and the most appropriate solution to be selected from the alternatives.

The type of problem to be solved may concern the sequence of decisions necessary for the optimum operation of single or multiple reservoirs in a system, the design of water resources systems incorporating reservoirs and aquifers, and such requirements as the need to regulate rivers to maintain adequate water quality and the examination of alternative programmes for developing water resources and estimating their costs.

Many of the problems encountered are at present too complex to be solved analytically; therefore simulation methods must be applied. This technique involves formulating the model by arithmetic and algebraic relationships together with non-mathematical logical processes[54]. It is likely to incorporate a large number of interacting components (e.g. reservoirs and aquifers) to which operating rules are assigned. All the parameters are fixed before the simulation is made at a chosen level of water demand; to obtain alternative results the parameters are changed and the programme recalculated.

Modelling processes move through a number of well-defined stages:

1. definition of the problem and the criteria that the solution must satisfy;
2. construction of the model following an investigation of the real system;
3. testing the validity of the model;
4. using the model to obtain a solution or, more commonly, alternative solutions to the problem;
5. implementation of the preferred solution;
6. updating the model and using it to control or manage the future behaviour of the system.

As experience grows one feature stands out: with an appropriate but simple model, a large measure of understanding of a problem can be gained. However, as the model escalates in complexity it becomes increasingly difficult to explain the residual part of the problem. This is particularly so on a catchment scale where the permutations of ground conditions, land use and rainstorm variation are enormous. Inevitably, many models will be 'lumped' ones for the foreseeable future; that is, no account will be taken of the spatial distribution of the input variable or of the spatial variation of factors influencing the physical processes[45]. However, attempts will continue to be made to overcome the inherently unsatisfactory nature of the lumped system for many problems, by identifying sub-systems of the whole. For example, if the system under consideration is a river basin, geographical sub-areas of similar characteristics can be recognised by taking into account types of vegetation, land use, geology, elevation of the ground etc. This is referred to as a 'distributed system' approach. It is extensively used to solve problems in the groundwater field.

Data extension

Two different approaches are proposed. The first uses an understanding of the statistical structure of a series of figures, say of runoff, as a basis for extending the record into the future. The second is based on the assumption that the past is a sound tutor for what is to come and seeks to collate a knowledge of the complete hydrological regime previously experienced at a site.

The former is commonly termed Time Series analysis[55] and lies at the heart of stochastic hydrology, i.e. the form of hydrology that views events as being essentially probabilistic in nature; for example, today's runoff is derived from yesterday's runoff, today's expected mean, and a random number. (Its apparent converse is deterministic hydrology, in which each event is determined by the preceding one, together with a newly added piece of data; for example, runoff today is derived from yesterday's runoff and today's rainfall.)

The second approach can involve the evidence of archaeology, newspaper records, flood-level stones, the memory of the 'oldest living inhabitant', fishing-level staff gauges, or reservoir record books. Often events have been described but not measured. The *British Rainfall* volumes include many such examples. The UK *Flood Studies Report*[56] contains examples of how, by Bayesian theory, qualitative information may be incorporated into an otherwise quantitative exercise. Naturally extension back by correlation with a nearby longer record falls under this heading too. Alternatively, modelling runoff by a catchment model based on rainfall, potential transpiration, soil moisture and aquifer storage may be successful enough over a known period to warrant its extension as far back in time as rainfall data permit[57].

Problems exist with both approaches. Extension by Time Series can only preserve those statistics of the original record that are envisaged at the outset of the study and that its mathematical equations permit. Thus preserving the mean and standard deviation of monthly flows, together with the daily flow duration curve, is now readily accomplished[58]. However, it requires a special model to represent the perfect drought recession interrupted but rarely by a modest spate. In fact preserving the characteristic sequential flow deficiency that produces major reservoir drawdowns is noticeably difficult[59]. Consequently, such hydrograph models may be more useful in water-quality models than in obtaining yield–storage relationships or droughts rarer than those in the original record.

Water resources development

The development of water resources can be based on one or more of a number of different approaches:

1. river intakes
2. storage of runoff in upland areas in reservoirs
3. lowland pumped storage reservoirs
4. development of storage in aquifers (supplemented where possible by artificial recharge)
5. storage behind estuarial barrages
6. reuse of water, including the direct use of effluent
7. desalination of sea water
8. treatment of brackish water

In England and Wales, the integrated development of the various water resource options is concerned particularly with the problems of regulating river flows by means of reservoirs and by the use of groundwater storage, and also with the difficulties of using off-channel storage fed from the lower reaches of rivers that may contain a significant proportion of effluent. The concentration of population in urban and industrial centres creates water quality problems, particularly in view of the need to reuse water and the intention to use more rivers as natural links in regional distribution systems. Aspects commonly associated with the integrated use of water resources of a river basin (for example flood control, irrigation and the development of hydroelectric power) are of less importance. In Scotland and Northern Ireland public water supplies rely to a greater extent on direct supply from upland storage. Water quality is only a problem in industrial areas such as the central valley of Scotland. Generation of electricity by water power is important in the mountainous areas of both northern and southern Scotland.

Types of resource

Reservoirs

These are of two main types: 'line of the valley' impounding reservoirs, usually at upland sites, and pumped storage reservoirs, which are filled by pumping from one or more rivers and are often constructed on lowland flood plains. The UK is well endowed with potential reservoir sites in south-west England, Wales, north-west England, the Pennines, northern and southern Scotland and the upland areas of Northern Ireland. Natural storage exists in the Scottish lochs and in lakes of the Lake District, north Wales and Northern Ireland. Many of the most favourable sites near centres of demand have already been developed. Manchester uses natural storage in the Lake District, Liverpool and Birmingham turned to north and central Wales, Glasgow to Loch Katrine, Belfast to the Mourne Mountains, and many centres in Yorkshire and the Midlands of England depend on storage in the Pennines. London relies to a large extent on pumped storage reservoirs in the Thames and Lee valleys.

Before 1960 most reservoirs were of the direct-supply type, being connected by pipelines to centres of demand. More recently reservoirs for river regulation have been constructed, such as Llyn Celyn and Clywedog. Regulating reservoirs have several advantages over those of the direct-supply type. Their yield is greater, for the runoff from the entire catchment above the downstream abstraction point is available; the water may be used several times as it flows down the river; and a much shorter length of aqueduct needs to be constructed from the abstraction point to the city where the water is required. However, this water usually needs more

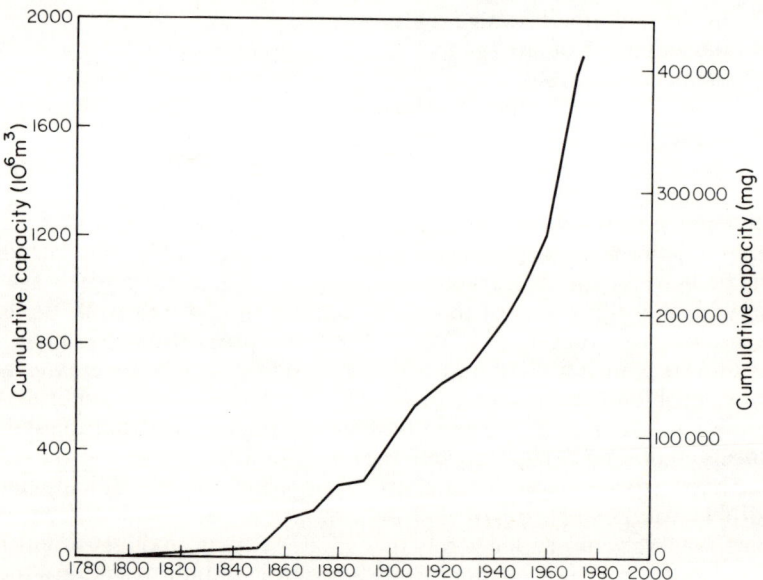

Figure 8.5   Cumulative capacity of water supply reservoirs in the UK (from Hetherington[60], courtesy Surveyor)

sophisticated treatment, while to operate a regulating reservoir to the maximum benefit, i.e. for both water supply and flood alleviation, requires a complicated management strategy. In fact the reservoir may be needed to regulate the flow of the river for only a small part of the year but this could be a vital period of intense drought or high flood.

The total reservoir capacity used for water supply in the UK in 1970 was just over 1800 million m³, having doubled since the mid-1940s. The average size of reservoirs constructed increased from 4.8 million m³ in the period 1901–1950 to 11.5 million m³ between 1951 and 1972 and 38 million m³ under construction in 1972 *(Figure 8.5)*. Two large reservoirs being built at present are Empingham in Rutland, which will have a surface area of 12.55 km², not far short of the area of Windermere (14.5 km², the largest expanse of water in England), and Kielder in Northumberland, which is designed to yield 955 Ml/d for the north-east of England. It should be noted that the reservoir capacity used for hydroelectric development far exceeds that used for water supply (Appendix 6).

The problem is often selecting the best site from the potential sites that are available. Examples are given later in the chapter of reservoirs constructed on river systems in series and in parallel. In these circumstances, the design problem is selecting the sites to obtain optimum storage capacity from the river system. The economic objective is to minimise total costs after taking into account the benefits that may be derived from the scheme. The final decision is unlikely to be on the basis of straightforward economic grounds, as social costs and political realities also have to be considered.

Estuarial storage

The possibility of creating freshwater storage in the intertidal areas of estuaries has some attraction in England and Wales in view of the strong objections often made to inland storage proposals. In recent years five estuaries have been examined in detail with this object in mind, namely Morecambe Bay[61], Solway Firth[61], the Dee[62], the Wash[63] and the Severn[90]. Solway Firth is remote from centres of water demand and is not favoured on that account, while the Severn scheme would be very expensive. The Dee proposal has the added attraction that it includes a road link, which would relieve traffic congestion and bring opportunities for industrial expansion in the area.

The original proposals in each case were for the construction of a barrage across the estuary, but the closure of such constructions under the tidal conditions encountered around the coasts of Britain would be costly and difficult. Closures on the scale that would be necessary and under these tidal conditions have not yet been attempted anywhere[64]. An additional problem is the limited depth of storage that can be created between an upper level low enough to permit drainage of adjacent land and a lower level sufficiently high to restrict seepage of saline water into the reservoir.

As a consequence, the construction of pumped storage reservoirs on intertidal areas is now favoured, particularly in the Dee and the Wash *(Figure 8.6)*. These would be filled by pipeline or tunnel from water held behind a short barrier across the upper part of the estuary or the tidal reach

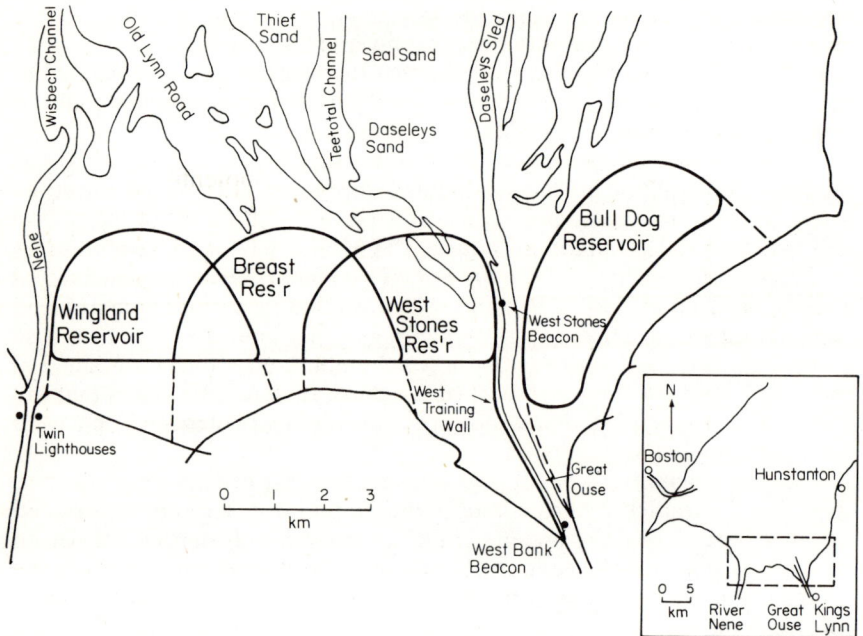

Figure 8.6   Possible form of bunded reservoirs in the Wash (courtesy Binnie and Partners)

of the river. Such pumped storage reservoirs have the following advantages[64]:

1. Quality control is easier, as water can be selected for pumping into storage.
2. Depth of water can be increased thereby reducing the greater tendency for algal blooms in shallow water.
3. Engineering problems are considerably reduced.
4. Siltation problems in the remaining part of the estuary are likely to be less severe.
5. Interference with fish and wild-life habitats and navigation is reduced.
6. Gradual construction of storage units to meet demand is possible, and therefore capital is invested gradually.

In the case of the Wash[63], a preliminary examination of the feasibility of a barrage led to an estimated yield of over 2700 Ml/d. The present view is that three or four pumped storage reservoirs would be preferable and these could yield ultimately 2400 Ml/d[63]. They would cover some 100 km² rising to a height of about 9 m. The first stage could cover an area of 7 km² with a storage of 65000 Ml supplied from the Great Ouse. The yield would be 225–450 Ml/d depending upon the need for a residual flow of fresh water downstream of the river abstraction point. Water storage in the Dee Estuary would probably take the form of three or four pumped storage reservoirs constructed as the need arose. The capacity would be up to 210 million m³. Storage in Morecambe Bay could combine reservoirs behind barrages across the Kent and Leven estuaries with bunded reservoirs on the shore of the bay. The schemes could yield up to about 2000 Ml/d from a storage of between

200 000 and 323 000 Ml, but they would be developed in stages yielding 300–1000 Ml/d for each stage.

Groundwater

The development of a major aquifer can be considered to pass through three stages:

1. Initially abstraction represents a small proportion of the average infiltration to the aquifer and wells can be sited where the water is required.

2. As the demand upon the aquifer increases and more wells are drilled, the abstraction–infiltration ratio increases and the reduction of groundwater discharge from the aquifer becomes apparent because of lower river flows during dry periods. During this stage steps have to be taken to maintain river flows to meet needs such as irrigation, navigation, fisheries and amenity.

3. When average abstraction exceeds average infiltration, 'mining' of water storage is occurring. If this is unacceptable, artificial recharge must be introduced to restore the balance between input and output.

In the UK it is now accepted that the use of aquifers to maintain acceptable river flows is important and that, where necessary, groundwater and surface water development should be integrated[65]. In such circumstances the development of groundwater resources depends upon using groundwater storage to control artificially the discharge from aquifers at a more even rate. In some situations the ultimate objective may be to provide a relatively constant flow approaching the mean discharge from the river basin, after taking into account variations in direct runoff. It may be necessary, however, to consider the effect of flow variations on the ecology of a river and on the sedimentological nature of the river bed.

There are several methods[66] of developing surface water resources in conjunction with groundwater:

1. A water supply can be obtained from a river intake, the flow of the river being supplemented by pumping from wells into the river at times of low flow. The river intake can be replaced by wells if the river is in hydraulic continuity with the aquifer, the wells deriving their yield from the river by induced recharge. This has the advantage that the water receives some degree of purification as it flows through the ground to the wells.

2. Water can be pumped directly to supply continuously, but additional wells are required to pump compensation water into the river system to maintain acceptable flows.

3. Seasonal abstraction can take place from a river intake (or riverside wells if conditions are favourable) when the flow exceeds that required to meet non-water–supply demands on the river. At other times pumping is from wells at sites remote from the river. Some groundwater abstraction may be required to maintain acceptable river flows.

Successful operation of these methods depends upon taking advantage of the large storage capacity of aquifers and the low rate of groundwater flow.

The effects of groundwater abstraction are time dependent, and wells have to be sited to take advantage of the delay between the abstraction of groundwater and the reduction in discharge at natural outlets. As already mentioned in Chapter 5, the delay between cause and effect, referred to as the aquifer response time, depends upon the aquifer's permeability, storage coefficient and dimensions. Siting of wells requires careful consideration of the aquifer properties and the nature of the river bed. The overall purpose is to develop storage in the aquifer, and to do so with the smallest number of wells necessary, by taking advantage of the aquifer as a transmission system. In catchments where river beds are permeable wells are usually sited away from the river, but if the river bed is impermeable there are economic advantages in siting near the river. In all cases environmental constraints may be an over-riding factor[66].

The success of a scheme for regulating a river with groundwater may be judged by the net gain to river flow during the abstraction period and its change with time. Net gain is defined as:

$$\frac{G - R}{G}$$

where G is rate of groundwater abstraction and R is reduction of river flow due to groundwater being intercepted by wells and any loss through the river bed due to recirculation. During the period when groundwater is being abstracted (i.e. generally during the summer) there is a positive net gain, but during the following winter, when pumping has stopped, infiltration preferentially replenishes the depleted groundwater storage and there is a net decline in river flow *(Figure 8.7)*. Thus in the long term 'gains' must balance 'declines' unless additional aquifer flow has been induced into the catchment as a consequence of a lowered water table.

The ideal aquifer for this type of development has a relatively low permeability and a high storage coefficient, which give a slow response time. The Triassic sandstones are generally more suitable than the Chalk and confined aquifers are usually unsatisfactory because of their fast response time. The change in net gain with time due to a simple intermittent abstraction regime is illustrated for fast and slow response situations in *Figure 8.8*.

The feasibility of regulating rivers with groundwater is being investigated in a number of river basins. The Thames Water Authority is promoting a scheme to provide 450 Ml/d by regulating the Thames with groundwater pumped from the Chalk and Jurassic limestones[67]. It is estimated that up to 440 Ml/d may be available for export to south Essex by regulating the Ely Ouse, again using storage in the Chalk. In the case of the Ely Ouse proposal, some 345 wells will be required in a 2462 km^2 development area[68]. The possibility of regulating the Severn and the Yorkshire Ouse with groundwater from the Triassic sandstones is being examined.

These examples involve the use of surface water in conjunction with groundwater storage in the same river catchment, but the aquifer could be in a different basin. For example the resources in the Triassic sandstones in Nottinghamshire and south-east Staffordshire could be developed in conjunction with rivers from the Peak District, and the Lincolnshire Limestone could be developed with the rivers Welland and Nene regulated

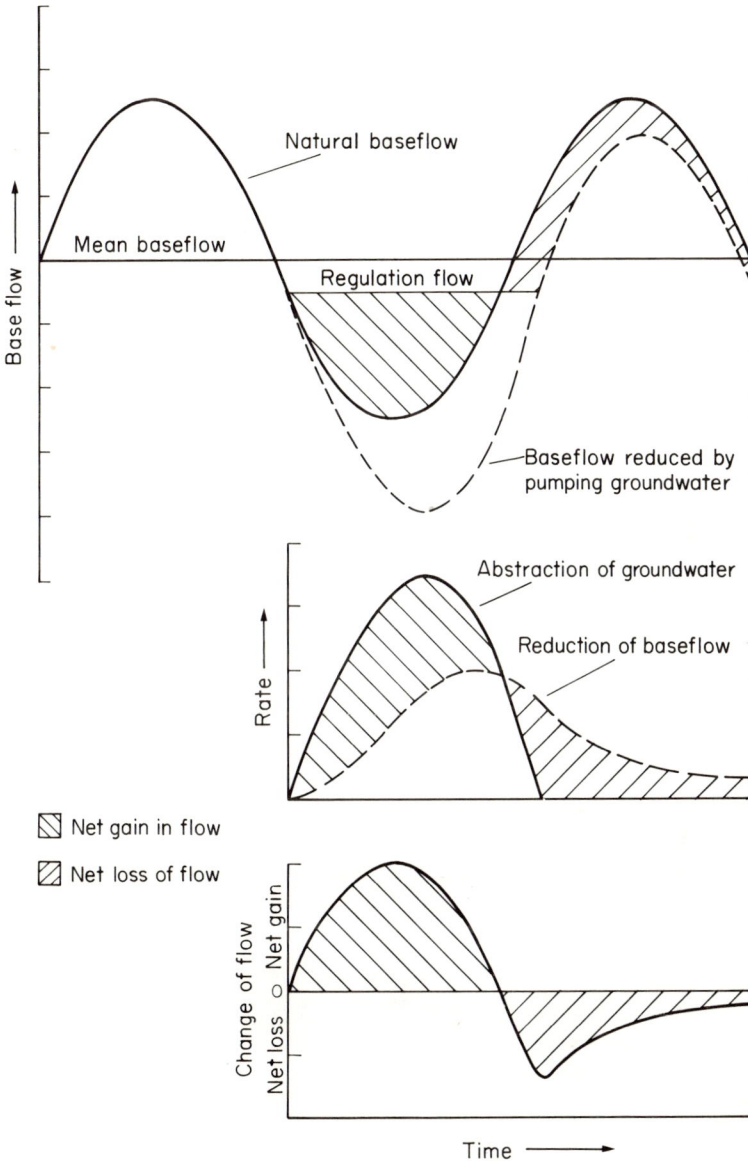

Figure 8.7 Schematic diagram illustrating the regulation of river flow by pumping groundwater into a river (with continuous seasonal regulation a new equilibrium would ultimately be attained and the winter baseflow would then be reduced to a greater extent than shown)

by the Empingham Reservoir. Further examples could be given, for there is still considerable potential in England for the development of groundwater storage in combination with surface resources. The geographical distribution of the important aquifers and of the areas where impermeable rocks crop out facilitates this type of development. In situations where the infiltration to an aquifer is already used for continuous direct supply from wells to distributions systems, increased yields could be obtained by switching to a seasonal use of the aquifer in combination with surface runoff.

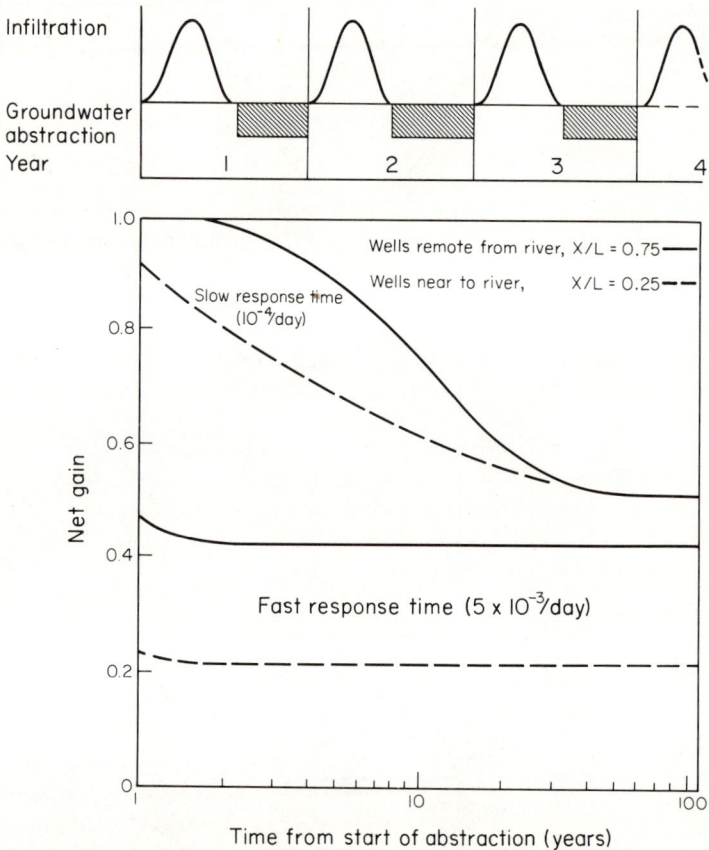

Figure 8.8   The change in net gain with time due to a simple intermittent groundwater abstraction regime; X is the distance of the well from the river and L the distance between the catchment boundary and the river, both measured at right angles to the river; the hydraulic gradient is maintained towards the river (from Downing et al.[66], courtesy Elsevier Scientific Publishing Co.)

It should be realised that the optimum development of groundwater resources is feasible without the need for artificial recharge, providing surface water and groundwater sources are interconnected to a regional distribution system allowing flexibility in their use according to availability.

Artificial recharge

The natural infiltration to an aquifer may be supplemented artificially; this process is referred to as *artificial recharge*[69]. The purpose is usually the storage in aquifers of surplus water for use at times when supplies are less plentiful; the principle is similar to that of pumped storage reservoirs. A secondary benefit is that the quality of a surface water is improved as it infiltrates through unfissured rocks; in some situations this improvement is the primary purpose. The source of water for recharge is usually surface runoff, treated effluents and cooling water.

Water may be introduced into an aquifer through basins or wells, or simply by surface spreading through ditches or controlled flooding. Wells are necessary if the aquifer is overlain by impermeable deposits. Recharge basins are typically about 3 m deep and contain perhaps 1 m of water, although this may be very much reduced during periods when algal blooms are likely. A filter is commonly, but not always, placed on the floor of the basin to avoid deterioration of the aquifer. A recharge basin acts as a slow sand filter and the chemical reactions that take place in the two processes are similar. The filter removes the suspended solids and part of the organic matter, but in doing so it ultimately becomes clogged and the surface layer has to be removed, a process that eventually leads to replacement of the filter. As water infiltrates the filter and the aquifer, organic matter and ammonia are oxidised. If there is insufficient oxygen available in the water and the aquifer to oxidise all the organic matter, any nitrates present will be used as a source of oxygen; once this has been used, however, anaerobic conditions will develop with undesirable consequences such as the reduction of sulphates and the solution of iron and manganese compounds.

An operational recharge works often incorporates rapid and slow sand filtration, aeration and chlorination treatment both before and after recharge. Some form of settlement is generally included as an initial pre-treatment phase, while activated carbon adsorption may be advisable during post-treatment. The extent to which treatment is necessary obviously depends upon the nature of the recharge source and the type of chemical reactions that occur in the filter and the aquifer.

If recharge is by wells they should ideally fully penetrate the aquifer and their diameter should be as large as possible. A well screen and gravel pack are usually essential and considerable pre-treatment is necessary to remove suspended solids and constituents that would promote bacterial growths on the well screen[70] or clog the pack. These precautions may be relaxed if the aquifer concerned is extensively fissured. Recharge water should not contain dissolved oxygen if it is to enter an oxygen-free environment containing iron and manganese salts in solution, and ideally it should be free of air as air bubbles reduce the permeability.

Artificial recharge has only been used to a limited extent in the UK. The Metropolitan Water Board[71] carried out experiments in the 1950s and 1960s, recharging the Chalk in the Lee valley by wells. At a number of localities cooling water is returned to aquifers after use and sewage effluents are recharged by surface spreading. The feasibility of recharge is being, or has recently been, examined in the London Basin, Nottinghamshire and Sussex. A large volume of Chalk and Tertiary sands has been dewatered in

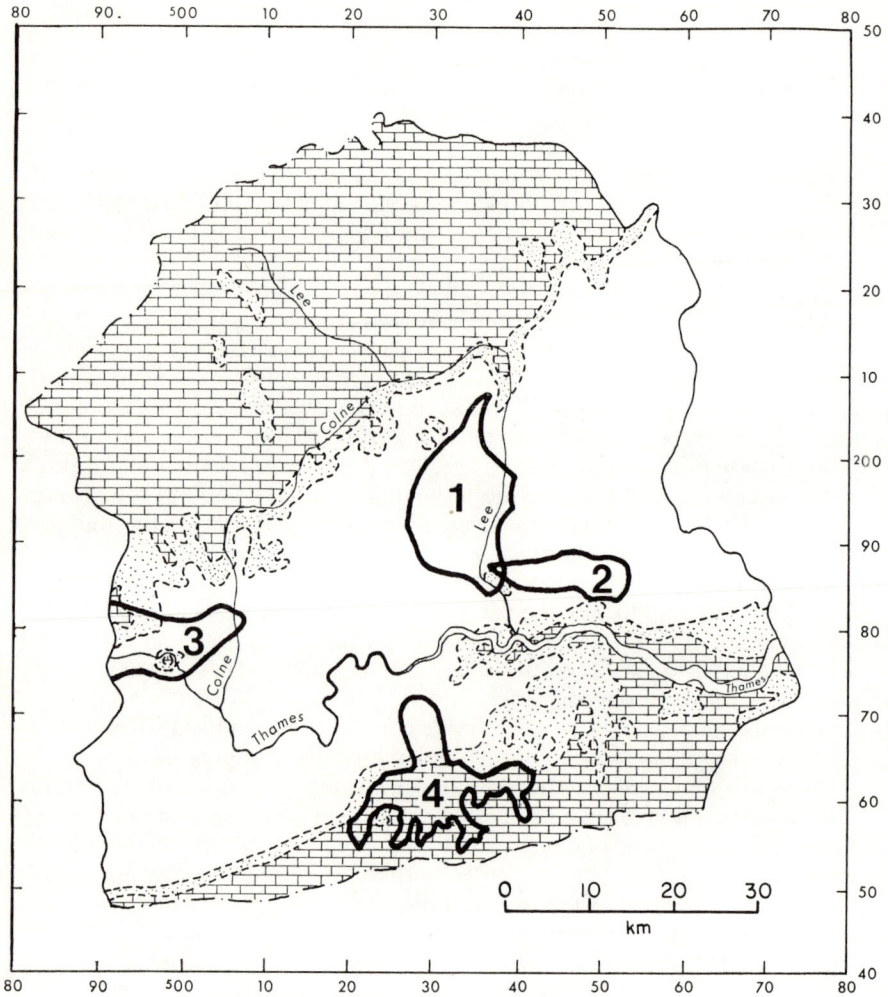

Figure 8.9   Central part of the London Basin showing principal potential recharge areas (after Water Resources Board[72])

the London Basin as water levels have fallen. In the Lee valley the storage is at least 90 million m³ and possibly 200 million m³. There is a further potential area in south-east London *(Figure 8.9)* where the storage is estimated to be over 110 million m³, and artificial recharge in this area would have the additional benefit of preventing further intrusion of saline water from the Thames[72,73].

At Hardham in Sussex, the possibility of recharging the Folkestone Beds by means of basins is being examined. During pilot studies an average recharge rate of 0.3–0.5 m/day per metre head of water in the basin has been achieved[74].

Pilot studies in Nottinghamshire have examined the feasibility of recharging the Triassic sandstones by wells and basins. The purpose of the basin study was to determine whether the aquifer could improve the quality of a polluted river such as the Trent. The only treatment prior to recharge was settlement and aeration, but nevertheless the entire recharge process substantially improved the water quality. Suspended matter, including bacteria and viruses, was removed, most of the organic matter was removed by adsorption and biodegradation, ammonia was completely oxidised and phosphate was largely eliminated; hardness and alkalinity increased and the small amounts of heavy metals were not affected, however. With the average recharge rate achieved of 0.35 m/d, land requirement for a prototype development was estimated to be 1 km² per 100 Ml/d[75].

Desalination

The methods available for desalting water are distillation, freezing, electrodialysis, reverse osmosis and ion exchange[76,77]. The application of the various processes to waters of different salinity is illustrated in *Figure 8.10.*

Distillation processes provide some 90 per cent of the world's supply of desalted water, but of these processes multi-stage flash distillation is the only one in common use for converting sea water to fresh water. Small plants exist in Jersey and Guernsey capable of producing 6.8 and 2.3 Ml/d respectively,

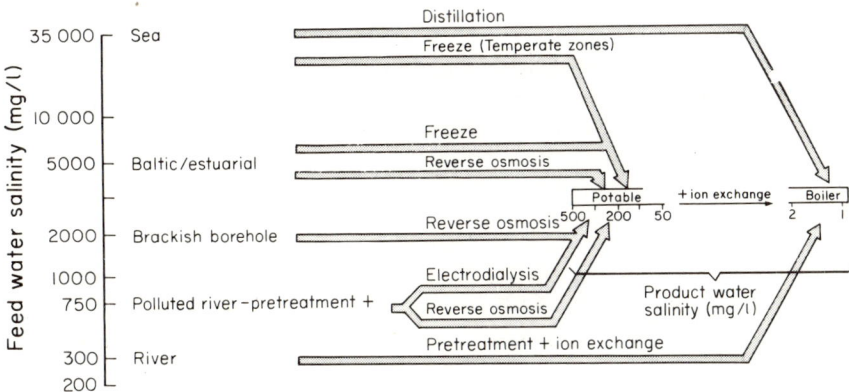

Figure 8.10   Schematic diagram of desalination processes and their applications (from Pugh[78], courtesy Surveyor)

but larger plants are in existence outside the UK; some 180 Ml/d will be provided by this British-engineered process in Hong Kong.

Most of the desalination plants that have been built are dual purpose, i.e. they produce both water and power, but, with the decline in oil reserves and hence the rise in the price of fossil fuels generally, the future for dual-purpose plants depends upon the stabilisation of costs following the introduction of nuclear energy generation[78].

Electrodialysis and reverse osmosis are both suitable for treating brackish water. Electrodialysis is an electrochemical process that separates salts from water using selective ion-exchange membranes. Selective cation and anion membranes arranged alternately divide the container where the process takes place into cells; eventually cells of increasing salt concentration alternate with those of decreasing concentration. The reverse osmosis process involves subjecting a brackish solution to a pressure in excess of the osmotic pressure across a semi-permeable membrane. Water is forced through the membrane, which retains the salts.

Electrodialysis is the only desalting process other than distillation in commercial worldwide use, commonly to reduce the salt concentration of brackish groundwater. A pilot-scale plant has been used successfully to desalinate brackish groundwater resulting from saline intrusion at Manning-tree in Essex, adjacent to the Stour estuary. Reverse-osmosis plants are in commercial use overseas for treating brackish groundwater, and these plants may prove to be attractive in certain parts of the UK. Both reverse osmosis and electrodialysis have potential for treating industrial and municipal effluents. Electrodialysis has a low tolerance to impurities and is less suitable than reverse osmosis for treating polluted water. Freezing processes depend on the basic fact that when sea water freezes pure ice is formed, the salt remaining as concentrated brine. However, plants using freezing processes are not yet in commercial use.

The Water Resources Board[79] studied the possible application of desalination to meet the demand for new sources of water in the UK and concluded that desalination is unlikely to make a substantial contribution this century. Water from a desalination plant operated to meet base load demand would cost 2½ to 15 times more than water from conventional sources even in south-east England, but the operating costs of a desalination plant are high and base-load operation is unlikely. Any introduction of desalination would probably be in conjunction with surface sources, e.g. to meet peak demands in coastal resorts or to increase the yield of a reservoir[79-82]. In these circumstances, the plant would be operated at a low load factor (i.e. at infrequent intervals during drought), but even when used in this manner the cost of the desalinated water would not be competitive with other more conventional sources.

The distillation processes produce a very pure water, which can cause problems during distribution[77]. If blending with another source is not possible, treatment is necessary to adjust the pH and introduce hardness to the water. One of the problems of desalination is the disposal of the concentrated brine effluent, hence the advantage of coastal sites for installations. There is, however, likely to be a reaction against unsightly 'water factories' and disposal of the concentrated brine could pose ecological problems in some situations.

Use of effluents

The effluent from sewage works is water that has been treated to some extent. It usually contains trade effluents and the composition varies considerably with time[83], daily and weekly cyclical variations being usual. Its importance as a water resource can be judged from *Table 8.4*, which gives an indication of the proportion of sewage effluent in the total flow of a number of rivers used for water supply. The values given in the table are a measure of the indirect use of such effluent both as a water resource and for maintaining river flows in dry weather. The proportion of effluent is likely to increase with time in rivers generally. For example, it is estimated that by the end of the century the flow of the Bedford Ouse may be more than 75 per cent effluent under normal low flow conditions at some existing water abstraction sites[84].

The increased use of rivers as natural aqueducts requires the maintenance

Table 8.4 PROPORTION OF SEWAGE EFFLUENT TO TOTAL RIVER FLOW AT WATER SUPPLY ABSTRACTION POINTS (FROM PACKHAM[83], COURTESY INSTITUTION OF PUBLIC HEALTH ENGINEERS)

River	Abstraction point	River flow (m^3/s) Average	Min.	Flow of sewage effluent (d.w.f.) (m^3/s)	Proportion of effluent to total river flow (%) Average	Max.
Blackwater	Langford	1.4	0.63[a]	0.14	10.0	21.7
Chelmer	Rushes Lock (above Langford)	1.7	0.95[a]	0.17	10.1	17.9
Dee	Chester	35[b]	8.8[c]	0.47	1.3	5.4
Derwent (Yorkshire)	Elvington	16.3	4.8	0.14	0.85	2.9
Eden	Bough Beech	2.3	1.3	0.14	6.1	10.8
Esk	Rushwarp	3.9	0.53	0.009	0.24	1.8
Great Ouse	Foxcote	0.26[d]		0.052	20[e]	
	Clapham	1.3[d]		0.47	36[e]	
	Offord	2.7[d]		1.26	47[e]	
	Brownshill	2.9[d]		1.37	47[e]	
	Denver	3.8[d]		0.95	24[e]	
Hull	Hempholme	4.1	1.1	0.099	2.4	9.0
Lee	New Gauge	3.1	1.7	0.59	19.3	34.4
	Chingford	5.1	2.7	1.26	24.6	47.0
Ouse (Yorkshire)	Acomb	45	8.1	0.56	1.23	6.9
Severn	Tewkesbury		11.5	2.43		21.0
	Sharpness Canal		15.6	5.32		34.0
Stour	Langham	3.0	1.32[a]	0.147	4.9	11.1
Thames	Buscot	7.4	2.0	0.66	8.8	32.8
	Staines	57.6	19.3	5.37	9.3	27.8
	Sutton Courtenay	26.0	6.6	1.61	6.1	24.3
	Swinford	13.0	3.8	0.78	6.0	20.4
	Teddington	76.4	26.3	10.0	13.1	38.0
Wharfe	Addingham	14.7	1.58	0.005	0.04	0.33

Note. a 1 in 10 year driest average flow
b High peak flood flows inflate average
c May be regulated at this level for extended periods
d River flow exceeded 90 per cent of the time
e Percentage based on d

and improvement of the quality of those rivers used for water supply. The extent to which reuse is advisable along a river depends upon the efficiency of the effluent treatment process and also the treatment available for the abstracted water. As discharges to rivers increase, the dilution ratio decreases, thereby requiring the application of higher treatment standards for both effluents and abstracted water. In this context parts of the Report of the Working Party on Sewage Disposal[85] have a particular relevance:

'The two standards proposed by the Royal Commission on Sewage Disposal (1898–1915)[86] for no more than 30 mg/l of suspended solids and 20 mg/l for BOD—a 30:20 effluent—are in general the normal minimum requirements of river authorities for sewage effluents. It may be noted that the Royal Commission in pronouncing these standards envisaged that the effluent would be diluted with eight volumes of clean river water with a BOD of 2 mg/l.'

'Examination by WPRL of Sewage Works Statistics for 1964–65, compiled by the Institute of Municipal Treasurers and Accountants, showed that the yearly average values for suspended solids and BOD in final effluents discharged by nearly 60 per cent of the works were outside the 30:20 consent conditions of the relevant river authority; if individual sample results had been used rather than yearly averages, the proportion would have been far higher.'

Because of the increasing reuse of water, more stringent and higher requirements are necessary where a dilution of 8:1 is not available. Water authorities will need to exercise great care in deciding what requirements will apply, having regard to the use to be made of the receiving water.

Effluent standards applied for the protection of fisheries impose limits for suspended solids, BOD and the presence of toxic substances including ammonia; but if the river is to be used for water supply attention should also be given to oxidised nitrogen, dissolved constituents, non-biodegradable organic matter, phosphates, detergents and oils, colour, and bacteriological aspects[87]. The principal methods[87] of controlling these constituents or characteristics are:

1. by augmenting streamflow, particularly during dry weather;
2. by special treatment of the effluent to acceptable standards;
3. by special treatment of the water at the river intake;
4. by mixing the treated water with water from another cleaner source before distribution.

The method or methods applied in any particular case depends upon the circumstances, and of course the treatment required depends upon the dilution factor and the nature of the reuse.

An approach to pollution control involves the modification of industrial processes so that the nature of the ultimate waste product is changed or the volume reduced. An example of this was the modification of the composition of synthetic detergents. The early detergents, the so-called 'hard' detergents, caused stable foams to form on rivers when present in concentrations as low as 0.1 mg/l. This also had the effect of reducing the uptake of oxygen by the river. The problem was overcome by changing the chemical composition of the detergents to make them more readily biodegradable, and this has reduced the concentration of detergent residues in rivers.

It is generally agreed that in the UK the direct reuse of sewage effluents for potable water supply is unacceptable and also that, if the proportion of sewage effluents increases above present values in rivers that are used for public supply, higher standards of treatment might be necessary[83]. However, the reuse of effluents for industrial process water[88] or for irrigation is likely to increase, together with the recycling of water by industry, thereby saving the cost of both a supply and in some cases the waste-water disposal.

Where a lowland river is used for water supply two factors have to be borne in mind. The first is the possibility of pollution of the river through the accidental discharge of a toxic substance; the second is the possible cumulative effect of small concentrations of hazardous substances, including synthetic organic chemicals, on human health. The first can be guarded against by having a limited amount of storage; it has been recommended[89] that at least seven days storage should be available where public water supplies are from an intake on a river. The second factor remains an unknown quantity. The effect on man of long-term exposure to small concentrations of many pollutants is simply not known.

Water resources in England and Wales

The use of water in England and Wales and the methods available to meet the deficiency in supply expected by the end of the century have been examined by the Water Resources Board[90]. The methods adopted in making the examination have been summarised by Rydz[91]. In 1971 the demand for water for public supply was 14 million m^3/d. Between 1964 and 1971 the growth rate was 3.03 per cent, but this is expected to decline to less than 2 per cent by the end of the century when public supply use will probably be about 28 million m^3/d. The direct industrial abstraction in 1971 was 28 million m^3/d, but this represents the quantity actually abstracted and not the net use, which is very much less because of reuse. By AD2000 the net direct industrial demand will be about 2 million m^3/d. To provide against the anticipated deficiency in AD2000, new resources to supply about 12.1 million m^3/d will be required; the new storage capacity could amount to about 2000 million m^3, of which perhaps 25 per cent will be in aquifers.

Parts of northern and eastern England, England south of the Thames, south-west England and west Wales have adequate resources within their boundaries to meet all anticipated demands until AD2000. These areas have been referred to as 'self-sufficient areas' *(Figure 8.11)*. New resources that can be developed to meet the requirements of these areas by AD2000, together with other local potential sources existing outside the self-sufficient areas, amount to some 2.6 million m^3/d. The remaining requirement of 9.5 million m^3/d is best provided by an integrated pattern of resource development.

In the short term, up to 1981, the programme recommended to satisfy the expected deficiency of 2 million m^3/d is summarised in *Table 8.5*. The total yield at full development from all these projects is about 5 million m^3/d, more than 50 per cent of the deficiency expected to arise by the end of the century. Interesting aspects of the programme are the strong dependence upon groundwater in the south-east, the considerable additional yields to be

Table 8.5 PROGRAMME OF STRATEGIC SOURCE DEVELOPMENT RECOMMENDED BY THE WATER RESOURCES BOARD FOR IMPLEMENTATION BY 1981 (AFTER WATER RESOURCES BOARD[90])

Source	River authority area	Regional water authority area	Assumed date of introduction	Yield with full development* (Ml/d) Individual source yield	Total increment to yield
New reservoirs					
1. Brenig	Dee and Clwyd	Welsh	1976	330	
2. Kielder	Northumbria	Northumbrian	1978	955	
3. Carsington	Trent	Severn/Trent	1978	180	1465
Existing reservoirs enlarged					
4. Lancs Conjunctive Use Scheme (Stocks)	Lancashire	North-West	1975	450	
5. Grimwith	Yorkshire	Yorkshire	1977	100	
6. Craig Goch	Wye	Welsh	1980	1200	1750
Tidal limit river abstraction					
7. Barmby Sluices	Yorkshire	Yorkshire	1975	145	145
Groundwater augmentation of river flows					
8. Chalk and Jurassic lsts	Thames	Thames	1974	455	
9. Chalk	Great Ouse	Anglian	1976	330	
10. Triassic sandstones	Severn	Severn/Trent	1978	225	1010
Existing reservoirs: change of use					
11. Gouthwaite	Yorkshire	Yorkshire	1972	70	
12. Usk	Usk	Welsh	1972	100	
13. Thirlmere	Cumberland	North-West	1981	180†	
14. Empingham	Welland and Nene	Anglian	1981	90‡	440
Existing reservoirs: compensation water redeployment					
15. Elan	Wye	Welsh	1972	120	
16. Ladybower	Trent	Severn/Trent	1974	50	
17. Vyrnwy	Severn	Severn/Trent	1976	25	195
Total yield of additional sources (rounded)					5000
Yield taken up by 1981 (rounded)					2000

*Part of the full yield will be developed after 1981 where schemes are staged
†Extra yield from Cumberland Derwent regulated by partial conversion of Thirlmere
‡Extra yield by regulation of R. Welland and conjunctive use with groundwater from Lincolnshire Limestone over and above authorised direct supply yield of 230 ml/d

obtained by enlarging existing reservoirs, and the change in the use of other reservoirs to river regulation *(Figure 8.11)*.

After 1981 the various options open range from complete reliance on inland storage on the one hand, to depending entirely on estuary storage on the other. The strategy preferred by the Water Resources Board[90] combines the advantages of both of these extreme choices. Acceptance of this view would involve the construction of a small number of large inland reservoirs, the enlargement of some existing reservoirs and the development of estuary storage at one, or possibly two, site(s). It would also require the maximum development of groundwater and the possible incorporation of artificial recharge in the London and Trent basins. Considerable emphasis was placed upon the movement of water from the wetter areas in the west to the drier regions of the east and south-east. This would require the large-scale transfer of water using rivers as carriers and the construction of aqueducts to facilitate transfer between rivers. Successive reuse of water along the entire length of a river would be an important aspect of this strategy. To derive maximum efficiency the use of the various sources would have to be integrated, thereby increasing the yield beyond the total of each source estimated independently.

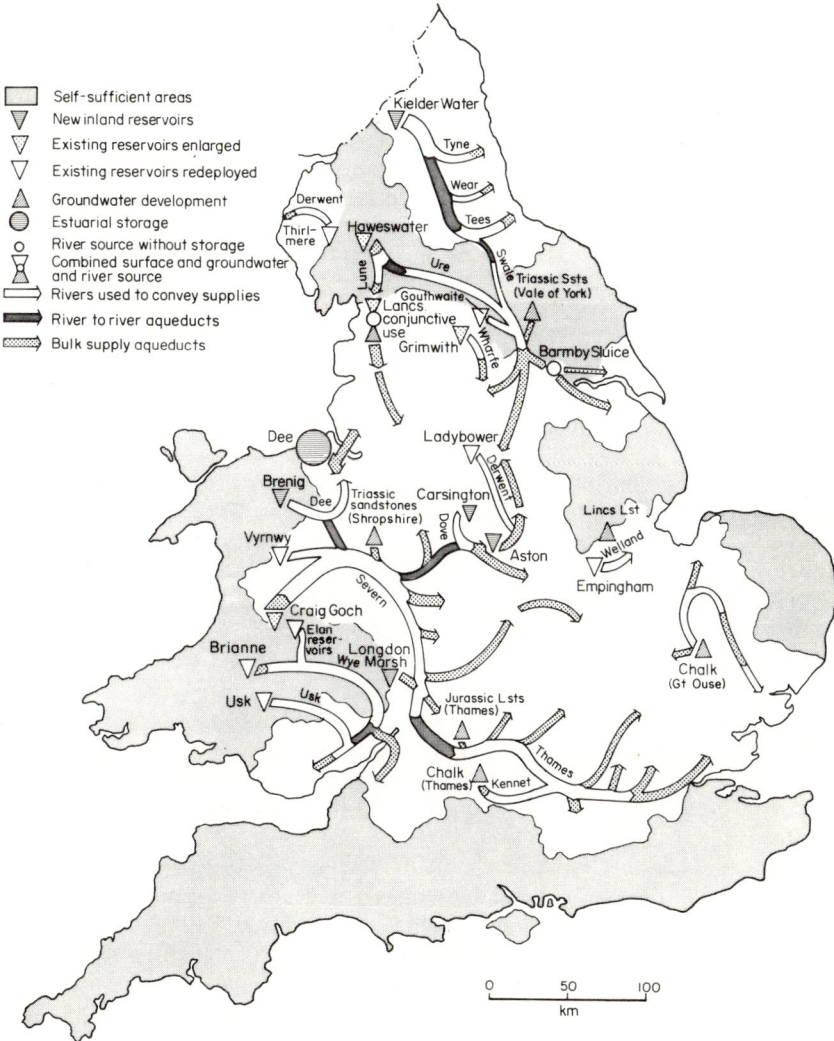

Figure 8.11 Strategy of water resources development in England and Wales in AD 2000 as preferred by the Water Resources Board

The projects recommended by the Water Resources Board[90] for introduction after 1981 and their estimated yields are:

1. enlargement of Hawes Water in the Lake District (820 ml/d)
2. enlargement of Llyn Brianne in central Wales (400 Ml/d)
3. storage in the Dee Estuary (up to 1080 Ml/d)
4. new reservoir at Longdon Marsh in Gloucestershire (660 Ml/d)
5. new reservoir at Aston near Derby (225 Ml/d)
6. redeployment of Lake Vyrnwy to regulate the Severn (110 Ml/d)
7. development of groundwater in the Triassic sandstones in the Vale of York (135 Ml/d)

These developments *(Figure 8.11)* could provide a total yield of about 3300 Ml/d which, together with the 5000 Ml/d provided by projects developed before 1981, and an estimated additional yield of about 1100 Ml/d from the reuse of effluents, would meet the estimated deficit of 9500 Ml/d at the end of the century.

An analysis of the recommended programme indicates that 3500 Ml/d could be derived from the Dee Estuary development and five new reservoirs (Brenig, Kielder, Carsington, Longdon Marsh and Aston), 3000 Ml/d from an enlargement of five existing reservoirs (Craig Goch, Grimwith, Stocks, Hawes Water, Llyn Brianne), 1000 Ml/d from the four groundwater schemes (in the Great Ouse and Thames basins, in Shropshire and the Vale of York) and 2000 Ml/d by the more efficient use of resources and the reuse of effluents. The total capital cost of the programme together with running costs until the end of the century would be about £1400 to £1500 million at 1972 prices. The proposals make imaginative use of the natural drainage patterns to transfer water from source to demand centre *(Figures 8.11 and 8.12)*. For example, advantage will probably be taken of the considerable scope for the conjunctive use of existing storage reservoirs on the Severn and the Wye to regulate these rivers. The key to this regulation would be the enlargement of Craig Goch reservoir to give a yield of 1200 Ml/d. The reservoir would regulate the Severn and the Wye. The next stage would be the enlargement of Llyn Brianne reservoir on the River Tywi, which would also regulate the Wye by tunnel to its tributary the River Irfon. This scheme would provide 400 Ml/d from the Wye or (with transfers from the Wye) from either the Usk or the Severn. Water from the entire system would supply the Midlands and South Wales and could also provide in part for south-east England by transfer from the Severn to the Thames if supported by a pumped storage reservoir at Longdon Marsh *(Figure 8.12)*.

Since the Water Resources Board's recommendations were published, it has been proposed that the Craig Goch reservoir could be enlarged still further to provide a yield of 2100 Ml/d from a storage of 550 million m³, which would make it the largest man-made reservoir in Europe.

There would be obvious advantages for water supply, amenity and recreation if the quality of the rivers that are at present unsatisfactory could be improved. For example, the polluted River Trent, which flows from south-west to north-east through the Midlands, has an average flow near Nottingham of 7 million m³/d; this flow represents a potential yield similar to that from an estuarial barrage. The feasibility of reclaiming the Trent has been the subject of an extensive research programme, which included the investigation of a number of methods for either improving the quality of the

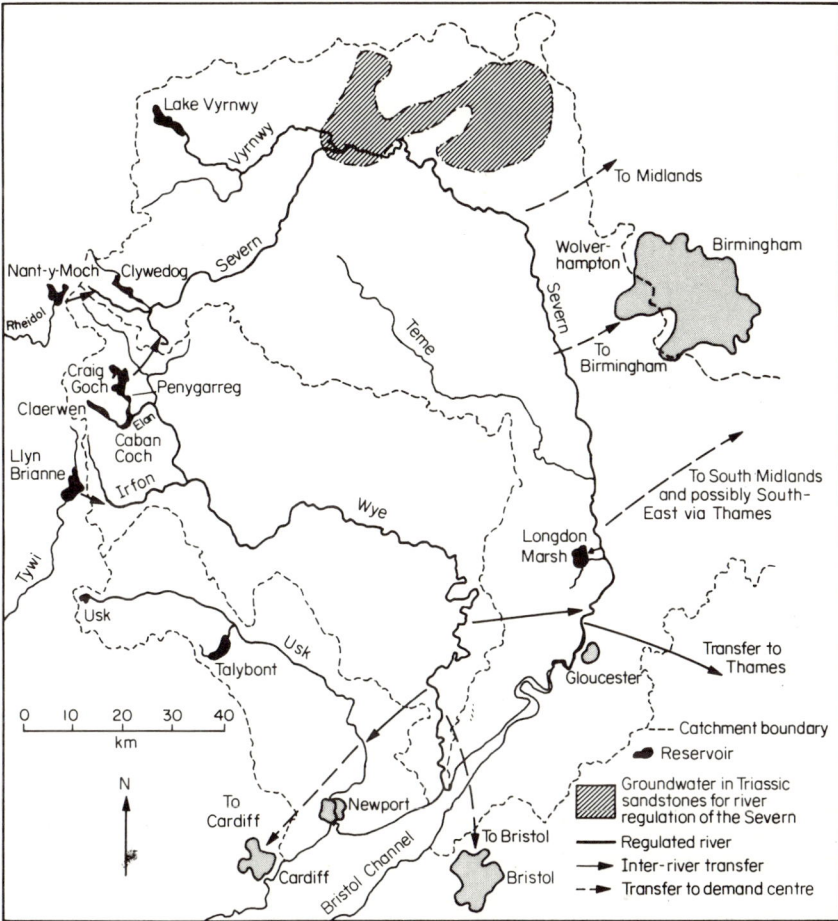

Figure 8.12   Regulation of the Severn, Wye and Usk from surface and groundwater storage (after Water Resources Board)

river or treating it to provide water suitable for supply purposes[92]. The aim of the programme was to estimate the cost of reclaiming the Trent for water supply and to compare this cost with those for other methods of obtaining the water.

Conventional treatment processes would not produce a potable water; the sequence considered necessary is storage, biological sedimentation (for ammonia removal), coagulation and softening, sedimentation and filtration, pH correction, granular activated carbon adsorption and disinfection by chlorine[93]. Although this sequence would produce a potable water, it would not necessarily be a 'wholesome' water, which has been described as 'water that could be consumed without risk from bacterial or chemical content'[77].

Purification lakes are to be used to protect the Trent from polluted water flushed from upper industrial tributaries during storms[94]. This will involve passing the River Tame through a lake system where the polluting suspended solids will settle and soluble biodegradable organic matter oxidise. If necessary the dissolved oxygen level would be maintained by artificial aeration. The pilot studies during the Trent Research Programme indicated that such a system on the Tame would protect the Middle Trent from storm pollution. Suspended solids would be considerably reduced, the BOD improved and the dissolved oxygen concentration in the river increased.

The quality of the Trent would be improved by treating both sewage and industrial effluents to higher standards[95], while reference has already been made to experiments designed to assess the feasibility of improving the quality of the river water by artificial recharge. A further alternative is to use the river as a lower-grade non-potable supply to selected industries[96]. At present some 185 Ml/d could probably be supplied in this way, rising to as much as 450 Ml/d by the end of the century. The Trent–Witham–Ancholme scheme[97] for Humberside industries makes use of Trent water; in a severe drought water would be passed from the Trent to the Witham via the Fossdyke canal and pumped from the Witham below Lincoln into the Ancholme's headwaters.

An economic model has been developed to find the cost of the alternative methods of meeting the future demand for water resources in the Trent basin, and deriving an optimum solution[52].

The Water Resources Board's national study[90] made use of models to indicate the cheapest programme of development[37] and to simulate the construction and operation of various programmes to obtain detailed cost analysis[98]. Models have also been used to evaluate the reliable yield of major water-resource sub-systems (such as the various sources and potential resources in the river basins draining into the Wash[99]; see *Figure 8.13*) and to derive the optimum operating conditions[99]. A method of analysis used to evaluate the design of such complex systems has been referred to as two-phase simulation[99]. The first phase involves the generation of synthetic river flow data extending over a long period of time (say 500 years). The second phase is the simulation of the system using models of the various components (e.g. reservoirs, aquifers) taking into account such factors as the capacity of reservoirs, constraints on developing groundwater storage and maximum pumping capacities.

Water resources in Scotland

Because of extensive areas of mountainous country that both induce precipitation and provide ideal sites for natural and artificial storage, Scotland is well endowed with resources of water to meet all foreseeable demands[100]. The average runoff is some 200 000 Ml/d, equivalent to 40 000 litres per head of the population. Public water supplies have tended to be derived from upland storage reservoirs by direct supply to individual centres of demand, involving considerable diversion of water from natural water courses.

In 1971 about 2130 Ml/d were used for public supply, but only 4 per cent of the estimated available yield from all sources was from underground storage and the available yield from springs was four times that from

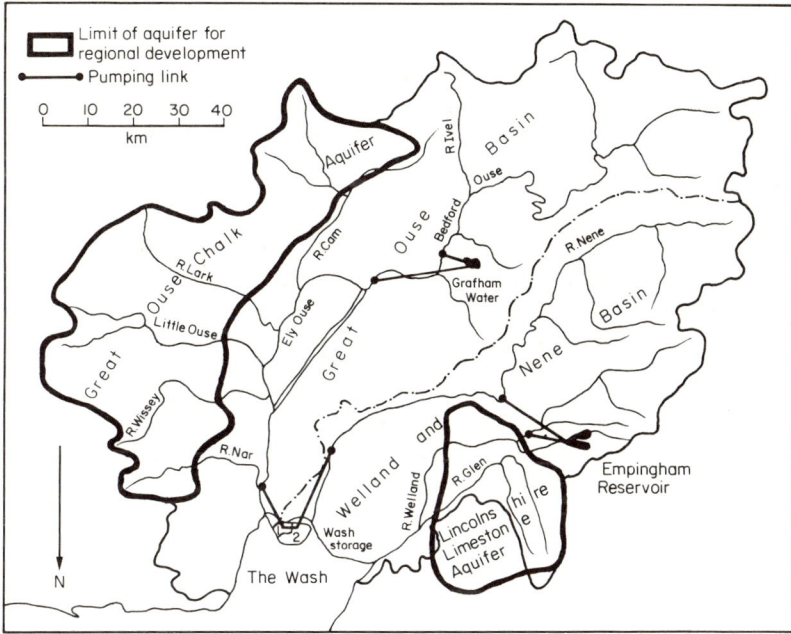

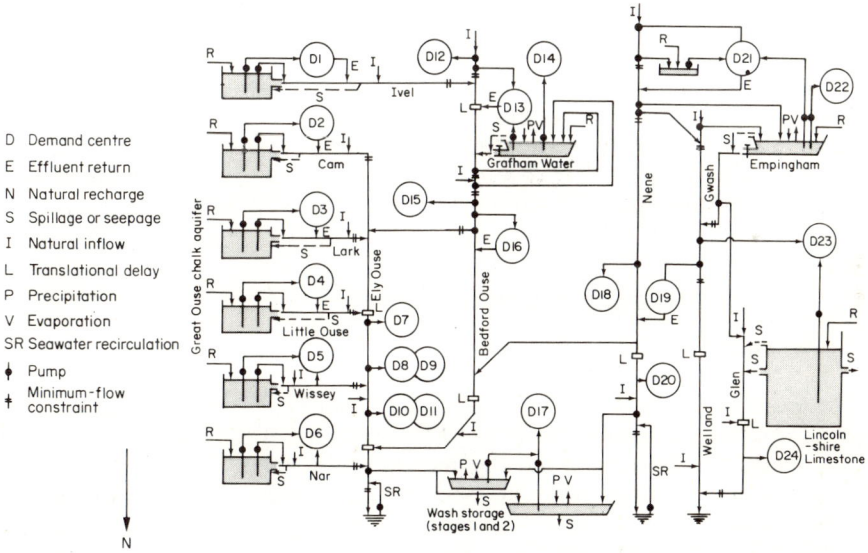

Figure 8.13 Schematic diagram of a complex water resource system incorporating pumped storage reservoirs (including estuary storage) and aquifers (after Jamieson and Sexton[99])

boreholes. Most of the groundwater is derived from glacial sands and gravels, river deposits, Carboniferous rocks and the Old Red Sandstone[101].

The gross use of water by industry in the more heavily populated central and south-east Scotland is about 10 000 Ml/d, but if hydroelectric usage is excluded this is reduced to about 2300 Ml/d, while the net use is only about

132 Ml/d. By the year 2000 the total demand for water in Scotland will be about 4400 Ml/d..

The future trend is towards larger water-resource developments serving larger areas which, together with a facility to transfer water, gives greater flexibility in use and distribution, and overcomes to some extent the difficulty of forecasting population growth and water demand in local areas—a problem in the past. With a fully interlinked water system the need for reserves can be reduced to a minimum. There are obvious advantages in having the reserve of water in a central position so that it can be transferred to any area if unforeseen industrial demands arise; the development of Loch Lomond to give a yield of 450 Ml/d is an example of this policy. In Scotland the problem is not a shortage of sources but the selection of those most appropriate to meet the need. Greater emphasis is likely to be placed on abstraction from low-lying lochs and rivers, nearer the demand areas, and upon the regulation of rivers from upland reservoirs. The yields from the larger potential schemes are between 500 and 900 Ml/d. For example, a river intake on the River Tay near the tidal limit at Perth would provide 900 Ml/d; regulation of the Clyde in south Lanarkshire by several reservoirs, 680 Ml/d; regulation of the River Doon in south Ayrshire by discharging water from Loch Doon, an existing hydroelectric reservoir, 400 Ml/d.

Considerable advantage is taken of the topography to store water for hydroelectric schemes. The total flow of water from catchments where such schemes exist amounts to about 30 000 Ml/d or 15 per cent of the total runoff from the Scottish mainland, indicating a significant use of available water resources for this purpose. Compensation water released from the storage generally maintains a greater-than-natural dry weather flow, and in the future benefits are likely to arise from the conjunctive use of water storage for both electricity generation and water supply. Most developments of hydroelectric power have been undertaken by the North of Scotland Hydroelectric Board[102-105], although major projects have been carried out in Galloway[106].

Water resources in Northern Ireland

Northern Ireland resembles Scotland in having adequate water resources, but with the exception of the Sperrin Mountains, the Mourne Mountains and the Mountains of Antrim most of the country is relatively low-lying. The country is essentially rural in character but with a growing industrial sector. The population is some 1½ million, about one third of whom live in Belfast. The use of water in the province for public supply amounts to about 450 Ml/d. Most is derived from river intakes, reservoirs and lakes, only about 6 per cent or 25 Ml/d being groundwater of which 16 Ml/d are from springs[107]. The storage capacity of the individual upland reservoirs tends to be small. Belfast derives supplies from the Mourne Mountains, the major scheme being the Silent Valley reservoirs with a storage capacity of about 21 000 Ml[108].

Northern Ireland contains Lough Neagh, the largest lake in the British Isles. The drainage basins of the lake and the Lower River Bann, which drains it to the sea near Coleraine, represent 43 per cent of the surface area of the province and include about 30 per cent of the population. The lough covers 385 km² and its capacity is about 3636 million m³. It has become the major new source for Belfast's water. Recently, a detailed study[109] has been completed of all aspects of the hydrology of the Lough Neagh and Lower River Bann drainage basins, including water supply, water quality, flood prevention and drainage, amenity and recreation. One conclusion was that up to 450 Ml/d could be exported with little adverse effect on other interests.

Groundwater resources, particularly in river gravels and the Permo-Triassic sandstones, are valuable, convenient assets in some areas for local supplies.

The future

Much of this chapter has been concerned with the future, particularly with the proposals for meeting the demands for water that are likely to arise up to the end of the century. These concluding remarks are meant to set these proposals and the subject matter of the preceding chapters in a broader context.

The UK is a relatively small, densely populated country. Management of the nation's natural resources in a manner that achieves greatest benefit to the community as a whole faces a number of constraints. Recent legislation for England and Wales, Scotland and Northern Ireland has provided a new framework for the overall management of water resources at regional and national levels. Some might argue that the pace of institutional change is too fast; on the other hand, a comprehensive approach to river basin management seems essential in a country where water resources are already highly developed and the range of options is somewhat restricted. With this framework, it should be possible to plan as a single process what resources to develop, where to provide storage for flood control and to regulate discharge for flow maintenance, what abstractions to permit and what discharges to allow to aquifers and water courses, and also to estuaries and coastal waters. It should also be possible to provide uniform opinions in those problems of planning where water is important; for example, where information about flooding, water supply and discharge of effluent might influence the siting of a new town. This is not to suggest that there can be one optimum solution for each region—in managing the water cycle there are many objectives even for a single basin. For instance, technical and economic factors have to be weighed against other criteria that are by their nature more subjective, such as the environmental aspects of development and the possible hazards to health resulting from the reuse of water.

In recent years, the trend has been towards increasing dependence on lowland river sources for water supply and the use of rivers as natural aqueducts, a use that will grow if the inter-basin transfer schemes come to fruition. Development of storage in estuaries is also planned. This trend represents a change from previous practice, which was to provide high-quality water from upland sources or from groundwater by direct supply. Present trends emphasise the importance of improving the quality of the nation's rivers, so that a potable supply of water can be maintained that is not prejudicial to good health. The improvement of rivers is also important to the aim of using all the nation's water space as an environmental amenity and a recreational asset, an aim that is likely to receive increasing support with the continued rise in water-based leisure activities.

There can be little doubt that the future has a considerable impact on today's decisions about water resources and their development. There is much more doubt about the ability to see into the future that is the basis of these decisions, and particularly about the forecasts of water demand that are fundamental to today's planning. There are, of course, considerable hazards to forecasting demand, stemming not only from the factors controlling demand but also from those determining the ability to meet it. For example, during the last ten years there has been a substantial decline in the population forecast for the end of the century. Forecasts published in 1973[110] gave totals for the UK for AD2001 of 59.0, 62.3 and 67.1 millions, based upon low, medium and high fertility natural-increase models. A slow growth of population, changes in the supply of money, inflation, high interest rates, coupled with a standard of living that shows little tendency to rise, are all factors important to the future demand for water. Between them they are likely to result in a weak consumer market, which could have consequences for the rate of growth of both the industrial and the domestic demand for water, and could affect investment in water-resources projects. In a parallel fashion, more realistic pricing of water and waste disposal services, taking into account marginal costs, in association with a more extensive use of meters, could bring about an appreciable change in the volume and pattern of demand. Any changes would reflect the public's attitude to the true value of water. Improvements to distribution systems together with increased efficiency in detecting leaks and stopping wastage would also alter demand. Greater use of non-potable water by industry and more recycling of industrial water could have similar results. Also of considerable significance to water resources are the measures likely to be taken to improve agricultural productivity: an extension of field drainage on a large scale, a widespread increase in the practice of irrigation, and a continuation of the high rates of application of nitrogenous fertilisers.

Other factors that may have effects on the future of water resources and their development are: increasing cost of energy and raw materials, climatic change, and the growing involvement of the public and various pressure groups in environmental matters. Of these factors, by far the most attention has been given to climatic change, particularly by Lamb[111,112]. Most of the evidence seems to indicate that during the remainder of this century, at least, the retreat from the climatic optimum of the 1940s will continue. Global temperatures will suffer a further decline and the intensity of the atmospheric circulation system will continue to decrease. For the UK, the lessening incidence of the westerlies points to a harsher climate with cooler winters and shorter summers. The amount of precipitation is expected to remain much as it is at present but rates of rainfall may increase and dry periods lengthen. Neither the magnitude of these changes nor how far they will continue into the 21st century is particularly clear.

The future of the science of hydrology in the UK is probably even less certain than the estimates of the demand for water in AD2000. It is also much more susceptible to influences from abroad, particularly the advances made in the United States. Probably the next twenty to thirty years will see a consolidation[113] of the gains made in the last ten years, the gains based on the increasing use of computers to control data-collection systems and to develop and operate mathematical models. This consolidation will involve improvements in the understanding of hydrological processes, including greater use of remote sensing techniques[114]. Orbiting satellites will probably be employed routinely as platforms for some remote sensing purposes, and geostationary satellites for relaying data from sensors on the ground to one or more data-collection centres. Because of the substantial rise in the volume of hydrological data, there will be a need to present information about the availability of data[115], where it is held, in what form, whether or not it has been analysed in any way, and so on.

The hydrologist is likely to become increasingly involved in planning and management, either through the provision of information on which decisions may be based or by his being a member of the decision-making team. In the first case, demands are likely for a widening range of information: water quantity and water quality data will need to be supplemented by data on water costs and water use, damage costs for floods and pollution and evaluation of the merits and demerits of water resources schemes. In the second case, the hydrologist will probably be faced with the problem of communicating his professional views on a project to others less well versed in the subject. Then he should attempt to express the alternatives to his scheme, the likely social, economic and ecological effects, and the benefits of his proposals to the community as a whole[116].

That it is uncertain seems to be the only certainty about the future. Even if current world trends continue there are those prophets of doom[117] who interpret these trends as showing that the collapse of our expansionist civilisation is imminent. Others[118] adopt a less pessimistic view; growth is likely to proceed in the future but at a reduced rate. A number of civilisations in the past have foundered when the basis of their economies was put at risk and altered, particularly where water was involved. Most hydrologists would agree that this is unlikely to be at the root of any calamity that might befall modern civilisation.

REFERENCES

1. Penman, H. L., 'The catchment area of the Institute', in *A view from the watershed*, Institute of Hydrology Report No. 20 (1973)
2. *A background to water reorganisation in England and Wales*, HMSO, 36 (1973)
3. Department of the Environment, 'Reorganisation of water and sewage services: government proposals and arrangements for consultation, Circular 92/71 (1971)
4. Road Research Laboratory, *Annual report 1969*, HMSO (1970)
5. British Waterways Board, *Annual report and accounts 1970*, HMSO (1971)
6. Department of the Environment, *Report of a River Pollution Survey of England and Wales 1970*, 1 (1971)
7. Water Resources Board, *Annual report for the year ending September 1972* (1973)
8. Penman, H. L., *Vegetation and Hydrology*, Technical Communication No. 53, Commonwealth Bureau of Soils, Harpenden, 124 (1963)
9. Rutter, A. J., 'The water consumption of forests', in *Water deficits and plant growth*, 2 (T. T. Kozlowski, ed.), Academic Press, 23–84 (1968)
10. Pereira, H. C., *Land uses and water resources*, Cambridge University Press (1973)

11. Lloyd, D., 'The effect of forest on stream flow with reference to afforestation at Vyrnwy', *British Association Meeting,* Liverpool (1950)
12. Law, F., 'The effect of afforestation upon the yield of water catchment areas', *J. British Waterworks Assoc.,* November, 489–494 (1956)
13. Law, F., 'Measurement of rainfall interception and evaporation losses in a plantation of sitka spruce trees', *Proc. IASH General Assembly, Toronto 1957,* **2,** 397–411 (1958)
14. Green, M. J., 'Calibration of the Brenig Catchment and the initial effects of afforestation', *Proc. IASH/UNESCO Symp. on the results of research on representative and experimental basins, Wellington,* IASH Pub. No. 96, 329–345 (1970)
15. Painter, R. B., Rodda, J. C. and Smart, J. D. G., 'Hydrological research and the planner', *Surveyor,* **140,** 22–25 (1972)
16. Walling, D. E. and Gregory, K. J., 'The measurement of the effects of building and construction on drainage basin dynamics', *J. Hydrology,* **11,** 129–144 (1970)
17. Walling, D. E., 'Streamflow from instrumented catchments in south-east Devon', *Exeter Essays in Geography* (ed. K. J. Gregory and W. L. D. Ravenhill), 55–81 (1971)
18. Law, F., 'Water yields from forested and non-forested catchments', *Forestry Commission course on forest hydrology,* Alice Holt Lodge, 18–19 May (1972)
19. Rodda, J. C., 'Progress at Plynlimon—problems of investigating the effect of land use on the hydrological cycle', *British Association Meeting,* Swansea (1971)
20. Institute of Hydrology, *Research 1971–1972* (1972)
21. Bleasdale, A., 'Afforestation of catchment areas: the physicists' approach to the problems of water loss from vegetation', *J. Instn Water Engrs,* **2,** 259–268 (1957)
22. Kitching, R., 'Water use by tree plantations', *J. Hydrology,* **5,** 206–213 (1967)
23. Binns, W. O., Letter to the Editor on 'Water use by tree plantations', *J. Hydrology,* **7,** 109–110 (1969)
24. Reynolds, E. R. C. and Leyton, L., *The water relations of plants,* Blackwell Scientific Publication, 127–141 (1963)
25. Reynolds, E. R. C. and Henderson, C. S., 'Rainfall interception by beech, larch and Norway spruce', *Forestry,* **40,** 165–184 (1967)
26. Leyton, L., Reynolds, E. R. C. and Thompson, F. B., 'Rainfall interception in forest and moorland', *Proc. Int. Symp. on forest hydrology* (ed. W. Sopper and H. Lull), 163–178 (1967)
27. Leyton, L. and Armitage, I. P., 'Cuticle structure and water relations of the needles of Pinus radiata', *New Phytology,* **67,** 31–38 (1968)
28. Rutter, A. J., 'Evaporation from a stand of Scots pine', *Proc. Int. Symp. on forest hydrology* (ed. W. Sopper and H. Lull), 403–417 (1967)
29. Stewart, J. B. and Thom, A. S., 'Energy budgets in pine forests', *Q. J. R. Met. Soc.,* **99,** 154–170 (1973)
30. Monteith, J. L., 'Evaporation and environment', *Symp. Soc. Expl Biology,* **19,** 205–234 (1965)
31. Stewart, J. B. and Oliver, S. A. 'Evaporation from forests', in *13th Aberystwyth Symp. on agricultural meteorology: 'Aspects of forest climate'* (1970)
32. Rutter, A. J., Kershaw, K. A., Robins, P. C. and Morton, A. J., 'A predictive model of rainfall interception in forests—a derivation of the model from observations in a plantation of Corsican pine', *Agricultural Meteorology,* **9,** 367–384 (1972)
33. Monteith, J. L., Szeicz, G., and Waggoner, P. E., 'The measurement and control of stomatal resistance in the field', *J. Agric. Engng,* **2,** 345–355 (1965)
34. Clarke, R. T., 'A review of some mathematical models used in hydrology, with observations on their calibration and use', *J. Hydrology,* **19,** 1–20 (1973)
35. Manabe, S. and Holloway, J. L., 'Simulation of the hydrologic cycle of the global atmospheric circulation by a mathematical model', in *UNESCO Symposium on world water balance,* Reading, IASH Pub. No. 93, 387–400 (1970)
36. Grindley, J., 'The estimation of soil moisture deficits', *Met. Mag.,* **96,** 1137, 97–108 (1967)
37. O'Neill, P. G., 'A mathematical-programming model for planning a regional water resource system', *J. Instn Water Engrs,* **26 (1),** 47–61 (1972)
38. Newsome, D. H., Bowden, K. and Green, J. A., 'Trent mathematical model', in *Symposium on advanced techniques in river basin management: Trent Model Research Programme,* 151–159, Instn Water Engrs (1972)
39. Edwards, K. A. and Rodda, J. C., 'A preliminary study of the water balance of a small clay catchment', *J. Hydrology* (NZ), **9** No. 2, 202–216 (1970)

40. Water Resources Board, *Eighth annual report for the year ending September 1971,* Mathematical modelling of groundwater systems, 79–84 (1971)
41. Water Resources Board, *Artificial recharge of the London Basin: II Electrical analogue model studies,* Publication 19, 38 (1973)
42. Hall, M. J. and Wolf, P. O., 'Design criteria for laboratory catchment experiments with particular reference to rainfall simulation', *Proc. IASH General Assembly, Berne,* 395–406 (1967)
43. Reynolds, E. R. C., 'Internal Water Balance of Trees', *Forestry (Suppl.),* 28–37 (1966)
44. Collinge, V. K., 'Mathematical models of water resource systems', in *Symp. on advanced techniques in river basin management: Trent Model Research Programme,* 13–32, Instn Water Engrs (1972)
45. Clarke, R. T., 'Mathematical models in hydrology', *Irrigation and Drainage Paper 19,* FAO (1973)
46. Dawdy, D. R. and O'Donnell, T., 'Mathematical models of catchment behaviour', *J. Hydraul. Div., Am. Soc. Civil Engrs,* **91,** HY4, 123–137 (1965)
47. Skibitzke, H. E., 'The use of analogue computers for studies in groundwater hydrology', *J. Instn Water Engrs,* **17,** 216–230 (1963)
48. Pinder, G. F. and Bredehoeft, J. D., 'Application of the digital computer for aquifer evaluation', *Water Resources Research,* **4,** 1069–1095 (1968)
49. Wilkinson, W. B. and Oakes, D. B., 'The use of digital models in the development of groundwater resources', *Proc. Instn Civil Engrs,* **55,** 523–530 (1973)
50. Rushton, K. R. and Tomlinson, L. M., 'Digital computer solutions of groundwater flow', *J. Hydrology,* **12,** 339–362 (1971)
51. Zienkiewiez, O. C. and Parekh, C. J., 'Transient field problems: two-dimensional and three-dimensional analyses by isoparametric finite elements', *Int. J. for Numerical Methods in Engineering,* **2,** 61–71 (1970)
52. Bowden, K., Green, J. A., Phillips, G. W. and Renold, J., *The Trent Research Programme: Vol. 11 The economic model,* Water Resources Board, Reading, 48 (1972)
53. Buras, N., *Scientific allocation of water resources,* American Elsevier, New York, 207 (1972)
54. Hall, W. A. and Dracup, J. A., *Water resources systems engineering,* McGraw-Hill, London, 372 (1970)
55. Bloomer, R. J. G. and Sexton, J. R., 'The generation of synthetic river flow data', Publication No. 15, Water Resources Board, Reading (1972)
56. *Flood Studies Report* Vol. I, Ch. 3, Natural Environment Research Council (1975)
57. MacDonald, Sir M. and Partners, *Report on River Stour Study III,* Kent River Authority, Maidstone (1974)
58. Hall, M. J. and O'Connell, P. E., 'Time series analysis of mean daily river flows', *Water and Water Engng,* April, 125–133 (1972)
59. Beard, L. R., 'Some problems in stochastic hydrology', *Water Resources Bulletin,* American Water Resources Association, 633–638 (1973)
60. Hetherington, R. G., Presidential address to the Institution of Civil Engineers, reported in *Surveyor,* **140,** 137–140 (1972)
61. *Morecambe and Solway Barrages: Report on Desk Studies,* Publication No. 4, Water Resources Board, HMSO (1966)
62. *Report on the Dee Crossing study—Phase I,* Binnie and Partners, London (1967)
63. *The Wash: Estuary Storage: Report of the Desk Study,* Publication No. 10, Water Resources Board, HMSO (1970)
64. Rydz, B., 'Estuary storage in the United Kingdom', *Symp. on enclosures of the sea,* British Assoc. Meeting (1971)
65. Ineson, J., 'Development of groundwater resources in England and Wales', *J. Instn Water Engrs,* **24,** 155–177 (1970)
66. Downing, R. A., Oakes, D. B., Wilkinson, W. B. and Wright, C., 'Regional development of groundwater resources in combination with surface water', *J. Hydrology,* **22,** 155–177 (1974)
67. *Report on the Lambourn Valley Pilot Scheme (1967–1969),* Thames Conservancy, Reading, 172 (1972)
68. *Great Ouse Groundwater Pilot Scheme—Final Report,* Great Ouse River Authority, Cambridge, 103 (1972)
69. *Proc. Artificial Recharge Conference,* Water Research Association, Medmenham, 481 (1971)

70. Marshall, J. K., Saravanapatan, A. and Spiegel, Z., 'Operation of a recharge borehole', *Proc. Instn Civil Engrs,* **41,** 447–473 (1968)
71. Boniface, E. S., 'Some experiments in artificial recharge in the lower Lee Valley', *Proc. Instn Civil Engrs,* **14,** 325–338 (1959)
72. *The hydrogeology of the London Basin,* Water Resources Board, Reading, 139 (1972)
73. *Artificial recharge of the London Basin: III Economic and Engineering Desk Studies* Water Resources Board, Reading (1974)
74. *Hardham recharge investigation—2nd progress report,* Sussex River Authority, Brighton, 16 (no date)
75. *The Trent Research Programme: Vol. 7 Artificial recharge—Bunter Sandstone,* Water Resources Board, Reading, 68 (1972)
76. Skeat, W. O. and Dangerfield, B. J. (editors), *Manual of British water engineering practice: Vol. III Water quality and treatment,* Instn Water Engrs, 407 (1969)
77. Holden, W. S. (ed.), *Water treatment and examination,* J. & A. Churchill, London, 513 (1970)
78. Pugh, O., 'What's new in desalination', *Surveyor,* **142,** 40–41 (1973)
79. *Desalination 1972,* Water Resources Board, HMSO, 46 (1972)
80. Burley, M. J. and Mawer, P. A., 'Desalination as a supplement to conventional water supply, Part 1—a technical and economic assessment of desalination processes', *Tech. Paper 50,* Water Research Association, Medmenham, 364 (1966)
81. Burley, M. J. and Mawer, P. A., 'Desalination as a supplement to conventional water supply, Part 2—desalination developments, conventional water supply costs and the future of desalination in the United Kingdom', *Tech. Paper 60,* Water Research Association, Medmenham, 250 (1968)
82. Mawer, P. A., 'Desalination as an interim alternative to conventional water resource developments', *Tech. Paper 76,* Water Research Association, Medmenham (1970)
83. Packham, R. F., 'Potable water from sewage effluent', *Symp. on sewage effluent as a water resource,* Instn Public Health Engrs (in press)
84. Billington, R. H., Livesey, J. B. and Taylor, N., 'Abstractions to water supply in the Great Ouse Basin', *Symp. on sewage effluent as a water resource,* Instn Public Health Engrs (in press)
85. *'Taken for granted', Report of the Working Party on Sewage Disposal,* Ministry of Housing and Local Govt/Welsh Office, HMSO, 65 (1970)
86. Royal Commission on Sewage Disposal, *8th Report Vol. 1,* Cmd 6464, HMSO, London (1912)
87. Fish, H., 'Effluent standards and water reuse', *J. Water Pollution Control,* **68,** 307–312 (1969)
88. Steel, P. H., 'The Bristol regional foul-water drainage scheme', *Proc. Instn Civil Engrs,* **38,** 595–620 (1967)
89. Department of the Environment, *First Annual Report of the Steering Committee on Water Quality,* Circular 22/72 (1972)
90. *Water resources in England and Wales,* Water Resources Board, HMSO, 67 (1973)
91. Rydz, B., 'Regional water resources analysis', *Proc. Instn Civil Engrs,* **49,** 129–143 (1971)
92. *The Trent Research Programme, Vol. 1,* Report by the Water Resources Board, HMSO (1973)
93. Miller, D. G., and Short, C. S., *The Trent Research Programme: Vol. 5 Costs of river water treatment,* Water Resources Board, Reading, 94 (1972)
94. Lester, W. F., Woodward, G. M. and Ravan, T. W., *The Trent Research Programme: Vol. 6 River purification lakes,* Water Resources Board, Reading, 69 (1972)
95. Porter, K. S. and Boon, A. G., *The Trent Research Programme: Vol. 3 The cost of waste water treatment,* Water Resources Board, Reading, 15 (1972)
96. Jackson, J. K., *The Trent Research Programme: Vol. 9 Dual supply systems,* Water Resources Board, Reading, 68 (1972)
97. Lindley, M., Bowyer, F. J. and Hawkins, H. S., 'The Trent/Witham/Ancholme scheme for the supply of non-potable water to the industrial area of Humberside', *J. Instn Water Engrs,* **28,** 272–298 (1974)
98. Armstrong, R. B. and Clarke, K. F., 'Water resource planning in south-east England', *J. Instn Water Engrs,* **26,** 11–32 (1972)
99. Jamieson, D. G. and Sexton, J. R., 'The hydrological evaluation of regional water resource systems in the United Kingdom', *Int. Symp. on water resources planning,* Mexico 1972 (in press)

100. Scottish Development Department, *A measure of plenty; water resources in Scotland—a general survey*, HMSO, 99 (1973)
101. Earp, J. R. and Eden, R. A., 'Amounts and distribution of underground water in Scotland', *Water and Water Engng*, **65**, 255–259 (1961)
102. Fulton, A. A. and Dickerson, L. M., 'Design and constructional features of hydroelectric dams built in Scotland since 1945', *Proc. Instn Civil Engrs*, **29**, 713–741 (1964)
103. Paton, J., 'The Glen Shira hydroelectric project', *Proc. Instn Civil Engrs*, **5**, 593–618 (1956)
104. Roberts, C. M., Wilson, E. B. and Thornton, J. H., 'The Garry and Moriston hydroelectric schemes', *Proc. Instn Civil Engrs*, **11**, 41–68 (1958)
105. Roberts, C. M., Wilson, E. B. and Wiltshire, J. G., 'Design aspects of the Strathfarrar and Kilmorack hydroelectric scheme', *Proc. Instn Civil Engrs*, **30**, 449–487 (1965)
106. Chapman, E. J. K. and Buchanan, R. W., 'Frequency of floods of "normal maximum" intensity in upland areas of Great Britain', *River flood hydrology symp.*, Instn Civil Engrs, 77–86 (1966)
107. Manning, P. I., 'The development of the water resources of Northern Ireland, progress towards integration', *Q. J. Engng Geol.*, **4**, 335–352 (1971)
108. Agnew, N., 'Water supplies in Northern Ireland', *J. Brit. Waterworks Assoc.*, **37**, 379–388 (1955)
109. *Lough Neagh Working Group*, Govt of Northern Ireland, 96 plus appendices (1971)
110. *Report of the Population Panel*, Cmnd 5282, HMSO, 135 (1973)
111. Lamb, H. H., *The Changing Climate*, Methuen, London (1967)
112. Lamb, H. H., *Climate, Present, Past and Future: Vol. 1 Fundamentals and climate now*, Methuen, London (1972)
113. Rodda, J. C., 'Data collection systems and their impact on the future development of hydrology', *Proc. Conference on the tercentenary of scientific hydrology, UNESCO/WMO/IAHS*, Paris (1974)
114. Painter, R. B., 'The potential application of satellites in river regulation', *Water and Water Engng*, **77**, 487–491 (1973)
115. Rowntree, Sir N., *Conservation and use of water resources*, 18th Graham Clark Lecture, Council of Engineering Institutions (2 Oct 1972)
116. Institution of Civil Engineers, *Council Statement* (2 Oct 1973)
117. Meadows, D. H., Meadows, D. L., Landers, J. and Behrens, W. W., *The limits to growth*, Universe Books (1972)
118. Cole, H. S. D., Freeman, C., Jahoda, M. and Pavitt, K. L. R., *Thinking about the future, a critique of 'The limits to growth'*, Chatto & Windus for Sussex Univ. Press (1973)
119. Dixon, J. W., 'Water resources', Section 26-1 of *Handbook of applied hydrology* (ed. Ven Te Chow), McGraw Hill (1964)

Appendix 1

Conversion factors

Metric to Imperial units		Imperial to Metric units	
Length			
1 mm	= 0.0394 in	1 in	= 25.4 mm
1 m	= 39.37 in	1 ft	= 0.3048 m
1 m	= 3.2808 ft	1 yd	= 0.9144 m
1 m	= 1.0936 yd	1 mile	= 1.6093 km
1 km	= 0.6214 mile		
Area			
1 m^2	= 10.7639 ft^2	1 ft^2	= 0.0929 m^2
1 m^2	= 1.196 yd^2	1 yd^2	= 0.8361 m^2
1 ha	= 2.471 acre	1 acre	= 4046.86 m^2
1 km^2	= 247.105 acre	1 acre	= 0.4047 ha
1 km^2	= 0.3861 mile^2	1 mile^2	= 2.59 km^2
Volume			
1 m^3	= 35.3147 ft^3	1 ft^3	= 0.02832 m^3
1 m^3	= 1.3079 yd^3	1 ft^3	= 28.3168 litre
1 m^3	= 219.969 gal	1 yd^3	= 0.7646 m^3
1 litre	= 0.21997 gal	1 gal	= 4.546 litre
1 Ml	= 0.21997 million gal	1 million gal	= 4546 m^3
		1 million gal	= 4.546 Ml
Weight			
1 kg	= 2.2046 lb	1 lb	= 0.4536 kg
1 kg	= 0.0197 cwt	1 cwt	= 50.8023 kg
1 tonne	= 0.9842 ton	1 ton	= 1016.05 kg
		1 ton	= 1.0161 tonne
Flow rates			
1 m^3/s	= 35.315 ft^3/s	1 ft^3/s	= 0.0283 m^3/s
1 m^3/s	= 19.005 million gal/d	1 gal/min	= 0.0758 l/s
1 l/s	= 13.20 gal/min	1 gal/h	= 0.00126 l/s
1 l/s	= 791.9 gal/h	1 million gal/d	= 0.05262 m^3/s
1 l/s	= 19 005 gal/d	1 million gal/d	= 52.617 l/s
Pressure			
1 m water	= 1.4223 lbf/in^2	1 lbf/in^2	= 0.7031 m water
1 kN/m^2	= 0.145 lbf/in^2	1 lbf/in^2	= 6.8948 kN/m^2
1 kN/m^2	= 0.3345 ft water	1 ft water	= 2.9891 kN/m^2
Hydraulic conductivity			
1 m^3 d^{-1} m^{-2} (m/d)	= 20.436 gal d^{-1} ft^{-2}	1 gal d^{-1} ft^{-2}	= 0.0489 m/d

Transmissivity

$1\ m^3\ d^{-1}\ m^{-1}$ (m^2/d)	$= 67.044$ gal $d^{-1}\ ft^{-1}$	1 gal $d^{-1}\ ft^{-1}$	$= 0.0149\ m^2/d$

Flow per unit area

$1\ l\ s^{-1}\ km^{-2}$	$= 0.09146\ ft^3\ s^{-1}\ mile^{-2}$	$1\ ft^3\ s^{-1}\ 10^{-3}$ acre	$= 6.9972\ l\ s^{-1}\ km^{-2}$
$1\ l\ s^{-1}\ km^{-2}$	$= 0.1429\ ft^3\ s^{-1}\ 10^{-3}$ acre	$1\ ft^3\ s^{-1}\ mile^{-2}$	$= 10.9332\ l\ s^{-1}\ km^{-2}$

Miscellaneous equivalents

Area

1 ha	$= 10\,000\ m^2$
$1\ km^2$	$= 100$ ha

Volume

$1\ m^3$	$= 1000$ litre
1 Ml	$= 1\ 000\ 000$ litre $= 1000\ m^3$
1 Imp gal	$= 0.1605\ ft^3$
$1\ ft^3$	$= 6.2288$ Imp gal
1 Imp gal	$= 1.201$ US gal
1 US gal	$= 0.8327$ Imp gal
$1\ ft^3$	$= 7.4805$ US gal

Weight

1 tonne	$= 1000$ kg

Flow

1 l/s	$= 86.4\ m^3/d = 86\,400$ l/d
$1\ m^3/s$	$= 1000$ l/s $= 86\,400\ m^3/d = 86.4$ Ml/d
$1\ m^3/d$	$= 0.01157$ l/s
$1\ ft^3/s$	$= 0.53817$ million gal/d

Pressure

$1\ lbf/in^2$	$= 2.3067$ ft water

Miscellaneous

1 mm rainfall/km^2	$= 1000\ m^3 = 0.220$ million gal
1 in rainfall/$mile^2$	$= 14.47$ million gal $= 65\,786\ m^3$

Appendix 2

Recent legislation in the UK affecting hydrology and water resource development

The water industry in the UK has been completely reorganised over the last few years by the various Acts of Parliament discussed briefly below. The administrative structure has been radically changed but many of the powers and requirements of earlier legislation still apply, particularly the Water Acts of 1945, 1948 and 1958 in England and Wales, and the Water (Scotland) Acts of 1946, 1949 and 1967 in Scotland[1]. There has been a general extension of the scope of the provisions, from the Water Acts of the 1940s through the Water Resources Act of 1963 to the more recent legislation. This was a direct consequence of the steady increase in the demand for water, the need to make more efficient use of water resources and the need to protect these resources from pollution.

Control of Pollution Act 1974

This Act applies to England, Wales and Scotland but only to a limited extent to Northern Ireland. It introduces provisions with respect to waste disposal, water pollution, noise, atmospheric pollution and public health, which it is anticipated will be implemented, in the main, by 1976. Local authorities are made responsible for the adequate disposal of household, commercial and industrial wastes and are given powers to licence waste-disposal sites; licences may be granted subject to conditions. Except in certain limited cases, the unlicensed disposal of waste is prohibited. Under Section 31 it is an offence to pollute streams or coastal waters or 'specified' underground waters, to impede the flow of water in streams, leading to or aggravating pollution, and to allow any solid waste to enter streams or coastal waters. (A 'specified underground water' refers to groundwater that a water authority declares is being used or is expected to be used for any purpose.) There are certain exceptions to Section 31; the discharge may be authorised by, for example, a disposal licence or a consent given by the Secretary of State for the Environment or a water authority. Exceptions also apply if the pollution is due to an act or omission in accordance with good agricultural practice (i.e. recognised as acceptable practice), or if it is caused by water from an abandoned mine entering inland, coastal or specified underground waters.

376

Solid waste from mines and quarrying is also excluded, providing the consent of a water authority has been given to deposit the waste and reasonable steps are taken to prevent the refuse entering streams. All discharges of trade and sewage effluent whether to the land, underground strata, rivers, lakes or the sea (including discharges made by pipeline beyond the territorial limit) will require the consent of the appropriate water authority. Section 52 allows water authorities in England and Wales and Section 53 allows river purification boards in Scotland to make a charge, in some circumstances, in respect of discharges of trade and sewage effluents.

The provisions of the Act, with respect to water, represent a strengthening of the legislation in the Rivers (Prevention of Pollution) Acts 1951 and 1961, and the Rivers (Prevention of Pollution) (Scotland) Acts 1951 and 1965. The Control of Pollution Act, when implemented by the Secretary of State for the Environment, will fully repeal the Deposit of Poisonous Waste Act 1972 and most of the Rivers (Prevention of Pollution) Acts.

Local Government (Scotland) Act 1973

The law relating to water supply in Scotland is contained in the Water (Scotland) Acts of 1946, 1949 and 1967. The 1946 Act empowered town councils, county councils and joint water boards (which were combinations of two or more local authorities) to provide public water supplies. This function was transferred to thirteen regional water boards by the 1967 Act. A Central Scotland Water Development Board, comprising seven of the regional boards, was also established under this Act. The board could develop new sources of supply for two or more of its constituent water boards and could also supply water in bulk. The Water (Scotland) Act 1949 provided for a uniform system of rating and charging for water throughout Scotland[1].

Pollution control was made the responsibility of twenty-one river purification authorities, consisting of nine river purification boards and twelve other local authorities in areas where the boards did not function, under the Rivers (Prevention of Pollution) (Scotland) Acts of 1951 and 1965.

Further reorganisation, with effect from 16 May 1975, resulted from the Local Government (Scotland) Act, 1973. The administration of local government became the responsibility of newly formed regional and islands councils (see *Figure 1.5*). The functions of the regional water boards were transferred to water authorities, but in fact the water authority for any area is the regional or islands council for that area, with minor exceptions detailed in Section 148 (3) of the Act. The Central Scotland Water Development Board was retained, but now comprises the limits of supply as water authorities of the Tayside, Fife, Lothian, Borders, Central and Strathclyde regional councils.

The river purification authorities were dissolved and their functions under the Rivers (Prevention of Pollution) (Scotland) Acts of 1951 and 1965 were taken over by seven new river purification boards covering the mainland of Scotland (see *Figure 1.5*).

Water Act 1973[2]

This Act required the inclusion in a national policy for water in England and Wales of provisions for the conservation, augmentation, distribution and proper use of water resources; the provision of water supplies; the provision of sewerage and the treatment and disposal of sewage and other effluents; the restoration and maintenance of the wholesomeness of rivers and other inland water; the use of inland water for recreation; the enhancement and preservation of amenity of inland water; the use of inland water for navigation; the provision of land drainage and the protection of fisheries in inland and coastal waters. Promotion of the policy is the responsibility of the Secretary of State for the Environment and the Minister of Agriculture, Fisheries and Food.

The Act establishes ten water authorities to cover England and Wales (nine regional water authorities and the Welsh National Water Development Authority). These authorities were given multi-purpose functions, being responsible for public water supply, sewerage and sewage disposal, control of pollution, abstraction of water, land drainage and fisheries.

A National Water Council was established. Its duties include advising the government on matters relating to the national policy, and advising the government and the water authorities on matters of common interest. Some of its duties relate to the water industry in the UK as a whole. A Water Space Amenity Commission was also established to advise on the use of water for recreation and amenity in England. Statutory Water Companies continue in existence, but they act as agencies for the water authority in whose area they are situated. Local authorities act as agents of water authorities in respect of sewerage functions and the provision and maintenance of local sewers, but not trunk sewers or sewage disposal works, with the exception of those constructed under the powers of the Housing Act. The water authorities are required to prepare a Water Development Plan surveying water resources and their use, and estimating future demand for water during the next twenty years (or such period as may be directed by the Secretary of State for the Environment or the Ministry of Agriculture, Fisheries and Food). The Plan has to be reviewed at least once every seven years.

Water Act (Northern Ireland) 1972

The provisions of this Act are similar in scope to the Water Act 1973 and the Control of Pollution Act 1974. Under the Act the Department of the Environment became responsible in Northern Ireland for the planning and management of water resources, the management of water supply and sewerage services, and the control of pollution, while the Ministry of Agriculture retained responsibility for drainage and fisheries.

Water Resources Act 1963

Under the Water Act 1945, Ministers were required to promote the conservation and proper use of water resources and to provide water supplies in England and Wales. The Water Resources Act 1963 extended these responsibilities to include measures for augmenting and redistributing water resources and for transferring water resources from one area to another. (These requirements were further extended, as indicated earlier, by the Water Act 1973.)

The 1963 Act established twenty-nine river authorities in England and Wales and made them responsible for the conservation and use of water resources in addition to land drainage, prevention of river pollution and fisheries (the functions of the former river boards). The Water Resources Board was also set up to advise the government and the river authorities and to ensure the proper use of water resources. Important provisions in this far-reaching act included the requirement that river authorities had to submit for the approval of the Water Resources Board hydrometric schemes for measuring and recording rainfall, evaporation, and the flow, level or volume of inland waters. Furthermore, river authorities could (and, if required by the Water Resources Board, had to) develop a network of observation boreholes to record changes in groundwater storage. Abstraction of water from both surface and groundwater sources was controlled by licences granted by the river authorities, except for minor abstractions specified in Section 24 of the Act. From 1 April 1969, all licensed abstractors of water were charged for the water abstracted in accordance with their licences, with the exception of agricultural abstractors abstracting from a groundwater source, unless it was for spray irrigation.

The administrative structure established by this Act has been replaced by the new structure required under the Water Act 1973, and many of the responsibilities have been transferred to the Regional Water Authorities.

REFERENCES

1. Skeat, W. O. and Dangerfield, B. J. (editors), *Manual of British water engineering practice,* Vol. 1, Instn Water Engrs, London, 319 (1969)
2. Telling, A. E., *Water authorities,* Butterworths, London, 152 (1974)

Appendix 3

Chemical standards for drinking-water

This summary is based almost entirely on *International Standards for Drinking-water* (third edition), World Health Organisation, Geneva (1971). For further details and information about standards of bacterial quality, and the virological, biological and radiological examination of water, reference should be made to this publication and *European Standards for Drinking-water* (second edition), World Health Organisation, Geneva (1970).

	Highest desirable level	Maximum permissible level
Total dissolved solids	500 mg/l	1500 mg/l
pH range	7.5–8.5	6.5–9.2
Anionic detergents	0.2 mg/l	1.0 mg/l
Mineral oil	0.01 mg/l	0.3 mg/l
Phenolic compounds (as phenol)	0.001 mg/l	0.002 mg/l
Total hardness (as $CaCO_3$)	100 mg/l	500 mg/l
Calcium (as Ca)	75 mg/l	200 mg/l
Chloride (as Cl)	200 mg/l	600 mg/l
Copper (as Cu)	0.05 mg/l	1.5 mg/l
Total iron (as Fe)	0.1 mg/l	1.0 mg/l
Magnesium (as Mg)	< 30 mg/l if sulphate > 250 mg/l, otherwise 150 mg/l may be allowed	150 mg/l
Manganese (as Mn)	0.05 mg/l	0.5 mg/l
Sulphate (as SO_4)	200 mg/l	400 mg/l
Zinc (as Zn)	5.0 mg/l	15 mg/l

Tentative limits for toxic chemical substances

	Upper limit of concentration
Arsenic (as As)	0.05 mg/l
Cadmium (as Cd)	0.01 mg/l
Chromium (as hexavalent Cr)	0.05 mg/l
Cyanide (as CN)	0.05 mg/l
Lead (as Pb)	0.1 mg/l
Mercury (total as Hg)	0.001 mg/l
Selenium (as Se)	0.01 mg/l

Specific chemical substances that may affect health

Fluoride (as F):

Annual average maximum daily air temperature °C	Recommended control limits for fluorides mg/l	
	Lower	*Upper*
10.0–12.0	0.9	1.7
12.1–14.6	0.8	1.5
14.7–17.6	0.8	1.3
17.7–21.4	0.7	1.2
21.5–26.2	0.7	1.0
26.3–32.6	0.6	0.8

Nitrate (as NO_3): Recommended that concentrations should not exceed 45 mg/l

Appendix 4

Principal rivers in the UK

A. Lengths

River	Length (km)	River	Length (km)
Severn	354	Bann — Upper	40 ⎫ 93
Thames	338	Bann — Lower	53 ⎭
Trent	274	Derwent (Yorkshire)	92
Great Ouse	251	Don	92
Wye	210	Stour (Hampshire)	90
Tay	195	Derwent (Derbyshire)	87
Spey	172	Ribble	87
Clyde	171	Ure	87
Tweed	156	Bure	80
Avon (Warwickshire)	155	Taw	80
Eden	145	Teifi	80
Nene	145	Waveney	80
Dee (Scotland)	140	Yare	80
Witham	130	Annan	79
Nith	127	Avon (Hampshire)	77
Aire	126	Nidd	76
Avon (Bristol)	120	Stour (Essex)	76
Dee (Wales)	115	Torridge	76
Medway	115	Lee	74
Mersey	115	Ouse (Yorkshire)	70
Tees	115	Weaver	70
Usk	115	Lune	68
Welland	115	Great Stour	65
Forth	105	Taff	65
Tywi (Towy)	105	Arun	60
Wear	105	Test	58
Wharfe	105	Colne (Essex)	55
Swale	103	Chelmer	55
Exe	95	Clwyd	55
Tamar	95	Tyne	50
		Dove	47
		Itchen	40

Figure A.1   Principal rivers; numbers indicate gauging stations referred to in Appendix 4B

B. Discharges

River	Gauging station (see Fig. A.1)	Drainage area to station (km²)	Start of continuous record	Mean flow (m³/s)	Highest flow (m³/s)	Date	Lowest flow (m³/s)	Date	Type of gauge
1. Spey	Boat o'Brig	2860	1957	61.10	970	30 Jul 56	9.32	16 Aug 55	Section
2. Dee	Woodend	1370	1935	36.17	1130	24 Jan 37	4.99	26 Aug 55	Section
3. Tay	Ballathie	4590	1956	155.6	1387	12 Feb 62	11.46	6 Aug 55	Section
4. Forth (Teith)	Bridge of Teith	518	1963	21.41	255	13 Dec 61	2.07	19 Sep 59	Section
5. Tweed	Norham	4390	1962	82.91	1186	6 Mar 63	5.63	7 Oct 59	Section
6. Tyne	Bywell	2180	1956	44.94	1316	8 Dec 64	3.42	often	Section
7. Wear	Sunderland Bridge	658	1957	10.48	236	7 Mar 63	1.08	4 Oct 59	B.C. weir
8. Tees	Broken Scar Weir	818	1956	18.44	478	6 Mar 63	0.02	16 Oct 59	Crump weir
9. Derwent ⎫ Yorks. Ouse	Stamford Bridge	1630	1961	15.90	123	22 Feb 66	3.34	4 Dec 64	B.C. weir/section
10. Nidd ⎬	Hunsingore Weir	484	1953	7.55	242	9 Dec 65	0.41	27 Sept 41	B.C. weir
11. Wharfe ⎭	Flint Mill Weir	759	1955	16.61	430	16 Feb 50	0.43	23 June 57	B.C. weir
12. Aire	Beal Weir	1930	1962	35.79	320	13 Dec 64	5.49	10 Apr 64	Weir
13. Don	Doncaster	1260	1960	15.06	192	5 Dec 60	3.06	6 Oct 64	Section
14. Trent	Colwick A	7490	1958	83.31	790	5 Dec 60	17.43	8 Oct 59	Section
15. Witham	Claypole Mill	298	1959	1.55	29	4 Dec 60	0.11	20 June 60	B.C. weir
16. Welland	Tixover	404	1963	2.1	24	26 Mar 64	0.85	often	Section
17. Nene	Orton	1630	1939	8.29	382	18 Mar 47	0.09	29 July 48	Weirs & sluices
18. Great Ouse	Brownshill Staunch	3030	1936	14.00	310	16 Mar 47	nil	often	Sluices
19. Cam	Bottisham	811	1936	3.51	73	14 Mar 47	0.23	Oct 49	Weir & sluices
20. Bure	Ingworth	165	1959	1.04	8.5	4 Dec 60	0.43	14 Sept 69	2 weirs
21. Yare	Colney	232	1960	1.51	21	11 Dec 65	0.22	10 Aug 64	Weir
22. Waveney	Needham Mill	373	1963	1.31	30	10 Dec 65	0.20	25 July 64	2 weirs
23. Stour	Stratford St. Mary	844	1928	2.29	34	3 Nov 60	0.04	13 Apr 60	Weirs, sluices, meters
24. Colne	Lexden	238	1959	1.03	22	16 Mar 64	0.03	30 Aug 65	S.W. flume
25. Chelmer	Rushes Lock	534	1932	1.43	18	Dec 65	nil	often	B.C. weir
26. Lee	Feildes Weir	1040	1932	3.48	118	16 Mar 47	nil	often	Weirs & sluices
27. Thames	Teddington	9870	1883	67.08	1064	18 Nov 1894	0.88	29 Oct 34	Weirs & sluices

No.	River	Station								
28.	Medway	Teston	1260	1956	11.41	294	4 Nov 60	0.68	30 Sept 59	Weirs, sluices, section
29.	Great Stour	Wye	230	1962	2.06	24	18 Nov 63	0.13	6 Oct 62	Weir & section
30.	Cuckmere	Sherman Bridge	131	1959	1.77	83	30 Jan 61	0.03	Oct 59	B.C. weir & section
31.	Itchen	Allbrook	360	1958	5.31	12	4 Dec 60	2.46	Oct 59	Weir & section
32.	Test	Broadlands	1040	1957	13.14	38	9 Oct 60	4.73	6 Jan 65	Section
33.	Avon	East Mills Flume	1450	1963	14.77	54	27 Feb 66	5.77	11 Nov 64	Weirs & flume
34.	Frome	East Stoke Mill	414	1960	6.60	22	24 Dec 65	2.29	14 Sept 64	Weirs & flume
35.	Exe	Thorverton	601	1956	16.06	495	4 Dec 60	1.18	31 July 62	Section
36.	Teign	Preston	381	1956	9.96	410	30 Sept 60	0.71	Sept 59	Section
37.	Tamar	Gunnislake	917	1956	23.54	410	26 Nov 59	0.69	Oct 59	Section
38.	Torridge	New Bridge	663	1962	14.20	314	19 Dec 65	0.76	5 Oct 64	Section
39.	Taw	Umberleigh	826	1961	16.25	351	19 Dec 65	0.79	30 July 62	Section
40.	Avon	Bath	1600	1953	19.59	365	5 Dec 60	1.42	often	Section
41.	Avon	Evesham	2100	1937	13.41	242	25 Jan 60	1.27	often	Section
42.	Severn	Bewdley	4330	1921	62.47	654	21 Mar 47	5.66	Sept 49	Section
43.	Wye	Cadora	4040	1937	69.9	906	20 Mar 47	6.91	Sept 59	Section
44.	Usk	Chain Bridge	912	1957	28.74	787	4 Dec 60	2.32	Oct 59	Section
45.	Taff	Tongwynlais	487	1961	19.72	481	18 Dec 65	2.83	13 Sept 64	Section
46.	Tawe	Ynys Tanglws A	228	1957	11.36	310	12 Dec 64	0.45	Sept-Oct 59	Section
47.	Tywi (Towy)	Ty-Castell Farm	1090	1958	38.23	526	13 Dec 64	1.19	Sept-Oct 59	Section
48.	Teifi	Glanteifi	894	1959	26.96	260	18 Dec 55	0.91	Oct 59	Section
49.	Ystwyth	Pont Llolwyn	170	1963	5.43	210	12 Dec 64	0.47	July 64	Section
50.	Dovey	Dovey Bridge A	471	1962	21.43	580	12 Dec 64	1.22	27 June 64	Section
51.	Conway	Cwm Llanerch A	344	1964	20.50	509	12 Dec 64	0.45	Sept 63	Section
52.	Clwyd	Pont y Cambwll	404	1959	5.73	72	13 Dec 64	0.45	24 Sept 63	Section
53.	Dee	Erbistock A	1040	1937	31.02	665	13 Dec 64	1.93	Oct 37	Section
54.	Weaver	Ashbrook	622	1937	5.44	212	8 Feb 46	0.65	2 Sept 61	Section
55.	Mersey	Irlam Weir	678	1945	13.25	266	23 Jan 44	0.65	26 Aug 55	B.C. weir
56.	Ribble	Samlesbury	1140	1964	39.68	779	17 Dec 65	5.41	4 Oct 64	Section
57.	Lune	Halton	995	1959	37.37	864	17 Dec 65	0.92	13 Sept 59	Section

continued

B. Discharges *continued*

River	Gauging station (see Fig. A.1)	Drainage area to station (km²)	Start of continuous record	Mean flow (m³/s)	Highest flow (m³/s)	Date	Lowest flow (m³/s)	Date	Type of gauge
58. Derwent	Camerton	663	1961	25.25	235	9 Dec 64	2.24	Oct 65	Section
59. Eden	Temple Sowerby	616	1964	14.95	206	4 Sept 66	2.24	Oct 65	Section
60. Esk	Canonbie	495	1963	17.63	580	7 Oct 64	1.83	28 Feb 63	Section
61. Annan	St. Mungo's Manse	730	1958	21.34	283	1 Dec 60	1.93	9 Sept 59	Section
62. Nith	Friar's Carse	799	1957	25.82	1275	16 Jan 62	1.46	2 Mar 63	Section
63. Cree	Newton Stewart	368	1963	15.13	230	14 Aug 66	0.59	25 July 66	Section
64. Girvan	Robstone	246	1963	6.36	92	14 Aug 66	0.43	6 Sept 64	Section
65. Kelvin	Killermont	335	1952	8.06	177	18 Oct 54	0.74	May 62	Section
66. Clyde	Blairston	1710	1958	38.8	577	14 Aug 66	4.50	11 Oct 59	Section
67. Leven	Linnbrane	786	1963	41.2	1029	17 Jan 65	11.52	Aug 66	Section

Note. The gauging record given refers to the lowest gauge on the river for which a relatively long record exists. On some rivers the gauging station is a considerable distance from the mouth of the river or, if a tributary, from its confluence with the main river

Appendix 5

Some principal lakes in the UK

Name	Location	Area (km²)	Height of lake above mean sea level (m)	Max. depth (m)
1. Lough Neagh	Northern Ireland	397.0	15	31
2. Lower Lough Erne	Northern Ireland	110.2	45	75
3. Loch Lomond	Strathclyde Region	71.1	7	190
4. Loch Ness	Highland Region	60.0	18	230
5. Upper Lough Erne	Northern Ireland	43.0	46	
6. Loch Awe	Strathclyde Region	38.5	36	94
7. Loch Morar	Highland Region	26.7	10	310
8. Loch Maree	Highland Region	23.2	5	112
9. Loch Shiel	Highland Region	23.1	10	128
10. Loch Lochy	Highland Region	21.5	30	161
11. Loch Tay	Tayside Region	20.6	106	154
12. Loch Rannoch	Tayside Region	20.3	203	130
13. Loch Arkaig	Highland Region	17.8	42	97
14. Lake Windermere	Cumbria	14.8	40	67
15. Loch Leven	Tayside Region	13.5	107	25
16. Loch Earn	Tayside Region	10.7	96	87
17. Loch of Harray	Orkney	9.6	1	4
18. Loch Lurgainn	Highland Region	9.5	54	45
19. Fionn Loch	Highland Region	8.9	170	44
20. Loch Langavat	Western Isles	8.7	33	30
21. Ullswater	Cumbria	8.7	145	63
22. Loch Sionascaig	Highland Region	8.3	47	51
23. Loch Assynt	Highland Region	7.9	65	85
24. Loch Lubnaig	Central Region	6.6	128	34
25. Lough Beg	Northern Ireland	6.4	16	3
26. Loch of Stenness	Orkney	6.4	1	5
27. Loch na Sealga	Highland Region	5.4	95	64
28. Bassenthwaite	Cumbria	5.4	68	21
29. Derwentwater	Cumbria	5.4	74	21

Note. This list does not include natural lakes that have had their levels raised by a major dam

Appendix 6

Reservoirs in the UK with a capacity greater than 10 million m³

The table on pages 390–392 is compiled from the *World Register of Large Dams,* published by the International Commission on Large Dams, 1973

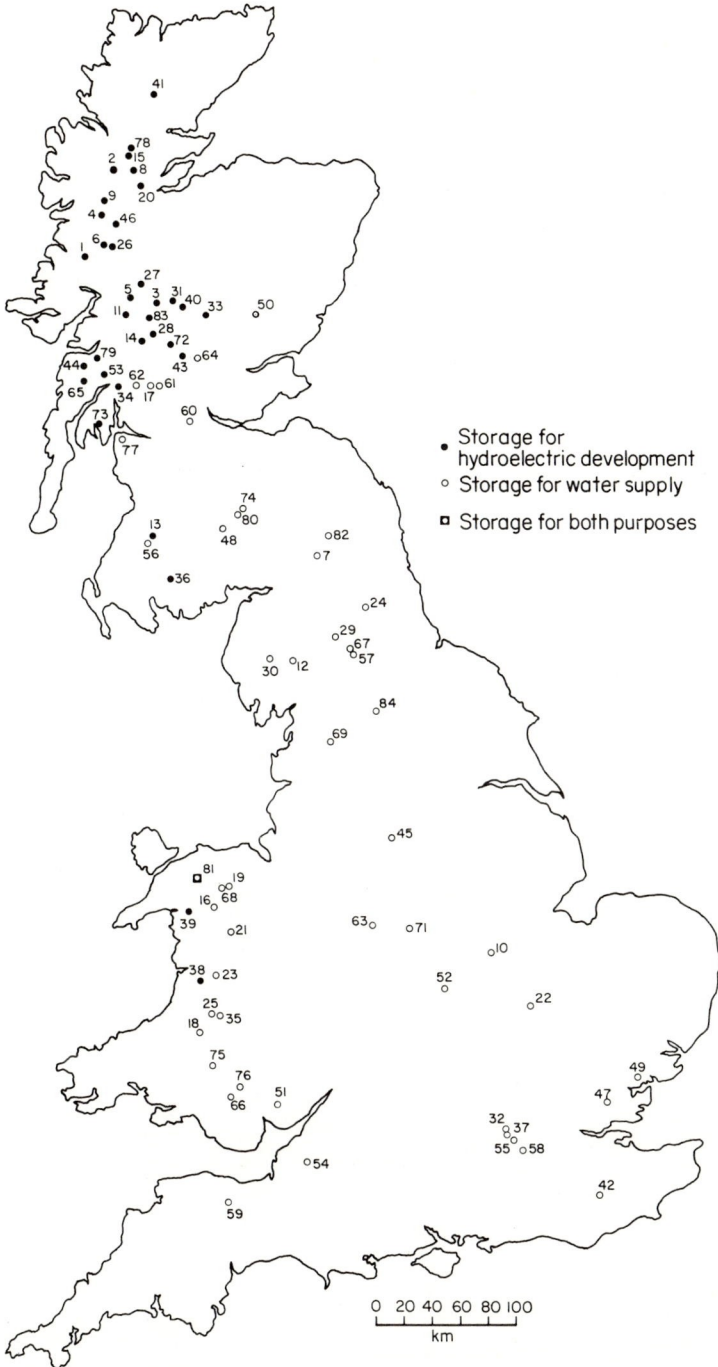

Figure A.2   Reservoirs with a capacity greater than 10 million m³

Name of reservoir (See Fig. A.2)	River	Gross capacity (10³m³)	Height above foundation (m)	Length of crest (m)	Year of completion (C = under constr.)	Type of dam*	Purpose†	Owner‡
1. Loch Quoich	Gear Garry	382 800	38	320	1956	ER	H	NSHEB
2. Fannich	Grudie	376 615	12	744	1957	PG/ER	H	NSHEB
3. Ericht	Ericht	229 367	17	404	1931	TE/PG	H	NSHEB
4. Mullardoch	Cannich	223 137	48	727	1951	PG	H	NSHEB
5. Treig	Treig	218 040	39	99	1934	TE/ER	H	British Aluminium Co. Ltd.
6. Cluanie	Moriston	203 032	41	675	1956	PG	H	NSHEB
7. Kielder Water	North Tyne	200 000	52	1150	C	TE	S	Northumbrian WA
8. Luichart	Conon	168 910	24	219	1954	PG	H	NSHEB
9. Loch Monar	Farrar	141 500	39	161	1963	VA	H	NSHEB
10. Empingham	Gwash	124 000	40	1200	1975	TE	S	Anglian WA
11. Blackwater	Leven	94 578	27	503	1909	PG	H	British Aluminium Co. Ltd.
12. Hawes Water	Haweswater Beck	84 123	38	469	1941	CB	S	North West WA
13. Loch Doon	Doon	82 969	16	299	1936	PG	H	SSEB
14. Loch Lyon	Lyon	82 968	39	530	1958	CB	H	NSHEB
15. Glascarnoch	Glascarnoch	78 155	43	534	1957	PG/TE	H	NSHEB
16. Lyn Celyn	Tryweryn	73 965	58	671	1965	TE	S	WNWDA
17. Loch Katrine	Achray Water	64 611	10	68	1859	PG	S	Strathclyde Regional Council
18. Llyn Brianne	Towy	61 000	91	274	1972	ER	S	WNWDA
19. Brenig	Afon Brenig	60 000	50	1200	C	ER	S	WNWDA
20. Orrin	Orrin	59 466	51	312	1959	PG	H	NSHEB
21. Vyrnwy	Vyrnwy	59 210	44	357	1891	PG	S	Severn-Trent WA
22. Graham Water	Diddington Brook	57 939	25	1761	1965	TE	S	Anglian WA
23. Clywedog	Afon Clywedog	49 700	72	183	1967	CB	S	Severn-Trent WA
24. Derwent	Derwent	49 500	36	975	1966	TE	S	Sunderland & South Shields Water Co.
25. Claerwen	Claerwen	48 307	67	355	1952	PG	S	WNWDA
26. Loyne	Loyne	45 307	22	549	1956	PG	H	NSHEB
27. Laggan	Spean	43 495	56	163	1934	PG	H	British Aluminium Co. Ltd.
28. Loch an Daimh	Allt Connait	42 758	35	463	1959	CB	H	NSHEB
29. Cow Green	Tees	41 000	30	600	1970	TE/PG	H	Northumbrian WA
30. Thirlmere	St. John's Beck	40 581	20	261	1894	PG	S	North West WA

No.	Name								
31.	Loch Garry	Garry	38 313	17	47	1956	PG	H	NSHEB
32.	Datchet	(Pumped storage)	38 200	22	5325	1975	TE	S	Thames WA
33.	Loch Tummel	Tummel	36 359	29	141	1950	PG	H	NSHEB
34.	Sloy	Inveruglas Water	35 679	56	357	1951	CB	H	NSHEB
35.	Caban Coch	Elan	35 531	44	186	1904	PG	S	WNWDA
36.	Clatteringshaws	Blackwater of Dee	35 396	26	448	1934	PG	H	SSEB
37.	Wraysbury	(Pumped storage)	34 500	20	5700	1970	TE	S	Thames WA
38.	Nant-y-Moch	Afon Rheidol	32 564	56	354	1962	CB	H	CEGB
39.	Trawsfynydd	Afon Prysor	32 564	31	209	1926	VA	H	CEGB
40.	Errochty	Errochty	32 480	49	501	1957	CB	H	NSHEB
41.	Loch Shin	Shin	31 800	20	448	1957	PG/TE	H	NSHEB
42.	Bewl Bridge	Bewl	31 300	29	780	1975	TE	S	Southern WA
43.	Lednock	Lednock	30 073	41	290	1958	CB	H	NSHEB
44.	Nant	Nant	30 000	28	372	1962	PG	H	NSHEB
45.	Ladybower	Derwent	28 600	43	381	1945	TE	S	Severn-Trent WA
46.	Benevean	Affric	27 467	37	137	1951	PG	H	NSHEB
47.	Hanningfield	(Pumped storage)	27 099	30	2088	1956	TE	S	Essex Water Company
48.	Daer	Daer Water	25 485	45	793	1956	TE	S	Strathclyde Regional Council
49.	Abberton	(Pumped storage)	25 274	17	733	1939	TE	S	Essex Water Company
50.	Backwater	Backwater	24 550	43	549	1969	TE	S	Tayside Regional Council
51.	Llandegfedd	Sor	24 408	39	351	1963	TE	S	WNWDA
52.	Draycote Water	Leam	22 700	21	1370	1969	TE	S	Severn-Trent WA
53.	Lochan Scron Mor	Shira	22 427	45	725	1959	CB	H	NSHEB
54.	Chew Valley	Chew	20 453	22	402	1953	TE	S	Bristol Waterworks Co.
55.	King George VI	(Pumped storage)	20 305	17	5271	1947	TE	S	Thames WA
56.	Bradan	Water of Girvan	20 000	30	438	1974	PG	S	Strathclyde Regional Council
57.	Balderhead	Balder	19 724	51	925	1965	TE	S	Northumbrian WA
58.	Queen Elizabeth II	(Pumped storage)	19 654	16	4351	1961	TE	S	Thames WA
59.	Wimbleball	Haddeo	19 500	60	280	C	CB/PG	S	South West WA
60.	Carron Valley	Carron	19 255	13	445	1939	TE/PG	S	Central Regional Council
61.	Glen Finglas	Turk	19 018	40	229	1965	PG	S	Strathclyde Regional Council
62.	Loch Arklet	Arklet Water	18 434	14	290	1912	PG	S	Strathclyde Regional Council
63.	Blithfield	Blithe	18 123	16	920	1953	TE	S	S. Staffordshire Waterworks Co.

continued

Name of reservoir (See Fig. A.2)	River	Gross capacity (10³m³)	Height above foundation (m)	Length of crest (m)	Year of completion (C = under constr.)	Type of dam*	Purpose†	Owner‡
64. Turret	Turret Burn	18 100	23	361	1964	TE	S	CSWDB
65. Awe	Awe	17 800	18	87	1962	PG	H	NSHEB
66. Taf Fechan	Taf Fechan	15 461	33	308	1927	TE	S	WNWDA
67. Selset	Lune	14 555	39	927	1960	TE	S	Northumbrian WA
68. Alwen	Alwen	14 527	46	137	1916	PG	S	WNWDA
69. Stocks	Hodder	13 568	31	354	1932	TE	S	North West WA
70. Silent Valley	Kilkeel	13 536	27	1374	1933	TE	S	Dept. of Environment (Northern Ireland)
71. Foremark	Foremark Bottom	13 300	39	1120	C	TE	S	Severn-Trent WA
72. Lochan Na Lairige	Allt A Mhoirneas	13 026	42	344	1958	CB	H	NSHEB
73. Loch Tarsan	Glentarsan Burn	13 000	30	344	1953	CB	H	NSHEB
74. Talla	Talla Water	12 714	25	320	1905	TE	S	Borders Regional Council
75. Usk	Usk	12 253	33	480	1955	TE	S	WNWDA
76. Talybont	Caerfanell	11 669	30	427	1938	TE	S	WNWDA
77. Loch Thom	Kip Water	11 347	20	427	1827	TE	S	Strathclyde Regional Council
78. Vaich	Vaich	11 213	38	257	1957	TE	H	NSHEB
79. Cruachan	Allt Cruachan	11 065	47	316	1964	CB	H	NSHEB
80. Fruid	Fruid Water	10 900	23	305	1968	TE	S	Borders Regional Council
81. Cowlyd	Afon Ddu	10 760	15	390	1920	TE	S/H	CEGB
82. Catcleugh	Rede	10 483	25	558	1905	TE	S	Newcastle and Gateshead Water Co.
83. Loch Eigheach	Gaur	10 194	17	110	1954	PG	H	NSHEB
84. Scar House	Nidd	10 036	63	460	1936	PG	S	Yorkshire WA

*Types of dam
CB Buttress
ER Rock fill
PG Gravity
TE Earth
VA Arch

†Purpose
H Hydroelectric development
S Water supply

‡Owners
CEGB Central Electricity Generating Board
CSWDB Central Scotland Water Development Board
NSHEB North of Scotland Hydroelectric Board
SSEB South of Scotland Electricity Board
WNWDA Welsh National Water Development Authority

Note. 1. The total storage capacity includes dead storage up to normal top or retention water level but not flood level (does not apply in the case of Loch Katrine, where storage quoted excludes dead storage)
2. The reservoirs included in the table are restricted to those maintained by dams higher than 15 m. The table therefore excludes some large reservoirs

Index